INTENSITY VARIATIONS IN TELEPHONE TRAFFIC

NORTH-HOLLAND
STUDIES IN TELECOMMUNICATION

VOLUME 10

NORTH-HOLLAND-AMSTERDAM • NEW YORK • OXFORD • TOKYO

INTENSITY VARIATIONS IN TELEPHONE TRAFFIC

CONNY PALM †

1988

NORTH-HOLLAND-AMSTERDAM • NEW YORK • OXFORD • TOKYO

ISBN: 0 444 70472 8

A Translation of:
Intensitätsschwankungen im Fernsprechverkehr
Ericsson Technics, No 44, 1943

Translated from German by Christian Jacobaeus

Published by:
ELSEVIER SCIENCE PUBLISHERS B.V.
P.O. Box 1991
1000 BZ Amsterdam
The Netherlands

Sole distributors for the U.S.A. and Canada:
ELSEVIER SCIENCE PUBLISHING COMPANY, INC.
52 Vanderbilt Avenue
New York, N.Y. 10017
U.S.A.

Printed in The Netherlands

PREFACE

Conny Palm's dissertation "Intensitäts-schwankungen im Fernsprechverkehr" was published in 1943. It was already from the beginning considered to be one of the most important contributions to teletraffic theory. It contains the solution of a great many problems in this branch of science. It also presents ideas on new problems which have since engaged other researchers. The work had a greater depth than was usual at that time and it is not an overstatement to say that is has given teletraffic theory a scientific status. It has meant much also indirectly in that it has drawn attention to the fact that human behaviour, i.e. telephony habits, follows its own laws in its interplay with the telephone system. Later researchers have in many cases been able to find in the work ideas for the solution of their own problems.

The dissertation was printed in German, a language which during the time between the two world wars was commonly used for articles on teletraffic research. Conny Palm also mastered German best of the world languages. The translation from the Swedish manuscript was done jointly by him and Nils Rönnblom.

At the International Teletraffic Congresses the dissertation has been discussed privately. The fundamental importance of this work for teletraffic theory has been pointed out. It has been stressed that the thesis is still of great interest and that this interest has actually grown during the past 20 years. Complaints have been made that it is only available in German, a language which nowadays is mastered only by a limited number of researchers outside the German-speaking countries. It has been hoped that someone would have the opportunity of making a translation into English. Several researchers in England and the USA have been seriously interested in undertaking such a task but without result so far. Finally I decided to make an attempt. I had several reasons for this; I got more time to spare than earlier, I know the subject to a certain extent and above all I am one of the few still living who knew Conny Palm well and have worked together with him.

Now when the result of my endeavours is presented, I hope that my colleagues will find the translation useful. The German original has been followed rather closely in the translation. Only evident mistakes have been corrected. The mathematical terminology has in some instances followed more modern practice.

The Swedish Telecommunications Administration and Telefonaktiebolaget LM Ericsson have jointly paid the costs. An agreement regarding printing and distribution has been made with the North-Holland Publishing Company, who have great experience of publications of this kind.

A great number of persons have been engaged in the translation work. I am very grateful for their interest and contributions. During the autumn of 1986 two of the most involved had to withdraw from active work. I also had to limit my partaking in the remaining work. Gösta Neovius then took over the responsibility as editor. He knew Conny Palm well from the time when they both worked for the Swedish Board for Computing Machinery. He also has a good common knowledge of the subject and has contributed with a paper to the first Teletraffic Congress, held in Copenhagen, 1955. Gösta Neovius has in turn engaged John Pyddoke and Gunnar Lind in the remaining work.

Conny Palm was a man of many (and rare) qualities. Those who wish to know more about him and his scientific work are recommended to read the paper *Conny Palm and teletraffic theory*, presented by Arne Jensen at the seventh Teletraffic Congress, held in Stockholm 1973. This paper has been published in the English version of TELE, No. 2, 1973.

The work of translation has been demanding for those involved. I am glad that by this work I have been able to make the contributions of a great scientist in the technical field more known, especially since he was one of my friends and fellow workers.

Stockholm, June 1987
Christian Jacobæus

Preface to the German edition

In presenting this work it is a special pleasure to express my thanks to all those who have assisted me in different ways in completing my task.

My respecful thanks go firstly to the Board of Directors of the Royal Swedish Telecommunications Administration, which with great generosity has placed at my disposal personnel, equipment and premises for the performance of the necessary and very time-consuming measurements, as well as to Telefonaktiebolaget LM Ericsson who enabled me to devote almost all of my time during the past five years to the research of which the results are presented in this work. I also express my warm thanks to Mr. N. Rönnblom, First Section Engineer, who has contributed in an extraordinary way to the success of the measurements and also in other ways has assisted me with advice and references. He has also scrutinized the manuscript. In the mathematical formulation of the theories, furthermore, I have been given very valuable advice by Dr. W. Feller, now at Brown University, Providence, RI, for which I warmly thank him. Mr. Gunnar Olsson, of the Telecommunications Administration, designed most of the measuring equipment and also supervised it. He has contributed an extraordinary number of suggestions in the course of the work.

I owe thanks besides to many others who have assisted me in different ways in the completion of this work.

CONTENTS

INTENSITY VARIATIONS IN TELEPHONE TRAFFIC

Investigations of the developement of stochastic processes applicable to telephone traffic problems

It is of special importance for the economy and proper functioning of a telephone plant that it has a suitable number of switches and circuits (in the following named devices). Very often there is a lack of empirical data that have a direct bearing on the conditions. The determination of the number of devices must therefore rely on theoretical investigations. Because of the general character of telephone traffic, the theories must be worked out in close relation to the methods of probability theory. It is primarily the theories of stochastic processes that have turned out to be of importance in this context.

The theoretical investigations of telephone traffic problems that have been made so far have as a basis the definite stochastic process, which corresponds to what technicians call *random traffic*. However, it has turned out in the course of time that in this way we do not get the precision necessary for satisfactory modelling of real traffic conditions in a telephone plant. In this work an effort has been made to arrive at a more reliable basis for the description of telephone traffic by considering more general stochastic processes than the one corresponding to random traffic.

In the first part, comprising Chapters 2 – 4, general stochastic processes have been developed in such a form that they can be used as a model for telephone traffic. Chapters 3 and 4 contain applications to some special problems that are somewhat outside the general scope of the work. They have, however, been included as they seem to offer valuable means of solving problems regarding random traffic that up to now have remained unsolved.

In the second part, Chapters 5 – 10, a special stochastic process has been investigated in detail, being a modification of the definite process and based on the assumption of what are called slow intensity variations. The process has turned out to be particularly suitable as a basis for modelling the properties of real telephone traffic. It also offers good means of a systematic study of the superposition problems, which are of great importance for many applications. A fundamental concept, traffic classes, emerges from these investigations and by studying them, certain "normal" forms for the mathematical models are arrived at. At the end of the second part various measurement methods are discussed in order to have the applicability of the theories confirmed.

In the third part, Chapters 11 – 14, a series of measurements are presented. These measurements were carried out in recent years in conjunction with the development of the theories put forward in this work. The results of the measurements suggest that these theories will be of great importance for the treatment of traffic questions in practice.

Chapter 1

INTRODUCTION

The problems that arise when studying the traffic conditions in a telephone plant are in principle as follows. In order to set up a connection between two subscribers a number of devices are used, which are often common to several subscribers. For technical as well as economical reasons the number of such devices cannot as a rule be so great that all subscribers served by these devices can have simultaneous connections; nor would that be necessary, as generally only a small proportion of the subscribers have calls at the same time. However, one must ensure that the number of devices is sufficient so that subscribers will not experience too often the inconvenience of not getting their wanted connections immediately. In order to estimate this inconvenience certain quantities are used, such as blocking, delay times, etc., and others which characterize the *grade of service* of the plant. Most of the traffic problems in telephony relate to the determination of such quantities. We have as starting point a certain knowledge of the call conditions of subscribers and the technical functioning of the switching devices.

The basic quantity used for modelling the afore-mentioned call conditions is the concept *telephone traffic*. The most appropriate course would be to define telephone traffic in relation to a certain group of devices. The groups we refer to are characterized by the fact that they comprise only similar devices. Further, any such device can set up a connection or a stage in a connection from certain subscribers and cooperate with the others in the group in such a manner that any of these subscribers can get their calls connected by any of the devices in the group. (In this respect certain limitations may, however, occur.) Each device can only handle one call at a time. It is occupied (busy) during the time of the call and cannot, until free, start handling another call.

By traffic in the considered group is understood a sum of the occupations of the different devices. Mathematically the traffic is represented by intervals along the time axis. These can be defined by the points of time for the arrivals of calls and the individual call durations. Thus they can be said to constitute the elements of the traffic. The occupations in traffic defined in this way need not be identical with any call durations in reality. They often correspond only to a stage in the setting up of the calls. A number of cooperating operators can thus be considered as a group of devices. The occupations of the operators are constituted by the times during which they are engaged in the connection and disconnection of the calls. It should also be observed that every single device in a telephone plant can be considered as a group of devices.

An important generalization of the traffic concept can be arrived at by using the call arrivals in the group, instead of the points of time for the start of occupations. The call arrivals comprise the points of time for the start of occupations as well as the points of time at which the setting up of a call – that is an occupation – by the devices in the group is wanted but cannot take place immediately as all devices in the group are busy. Such calls are said to be blocked for the group.

The simplest and most important quantitative property of the traffic is its magnitude, which means the number of occupied devices in the group at a given point of time. When an occupation starts or ends, the magnitude changes by jumps with the value 1. Often, however, we understand by traffic in quantitative terms, i.e. traffic intensity, the mean of the magnitude of the traffic during a certain length of time. A traffic intensity defined in this way does not need to have the value of an integer.

In an attempt to treat traffic problems by mathematical methods it is above all necessary, by studying the physical properties of the telephone traffic, to get an idea of the mathematical models that can illustrate these properties in the most precise and, at the same time, most simple way. We find then that traffic belongs to the type of physical phenomena of which the separate events clearly show a pronounced randomness. Their combined effect, however, very often exhibits a considerably more regular pattern. Such phenomena can appropriately be illustrated by means of the mathematical schemes that are common in probability theory.

According to probability theory, a process where a quantity varies with time in such a way that the instantaneous value of the quantity is not always wholly determined but can only be given with a certain probability, defines a *stochastic process*. The telephone traffic served by a group (of devices) can then be described as a stochastic process, the number of simultaneously busy devices constituting the value of the variable. In such a stochastic process the variable can only assume values of integers. One assumes, suitably, that a change in the number of busy devices in the group corresponds to a change in the value of the variable by jumps. Telephone traffic can thus be described as a *discontinuous* stochastic process. This means a simplification in comparison with general stochastic processes in which the variable can assume arbitrary values.

Clearly, it is possible and advantageous to specify further the mathematical scheme for the description of telephone traffic. The call arrivals are then looked upon separately, as are the call duration times following them. The traffic within the group is determined as a process originating from these two probability quantities. The advantage of this way of looking at the matter becomes especially clear if a certain independence between the two probability quantities can be assumed. This is generally the case. The call arrivals still constitute a discontinuous stochastic process. This is, however, a much simpler one than the one mentioned above, which is constituted by the traffic itself. One can always assume that the probability of more than one call arrival at one and the same point of time is zero. Thus the call arrival process can be modelled by a variable that can assume only two values, for instance 0 at the points of time when no call arrivals occur and 1 at the points of time when a call arrival takes place. Further, it can always be assumed that the probability of more than n call arrivals within a certain period of time goes towards 0 when n grows infinitely. The points where the variable has the value 1 can never form coherent areas. A stochastic process with these properties will be named a *point process*. Because of the special properties of these point processes, they can be treated in a completely different and considerably simpler way than is commonly the case in the theory of more general stochastic processes. In these we are, because of the mathematical difficulties, forced to confine ourselves to what is called *definite* processes. They are characterized by the assumption that the probability of an event is independent of all previous events. In point processes, on the other hand, it is possible to treat in a relatively simple way also those cases in which the previous events have an impact on the probabilities of the subsequent events. In this respect many of the processes that are treated below are considerably more generally valid than most processes that are treated in probability theory.

In a point process the definite case corresponds to the phenomenon that the points – in telephone traffic the call arrivals – are distributed fully at random over the time axis. This assumption has formed the starting point for most of the theoretical investigations of traffic problems in telephony. One speaks of *random traffic* and assumes that, for the duration of the occupations, some probability law is valid which is independent of the points of time for the call arrivals. Under these conditions a series of valuable results could be derived by relatively simple means regarding the quantities that determine the grade of service in the telephone plant. Gradually it became clear, however, that these theoretical results often do not reproduce the phenomena that occur in real life with satisfactory accuracy. Therefore they are not useful as a starting point for the technical treatment of traffic questions. This seems to be due to the fact that the pure random character of the traffic – which hitherto has been the basis for the theoretical investigations – does not correspond to its real properties to a sufficient degree. In the present work an attempt has been made to come to grips with the problem, by introducing more general point processes than the definite process. This attempt leads to the theory of *slow intensity variations* which is treated in the second part. The results arrived at are related to the theories applying to the definite point processes in such a way that the latter still remain fundamentally important for the calculations. In the treatment of the general properties of point processes that was necessary for the development of the theories of slow variations of the traffic intensity, however, it has become apparent that the more general point processes could also be of great importance for problems based on the assumption of definite processes. A treatment of problems that in this field have so far been considered too complicated seems to be possible. These uses are somewhat outside the general scope of this work, but nevertheless it has been considered appropriate to touch upon these questions in Chapters 3 and 4.

It should be remarked regarding the formal part of the presentation that the author has endeavoured to present the mathematical deductions in as simple a shape as possible. This has been done also when the area for the validity of the results has thereby been somewhat reduced. These reductions are, however, without any importance for the applications. For the use of some concepts, less well known to technicians than to mathematicians, illustrative explanations are given, for instance the *Stieltjes* integral concept. Finally it should be remarked that it has turned out to be advantageous in both definitions and deductions to replace probability concepts by the corresponding relative frequencies. This method should be taken purely as a matter of practicability. It does not mean that the author takes a stand regarding the question of the basic definition of the probability concept. However, there are many cases where a more illustrative picture of the probability quantities is in this way given to the layman. Thereby an understanding of the deductions will be facilitated.

Because of the special character of the subject treated in this work, it has proved necessary to introduce a number of new designations. It should be presumptuous to hope that the choice throughout can be considered successful. It was difficult to find adequate expressions in a number of cases.

TELEPHONE TRAFFIC AS A STOCHASTIC PROCESS

Chapter 2

General characteristics of the point processes

In order to model the arrival events in telephone traffic, point processes are considered that are characterized by the appearance of discrete points on the time axis. They should correspond to the call arrivals. The actual traffic problems in telephony are now mainly of such a nature that the point processes considered for their treatment can be specified as follows. For every point process, in relation to any arbitrarily chosen point of time, the process has gone on for an unlimited time and will continue for further unlimited time. Furthermore, all processes considered here are stationary in the sense that the probabilities that define them cannot show any unlimited continuing unilateral change as functions of absolute time. A more stringent definition of this kind of stationariness is given below. Finally we can generally assume that all points belonging to one and the same process are of the same kind. As a rule it is then not necessary to introduce a separation of the points into classes. In cases where such separation seems to be necessary, it is generally preferable to split the process into a number of cooperating processes. These will then each contain only points of the same kind.

Regarding the technical applications, it seems to be suitable in what follows to denote the point processes in question call arrival processes and, accordingly, to speak of the processes of call arrivals and not of their points. However, it should be observed that it is not a question of real traffic with real calls but of a mathematical model for describing the conditions that exist in reality.

A call arrival process may now be considered to be totally defined when the probability of a call arrival at any arbitrary point of time is known. It appears, however, that it is not necessary to know this probability. It is sufficient to assume that the probability exists. We need not have a closer description of it. The possibility then arises of stating a series of general propositions for the call arrival processes.

Consider now a call arrival process and, in it, a course of events consisting of $n + 1$ consecutive call arrivals with the conditions:

- the distance in time between the arrivals of the first and the second calls is between the limits x_1 and $x_1 + \Delta x_1$
- the distance in time between second and third call arrivals is between the limits x_2 and $x_2 + \Delta x_2$, etc., thus
- the distance in time between the call arrivals n and $(n + 1)$ is between the limits x_n and $x_n + \Delta x_n$.

With the point of time T as starting point we investigate the probability that an arbitrarily chosen call arrival, among the N next call arrivals, will start a course of the indicated kind. This probability is completely determined by the said definition of the process. It is generally a function of N and the point of time T. If, then, N is allowed to grow without limit, it may happen that the probability considered converges towards a definite limit. This is represented by

$$\mathfrak{F}_n (x_1, x_2, \ldots x_n) \triangle x_1 \triangle x_2 \ldots \triangle x_n \tag{1 a}$$

It is clear that this limit must be independent of the position of the point of time T, since between two call arrivals within a finite distance only a finite number of call arrivals can appear. For in Chapter 1 it is assumed that the probability of p or more call arrivals within a limited time tends towards zero with an infinitely growing p.

We now start from the same point of time T as before and investigate the probability that an arbitrarily chosen call among the next N call arrivals finishes a course of the indicated kind. It may then happen that also this probability converges towards a certain limit as $N \to \infty$. When the two limits exist and are equal, and this is valid for courses of every kind, i.e. also for all values of the parameters x_σ, Δx_σ and n, then the call process is described as *stationary*.

For all call processes treated below it is assumed that they are stationary in this sense.

In order to illustrate the probability (1 a) one can set up the following frequency representation for a stationary process. N consecutive calls are considered and the ratio is taken between the number of these calls that start a course of the kind considered, and the total number of calls N. When N grows infinitely, this ratio must converge in probability towards (1 a). Clearly, one need not choose those calls that begin courses of the indicated kind, but one gets the same result, naturally, if one takes the ratio between all calls that have a certain position in relation to a course of the above mentioned kind and the total number N of the calls. Thus (1 a) expresses also, for instance, the relative frequency of calls that are positioned p calls before the first call in a course of the indicated kind.

It immediately appears from the reflections concerning the relative frequency that (1 a) also expresses the probability that a call chosen at random starts a course of the considered kind or has a certain position in relation to this course.

We now let the quantities $\Delta x_1, \Delta x_2, \ldots, \Delta x_n$ all tend to zero and designate the limit value of the function $\mathfrak{F}_n (x_1, x_2, \ldots x_n)$ so derived by

$$F_n (x_1, x_2, \ldots x_n) \tag{1 b}$$

It can always be assumed that this limit is independent of the order of the limit transitions. It is not necessary, however, that the limit (1 b), which naturally is never negative, is limited in all points.

From what has been said above it follows that

$$F_n (x_1, x_2, \ldots x_n)\, dx_1\, dx_2 \ldots dx_n \tag{1 c}$$

expresses the probability that, after a call chosen at random, n calls follow that are so positioned that their mutual time distances are $x_1, x_2, \ldots x_n$ respectively. (In what follows this easily understandable and simpler form will be used instead of the more correct "intervals within the limits x_1 and $x_1 + dx_1$ etc.".) It is clear that (1 c) also expresses the probability that an arbitrarily chosen call finishes a course of the indicated kind or has any other determined position in relation to this course. The function (1 b) is called the *frequency function* and n the order of the course.

The probability that after an arbitrarily chosen call $n - 1$ calls follow that are so positioned that their mutual intervals are $x_1, x_2, \ldots x_{n-1}$ respectively, can clearly be derived by a summation of all probabilities (1 c) for all possible values of x_n. We get

$$\int_{x_n=0}^{\infty} F_n (x_1, x_2, \ldots x_{n-1}, x_n)\, dx_n = \\ = F_{n-1} (x_1, x_2, \ldots x_{n-1}) \tag{2}$$

For reasons of simplicity we assume that the frequency functions in all points are limited. This is always the case for the practical applications in this work. The integrals which occur are then always convergent. It should be noted that these limitations are not necessary but are made in order to simplify the presentation. The integrals occurring here can namely be transformed with the aid of *Stieltjes'* integral concept. Then no limitations are necessary in the respect mentioned. The presentation will, however, become more complicated formally.

When a frequency function of a course of a certain order is known, (2) makes it possible to calculate frequency functions of all courses of lower order. The calculation of frequency functions of higher order from functions of lower order does not seem to be generally possible. It is, however, possible to set up certain conditions that the frequency functions always must satisfy. These condition emerge from the above-mentioned fact that (1 c) expresses the probability of the occurrence of a certain course provided that either the first or the last call is chosen at random. Then, as is readily realized,

$$\int_{x=0}^{\infty} F_n (x, x_1, x_2, \ldots x_{n-1})\, dx = \\ = \int_{x=0}^{\infty} F_n (x_1, x_2, \ldots x_{n-1}, x)\, dx \tag{3}$$

Each of these integrals is equal to the frequency function of the right side of (2). For this function a similar condition can then be set. Totally n integral expressions are obtained which will satisfy each frequency function F_n. The nth expression so obtained is seen below in (4).

The frequency function of first order is of special interest. It is *the frequency function of the intervals between call arrivals* $F_1(x)$. The product $F_1(x)\, dx$ is the probability that the interval between an arbitrarily chosen call and the next one (or immediately preceding) is x. The sum of the probabilities of all these possibilities must be equal to 1. We obtain the expression

$$\int_0^{\infty} F_1 (x)\, dx = 1 \tag{4}$$

The probability that, after an arbitrarily chosen call, a time interval to the next call arrival follows that has at least the length t is clearly obtained by summing all $F_1(x)\, dx$ for $x \geqq t$. It will then be

$$\varphi (t) = \int_{x=t}^{\infty} F_1 (x)\, dx \tag{5 a}$$

Inversely

$$\frac{d\varphi (t)}{dt} = -F_1 (t) \tag{5 b}$$

The function $\varphi(t)$ is called *the distribution function of the intervals between call arrivals.* It is a function that never increases with increasing t. From the definition and from (4) it follows that $\varphi(0) = 1$. Further we see that $\varphi(\infty)$ must always equal zero.

We see from the definition of $\varphi(t)$ that this function also expresses the probability that the distance between a call chosen at random and the immediately preceding call is at least t. This is fully natural as $\varphi(t)$ gives the relative frequency of the intervals between call arrivals of a length of at least t.

The mean value of the number of calls per time unit is usually represented by y. The mean length between calls is then $1/y$ and is determined by

$$\frac{1}{y} = \int_0^{\infty} x\, F_1 (x)\, dx \tag{6 a}$$

From (5 a) it is now evident that

$$\int_{t=0}^{\infty} \varphi(t)\, dt = \int_{t=0}^{\infty} \int_{x=t}^{\infty} F_1(x)\, dx\, dt$$

As the integrand in the double integral can never be negative, the order of the integrations can be reversed.[15] We then obtain

$$\int_{t=0}^{\infty} \varphi(t)\, dt = \int_{x=0}^{\infty} \int_{t=0}^{x} F_1(x)\, dt\, dx = \int_{x=0}^{\infty} x\, F_1(x)\, dx$$

and thus

$$\frac{1}{y} = \int_{0}^{\infty} \varphi(t)\, dt \tag{6 b}$$

Generally we obtain from

$$\int_{t=0}^{\infty} t^{n-1} \varphi(t)\, dt = \int_{t=0}^{\infty} \int_{x=t}^{\infty} t^{n-1} F_1(x)\, dx\, dt$$

by reversal of the order of integrations

$$M_n = \int_{0}^{\infty} t^n F_1(t)\, dt = n \int_{0}^{\infty} t^{n-1} \varphi(t)\, dt \tag{7}$$

In the applications, especially in the compounding of traffic, a further function of greatest importance occurs, besides the frequency and distribution functions of the interval between call arrivals and is also associated with them. This function $\vartheta(t)$ expresses the probability that after *a point of time chosen at random* no call arrival will occur during the time t.

To derive this probability we can regard a time interval T and establish in which parts of it such points of time are present for which no call arrival occurs within the time t thereafter. The ratio between the sum of the lengths of these favourable time intervals and the total length of the time interval T then gives the probability that a point of time which is randomly chosen within the time interval T has the property in question. We then obtain the value of the function $\vartheta(t)$ by determining the limit of this ratio for an unlimited increase of T. In a stationary process this limit is obviously equal to the mean of the sums of the favourable time intervals referred to the time unit. This mean value is easy to determine. The favourable time intervals can only be present within call arrival distances of length at least t. On an average there are y call arrivals per time unit. Of these, on an average $y\,\varphi(t)$ have at least the length t. At a call arrival distance of length $x + t$ there is a region of length x that is favourable, so that the distances of the points of time in this region to the next call arrival are $\geqq t$. Now there are, on average, $y\,F_1(x + t)\,dx$ call arrival distances of length $x + t$ per time unit. Each of these distances contributes an amount x to the sum of favourable time intervals. On an average

$$y \int_{x=0}^{\infty} x\, F_1(x + t)\, dx$$

occurs per time unit as a favourable region. This expression obviously gives the wanted probability $\vartheta(t)$ immediately. In the same way as in the derivation of (6 a) and (6 b) we now obtain the simpler form

$$\vartheta(t) = y \int_{x=t}^{\infty} \varphi(x)\, dx \tag{8 a}$$

From (6 b) it now follows that $\vartheta(0) = 1$, as it should be. The function never increases with increasing t. By formation of the derivative we get

$$\frac{d\,\vartheta(t)}{dt} = -\,y\,\varphi(t) \tag{8 b}$$

From this we find that $y\,\varphi(t)\,dt$ expresses the probability that after a point of time chosen at random, the next call (or the immediately preceding) will be situated at the distance t.

The mean distance between a randomly chosen point of time and the next call arrival is obtained from

$$y \int_{0}^{\infty} t\, \varphi(t)\, dt \tag{9}$$

which with the help of the expression (7) can be written as $M_2/2\,M_1$. In Chapter 6, p. 53, it is shown that this is always $\geqq \frac{1}{2} M_1$ and that, even for an important class of distribution functions, the completely monotone functions, which most frequently occur in the sequel, the mean distance is $\geqq M_1$. Since a randomly chosen point of time always divides the distance between two call arrivals into two parts, it may seem paradoxical that the mean of the distances between a randomly chosen point of time and the next call arrival can be greater than the mean of the distances between two call arrivals. The explanation is, of course, that the probability that a randomly chosen point of time appears in the greater distances between call arrivals is greater than the probability that it appears in the shorter distances.

The function $\vartheta(t)$ can be described as *the distribution of the distance to the next call arrival* from a randomly chosen starting point. For the sake of simplicity the expression *distribution of the next call arrival* will be used in the sequel. From the above derivation it is clear, however, that $\vartheta(t)$ also expresses the probability that, in the time t *before* a randomly chosen point of time, no call arrival occurs.

Two functions of value for the applications has thus been derived from the frequency function of the distances between call arrivals, viz. the distribution of the distances between call arrivals and the distribution of the next call arrival. By corresponding integrations we can easily find similar functions and also general frequency functions of type (1 b). In this way we can present probabilities of courses that begin at a randomly chosen point of time, then exhibit call arrivals positioned at certain distances, and terminate with or without call arrivals. Such expressions will be derived in what follows insofar as they are needed. It will be unnecessary at this point to go further into these expressions, as their derivations do not essentially differ from what has been shown above.

We are sometimes interested in the probabilities of the appearance of certain courses together with other courses. Such conditional probabilities may, for example, have the following form. Among all occurring courses of a particular kind, a course is chosen by a random procedure and one seeks the probability of a course of another particular kind coming immediately thereafter. Such probabilities can generally be derived from the frequency functions.

The introduction of point processes of the general form that has been discussed above offers – at least in principle – good possibilities of the adaptation of the theoretical assumptions to the real conditions. The frequency functions may be supposed to be so chosen that they reproduce with any wanted degree of accuracy the properties that appear in measurements of the call conditions in telephone traffic. In order to get a complete description of the traffic, we need only find a suitable probability scheme that can also take into account the varying occupation durations and their possible connection with call arrival conditions. In principle this can be accomplished by defining the probabilities of different call arrival patterns in which not only the distance between call arrivals but also the length of the occupations following the call arrivals have definite values. In this way we could construct probability schemes that represent the measurable statistical properties with any wanted degree of precision. In practice, however, such a procedure for the treatment of traffic problems would be useless, not only because of the extreme complications that would arise in the mathematical treatment but also because the empirical determination of the frequency functions would involve great difficulties.

With the help of the methods that will be further developed in what follows, we can bring back the theoretical treatment of traffic problems to point processes which, according to certain fundamental rules, can be derived from the simple definite process. For this reason the point processes which occur here acquire a common mathematical structure which often makes the treatment easier. As regards the durations of the occupations it appears that we can generally be satisfied with a simple probability scheme which requires no connection with the positions of the call arrivals.

In the treatment of traffic problems two kinds of operations are mainly pertinent to the arrival processes in question. In one case a second process is derived from a given process by *selection* of certain calls. A *deformation* of the process may then occur due to a displacement of the calls according to certain rules. In the second case the properties of a process consisting of calls coming from several given processes will be studied. The derived process is then formed from the given ones by superposition. It appears now that for determination of the frequency functions of a process that is derived from another process by selection or deformation, we only need to know the frequency functions of the latter process. By the superposing of processes, however, it is generally impossible to derive the properties of the resulting process only from the frequency functions of the subprocesses. An arrival process is therefore not at all fully defined by its frequency functions. It is all the more remarkable that, for the solution of many important problems, we only have to know these functions. The process need not be defined in detail.

It seems to be appropriate to introduce a common designation for the probability densities of a process and of all the quantities that can be derived from it. We can call them, by way of suggestion, *elementary functions*. When the time axis has no firm division, when thus absolute coordinates of the call arrivals cannot be fixed, the elementary functions are obviously the only quantities that can be defined for an arrival process, inasmuch as no possibility exists to compare it with another simultaneous process. Besides, a fixed division of the time axis can be considered as an arrival process. The elementary functions are then the only quantities that can be defined in an arrival process that is not related to any other process.

The common structure just mentioned of most of the arrival processes in this work expresses itself in an occurrence of something that we suitably could present as *equilibrium points*. In such a point the probability of the further development of a course is independent of the positions of earlier call arrivals. The knowledge of the position of the equilibrium points offers considerable possibilities of simplifying the description of the processes. Occasionally, the equilibrium points coincide with a call arrival. If this is not the case, a comparison with other processes is necessary for the determination. Generally, the equilibrium points appear in regions where every point is an equilibrium point. These intervals are called equilibrium regions.

A special kinds of process, of great importance for further developments, constitutes the next generalization of the definite process. This process is characterized by the quality of each call arrival being an equilibrium point. That, however, does not exclude other points from also being equilibrium points. The probability of the further course after an *arbitrarily* chosen call arrival is then independent of the positions of the previous calls. Such processes can suitably be denoted *processes with limited after-effects*. We can justify this expression by the fact that for all other processes a kind of unlimited after-effect appears, since the probabilities of the conditions after a call arrival, which is not an equilibrium point, can be shown to be dependent on the position of a previous call which is unlimitedly far away.

If $F(x)$ is the frequency function of a process with limited after-effect then, because of the assumed independence the frequency function of the nth order will be

$$F_n(x_1, x_2, \ldots x_n) = F(x_1) F(x_2) \ldots F(x_n)$$

For a process with limited after-effect all elementary functions can obviously be derived from $F(x)$. A special kind of processes with limited after-effect is the *definite process*. It is characterized by the quality of every point on the time axis being an equilibrium point. For this distribution function we obtain the expression

$$\varphi(x + t) = \varphi(x)\,\varphi(t)$$

From this and from (6 b) follows that $\varphi(t) = e^{-yt}$. The frequency function becomes $y e^{-yt}$ and the distribution of the next call arrival e^{-yt}.

As has been said already, at the superposing of call arrival processes the elementary functions of the resulting process are generally not determined by the elementary functions of the subprocesses. There is, however, a very important exception from this rule: the case where the subprocesses do not have any mutual correlation. This case will here be further investigated.

If it appears that the probabilities, valid for the course of a process, are independent of the positions of the call arrivals of another simultaneous process, then the first process will be said to be *independent* of the latter. We now look at N simultaneous processes that are all independent of one another. We say that the process that is built up by the call arrivals that come from the N given processes originates from *random superposing* of the original processes.

The distribution function for the next call arrivals of the given processes are $\vartheta_1(t)$, $\vartheta_2(t)$, ... $\vartheta_N(t)$. The distribution function for the next call arrival of the resultant process, that is the probability for the event that in none of the given processes during the time t after a randomly chosen point of time a call arrives, is then because of the assumed independence

$$\Theta(t) = \vartheta_1(t)\,\vartheta_2(t) \ldots \vartheta_N(t) \tag{10}$$

Thus in order to derive the distribution of the next call arrival for the resulting process and thereby also the frequeny function and the distribution function for the distance between call arrivals, we need to know only the corresponding functions of the original process. This is valid in a similar way also for courses of higher order. Because of the assumed independence, we can always directly multiply the probabilities of different courses of the subprocesses that start at arbitrarily chosen points of time. We can then easily see how we can in prinicple determine the frequency functions for the courses of higher order.

It is obviously to be expected that the mean value per time unit of the number of call arrivals of the resulting process equals the sum of the corresponding quantities of the subprocesses. This follows immediately from (10). The general relation $\vartheta'(0) = -y$ can be obtained from (8 b). By forming the derivative of (10) we get

$$\Theta'(t) = \Theta(t)\left\{\frac{\vartheta_1'(t)}{\vartheta_1(t)} + \frac{\vartheta_2'(t)}{\vartheta_2(t)} + \ldots + \frac{\vartheta_N'(t)}{\vartheta_N(t)}\right\}$$

If we here insert $t = 0$, we get after insertion of notations that are easy to understand

$$Y = y_1 + y_2 + \ldots + y_N.$$

We consider then the case when the number of subprocesses N increases without limit, while all mean values of the call rates y_σ approach zero in the way that their sum Y always has a finite value different from zero. To derive the resulting limit we write the distribution function for the next call arrival for the subprocess in the form

$$\vartheta_\sigma (t) = 1 - y_\sigma t + y_\sigma^2 K_\sigma (t, y_\sigma)$$

Because of (8 b) we obtain

$$\varphi_\sigma (t) = 1 - y_\sigma K_\sigma' (t, y_\sigma)$$

It is then assumed that for all subprocesses

$$\lim_{y_\sigma \to 0} K_\sigma' (t, y_\sigma)$$

has a limited value for all finite values of t. This assumption obviously means that the probability of the distances between call arrivals being of limited length approaches zero when the mean call rate approaches zero. When now the derivative of $K_\sigma (t, 0)$ in respect to t is limited for all limited values of t then within this region $K_\sigma (t, 0)$ cannot increase from a limited to an unlimited value. Now $K_\sigma (0, 0) = 0$ for example, so that $K_\sigma (t, 0)$ must always be limited for all limited values of t.

As the probability of the call arrival distances that are greater than an arbitrary limited number t never can be zero when y_σ, the mean number of calls per time unit approaches 0, then for limited t-values and sufficiently small y_σ-values we must always have $\varphi_\sigma (t) > 0$ and hence also according to (8 a) $\vartheta_\sigma (t) > 0$. Because of that we have the inequality

$$y_\sigma t - y_\sigma^2 K_\sigma (t, y_\sigma) < 1$$

where the left side can of course never be negative. We can then develop $ln\ \vartheta_\sigma (t)$ in terms of powers of the left side of the inequality. We then obtain a series of the form

$$ln\ \vartheta_\sigma (t) = - \{y_\sigma t - y_\sigma^2 K_\sigma (t, y_\sigma)\} - $$
$$- \frac{1}{2} \{y_\sigma t - y_\sigma^2 K_\sigma (t, y_\sigma)\}^2 + \ldots$$

From (10) we then obtain

$$ln\ \Theta (t) = - (y_1 + y_2 + \ldots + y_N) t - R$$

where the residual term R only contains powers of the different y_σ of higher degree than the first with coefficients limited after the above assumption also when the different $y_\sigma \to 0$.

If we now let all $y_\sigma \to 0$ and simultaneously $N \to \infty$ and then in the way that $Y = y_1 + y_2 + \ldots + y_N$ retains a constant, non-zero, limited value then the limit of the residual term R tends to zero and we obtain

$$\Theta (t) = e^{-Yt}$$

That the distribution function for random superposing converges towards the simple exponential function is naturally the reason for the fact that for investigation of the call arrival processes this function has a dominant position. Indeed this corresponds to the basic importance of the normal distribution in the theory of errors and in other fields. The reason for the different forms of the distributions in the cases mentioned is to be found in the different ways in which superposing occurs. In the case of problems that lead to the *Gaussian* distribution function the superposition is additive, while in all arrival processes we have to deal with a superposition rule of factorial form (10). In the treatment of superpositions of additive form we usually change from the distribution functions to the characteristic functions, whereby the superposition rules take the factorial form. As this form is here already present in the distribution functions, the characteristic functions have not the same importance for the theory.

In the above derivation it was assumed that all $y_\sigma \to 0$. This is, of course, necessary in order that none of the subprocesses shall dominate. This condition is comparatively trivial and corresponds to a fully analogue condition in additive superposition. In the present case this assumption is not sufficient but must be supplemented by the condition that the probability of the distances between call arrivals being of limited length approaches zero when $y_\sigma \to 0$. This is not a condition of a merely formal kind. Further on we shall show that processes that do not satisfy this condition are of great importance for the treatment of many problems. We can say about these processes that they converge towards *clustering*. By this is meant that at decreasing call intensity, the call arrivals cluster in certain regions that are separated from each other by long intervals. Only in this way, can we explain why the probability of comparatively small distances from zero still maintain values differing from zero although the mean of the call arrival distances becomes increasingly greater. This explains also why such processes must not be present if the distribution of the resulting processes are to converge to an exponential function. A clustered process can namely in some regions gain a dominating position compared with the other subprocesses.

Above was shown that, under certain conditions with random superposition, the distribution function of the resulting process for the distance to the next call arrival and thereby, according to (8 b), also the distribution function for the distance between call arrivals converge towards an exponential function. From that, however, it does not necessarily follow that the resulting process converges towards a definite process. To do so, the probability of each course would have to be independent of the preceding course as well as of the conditions of every other process. That this really is the case can be proved in the following way. We look at a course in the resulting process that contains n call arrivals. The calls arriving within a limited time T before the course must then belong to subprocesses that are different from the processes that contain the n call arrivals. The probability that two call arrivals appear from one and the same subprocess within a limited time interval is zero according to the conditions. As the subprocesses are independent of each other, the positions of the n call arrivals must be independent of the positions of the preceding call arrivals during the time T. This is valid for every limited, arbitrarly great value of T.

This reasoning shows clearly the basic importance of the assumption that the subprocesses must not be clustered. It remains then to be shown that the resulting process cannot be dependent upon any other process. If such dependence exists, an unlimited number of subprocesses must also be dependent on this other process. This is not possible however, as these subprocesses cannot be mutually independent. This reasoning can be considered as sufficiently binding. It could probably without any great difficulty be deduced mathematically but this might be rather cumbersome.

Chapter 3

Blocking of traffic with limited after-effects

In the attempts made below to use the theories of point processes for problems regarding telephone traffic it seems appropriate, in considering the technical applications, to speak of traffic instead of processes. It must be noted, however, that it is not a question of any traffic in reality but of mathematical presentations by which attempts are made to describe the statistical properties of real traffic. In what follows, therefore, we shall talk about traffic with limited after-effects and definite traffic. For the latter the usual designation *random traffic* is often chosen.

The traffic concept now comprises not only the call arrivals but also the call durations. Accordingly, in this work, the expression traffic refers not only to the process constituted by call arrivals but also to the probability scheme for the duration of calls. It is here assumed that call durations are always independent of the actual times of the call arrivals. The former can then be treated without any close connection with the latter. When a call cannot be dealt with because of the conditions of the group of devices in question, that is, the wanted connection cannot be established, then the call duration is zero. The occurrence of such an event is, however, dependent upon the position of the previous call and thus contradicts the assumption that call durations are independent of call arrivals. In order to overcome this difficulty, such calls will be considered as nonexistent. Consequently the probability scheme of call durations should not comprise call durations which have the value zero. This is of the same significance as the assumption that the calls in question totally disappear or cause occupation of devices that do not belong to the group considered.

The concept *traffic intensity* was defined above as the average number of simultaneous occupations in a group during a certain time. This definition has the disadvantage that it is tied to the group of devices considered. It is possible, however, to introduce a more general definition of the traffic intensity that is more suitable for common use. The traffic intensity is then defined as the product of the mean of the number of calls during a certain time and the mean of the occupation times, the occupations of zero duration that correspond to blocked calls thus not being included in these occupation times. The advantage of this definition is that it permits the introduction of certain traffic quantities that are fictitious in the sense that they correspond to no direct physical reality but, however, possess a great theoretical and practical importance.

The simplest example of dealing with traffic in a telephone plant is offered by a *full availability group* in a *loss system*. The handling of the traffic in such a group can be described in the following way. The *group* consists of a number of devices n. The starting point is a call arrival process that on the average has y calls per unit time corresponding to the *incoming traffic* to the group. When a call arrives i.e. *comes in to the group* or hunts for the group, an *occupation* is established at the same time. As long as this occupation lasts, the device is *busy*. When the occupation ceases and the device becomes free it can be seized by another incoming call. The group will be defined as a *full availability group* if each call will cause an occupation of one of the devices, provided all devices are not already busy. When this is the case we say that the group is *blocked* (busy) for incoming calls. We say that the group is arranged as a *loss system* when the blocked calls in themselves cause no occupation. This means that, when the blocked calls cause no occupation of a device in another group, the subscribers must repeat their call attempts in order to obtain a connection. The blocked calls constitute a call process that is designated *traffic blocked for the group*. The calls that cause occupations in the group constitute a process that is designated *outgoing or carried traffic* in the group. One also speaks of the *traffic handled* in the group.

The mean duration of the occupations in the group is s. The product of s and the mean number per unit time of the call arrivals which are not blocked is designated the *traffic carried* by the group. The product of s and the mean number per unit time of calls blocked is designated the *blocked traffic*. This is an example of a fictitious traffic quantity of the above-mentioned type that is of great importance for estimating the traffic capacity of the group. Another traffic quantity that is usually at least partly fictitious is the product $s\,y$ where y stands for the mean per unit of time of all incoming calls to the group. This traffic will be designated as *offered* or *incoming* to the group. Clearly it must always be equal to the sum of the carried and blocked traffic.

The *delay system* is the opposite of the busy system. In a group that works as a delay system the blocked calls, that is those calls that arrive when all devices are occupied, can cause occupation of devices at a later time when free devices are available.

To find the blocked traffic for a full availability group in a loss system is one of the simplest, but at the same time the most important, tasks for traffic research. This problem was solved for the first time by *Erlang*[3] on condition that the incoming traffic was random (pure chance) and that the distribution of the occupation times follows a simple exponential function. Later it was proved that the same result is valid regardless of the distribution function of the occupation times, provided however that this distribution is independent of the position of the calls.[9] In this chapter the problem will be reconsidered. By introducing in the analysis the process with limited after-effects a far more penetrating description of the traffic conditions is possible. This offers the possibility of treating more complicated problems relating to the traffic conditions in groups where the incoming traffic consists of blocked or outgoing traffic from other groups.

For a full availability group with n devices in a loss system with incoming pure chance traffic and a traffic intensity $A = sy$, the mean fraction of the total time when all devices are occupied is given by the *Erlang* formula

$$E_{1,n}(A) = \frac{\dfrac{A^n}{n!}}{1 + \dfrac{A}{1!} + \dfrac{A^2}{2!} + \ldots + \dfrac{A^n}{n!}} \qquad (11)$$

(the index 1 indicates that the formula applies to loss systems). The quantity $E_{1,n}$ is often called *group busy time*. The amount of blocked traffic is $A\,E_{1,n}$. The formula (11) is valid independent of the manner of *hunting* (occupation pattern) for the devices in the group, by which is meant the rules governing the seizure of free devices by the calls. This selection often occurs more or less at random, but the devices in the group are sometimes so arranged that they are always hunted for in the same order. If the devices are numbered in order 1, 2, ... n, *sequential hunting* means that each call occupies the free device with the lowest number. It is evident that with sequential hunting the r lowest devices, r being an arbitrary chosen number not greater than n, together constitute a full availability group arranged as a loss system. The blocked traffic for this subgroup is then $A\,E_{1,r}$. Further, each device in itself can be considered as a full availability group with an incoming traffic $A\,E_{1,r-1}$, where r is the number of the device. The blocked traffic quantity is $A\,E_{1,r}$. The traffic quantity that is carried by the device, *the occupation of the device*, is then $A(E_{1,r-1} - E_{1,r})$. From (11) we can now derive the recursive formula

$$E_{1,r} = \frac{A\,E_{1,r-1}}{r + A\,E_{1,r-1}} \tag{12}$$

This shows that the traffic that a device can carry (for the same incoming traffic) becomes smaller the higher the number, r, of the device in the group. Later it will be shown that this decrease, which in some cases can be very considerable, is due to the fact that the incoming traffic is no longer random. The departures from the properties of random traffic are all the more pronounced, the more devices the incoming traffic has previously passed.

In this chapter only such traffic will be considered for which the occupation times are independent of the previous occupation times and of the position of the previous call arrivals, and for which the probability that an arbitrary occupation continues at least for a time t is expressed by

$$e^{-\frac{t}{s}}$$

Here s is obviously the mean occupation time. Because of the assumed exponential distribution of the occupation times, the probability that an occupation which has already lasted for a time x will continue at least for a further time t is likewise

$$e^{-\frac{t}{s}}$$

and therefore independent of the value of x. Introduction of this condition seems to be necessary in order that the call arrival process of the traffic in the group can be expressed in a fairly simple mathematical form. On the other hand it was already known that the exponential distribution of the occupation times is not a necessary condition for the validity of certain results, namely those concerning the mean values of states and occupations in loss systems. In spite of the special character of this condition, the investigations based thereon are important for judging the cases with an arbitrary distribution of occupation times.

We now consider a full availability group of n devices in a loss system where the hunting goes in a determined order, the devices being numbered 1, 2, ... n. It is assumed that the traffic arriving in the group has an arrival process with limited after-effects. The probability of a call arrival can thus be dependent upon the interval from the immediately preceding call but independent of other previous conditions. It is further assumed that at a certain point of time T_0 a call to the group is blocked. At this point of time all devices must then be occupied. The probability that during a certain time thereafter another call is blocked is then dependent upon two circumstances: firstly whether further calls arrive in the group or not and secondly how long the n occupations at the point of time T_0 continue. As it was assumed that the arriving traffic has limited after-effects, the distribution of the arrivals of calls to the group after the time T_0 is independent of the conditions prevailing before that time. As it was also assumed that the occupation times have an exponential distribution, the probability of their further duration is independent of how long they have lasted before the point of time T_0. Consequently the probability that a blocked call does not occur during a certain time after the point of time T_0 (that is, after the immediately preceding blocked call) is independent of the conditions prevailing before that time. This means that the traffic that is blocked for the group has limited after-effects.

As each particular device can be considered as a group in itself, we realize that the arriving as well as the blocked traffic of each device are processes with limited after-effects. By similar reasoning we can easily prove that the traffic handled by each device also has limited after-effects.

This fact, that all traffic in a group with full availability in a loss system has limited after-effects provided the arriving traffic has a limited after-effect and the occupation times have exponential distribution, is the reason why the traffic conditions can be determined comparatively easily. It should be remarked that we need not restrict our attention to the special case in which the incoming traffic is fully random in spite of the fact that the conditions in these cases will be very much simplified.

As all elementary functions in traffic with limited after-effects are determined by the distribution function of the intervals between call arrivals, it is sufficient to investigate this function in the present case. The distribution function for the interval between call arrivals for traffic blocked by the rth device, which is identical with the traffic offered to the $(r + 1)^{\text{th}}$ device, is $\varphi_r(t)$. The distribution function for the interval between call arrivals for the whole group is $\varphi_0(t)$ and the distribution function for the intervals between call arrivals of the blocked traffic for the whole group is $\varphi_n(t)$. The mean values of the call intensities of the corresponding traffic flows are represented by y_r.

We now consider the rth device at the point of time T_0 at which a call to this device is blocked. At this point of time the device must therefore be busy and a call must arrive at the device. The probability $\varphi_r(t)$ is sought that during the time t after the point of time T_0 no new call to the device is blocked. This can only happen in the following two cases which are mutually exclusive.

1 No new call to the device arrives during the time t considered.

2 During the time t at least one call to the device will arrive. However, no call to the device is blocked.

The probability of case 1, viz. that during the time t after a call arrival no new call arrival occurs, is $\varphi_{r-1}(t)$. The probability of case 2 can be deduced in the following way. We assume that the first call that hunts for the device after the point of time T_0 arrives at the point of time x after the point of time T_0. The probability of this event is obviously $-\varphi'_{r-1}(x)\,dx$ according to the general expression (5 b) from Chapter 2. In order not to have the call blocked, the device must be free at that point of time. The condition for this is that the preceeding occupation at the point of time T_0 has terminated during the time x. The probability of this is

$$1 - e^{-\frac{x}{s}} \quad \text{since} \quad e^{-\frac{x}{s}}$$

is the probability that the occupation still continues after the time x. The probability that a call to the device arrives at the point of time x after T_0 and then not only arrives but is also not blocked will then be

$$-(1 - e^{-\frac{x}{s}})\,\varphi'_{r-1}(x)\,dx$$

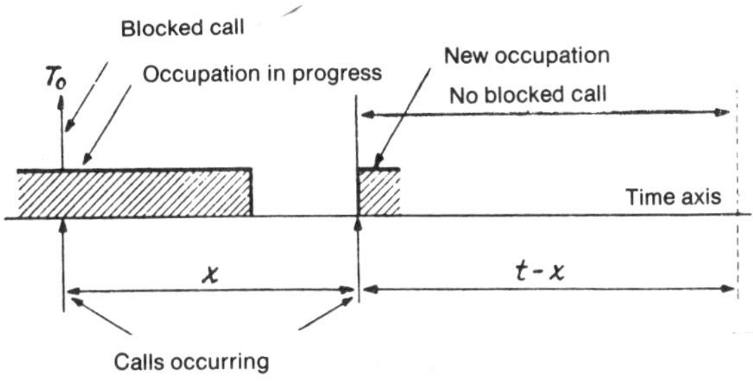

Fig. 1. Illustration of case 2 in the deduction of the integral equation (13).

For case 2 to occur, it is further necessary that no call to the device is blocked during the subsequent time $t-x$. The probability for this is $\varphi_r(t-x)$. It is independent of the probabilities deduced above and can be derived in the following way. According to the definition $\varphi_r(z)$ is the probability that, during the time z after a blocked call, no call to the rth device is blocked. $\varphi_r(z)$, however, also expresses the probability that, during the time z after an *incoming* call, no call to the device will be blocked. As the further duration of an occupation is independent of its previous duration, the conditions after a call arrival are independent of whether this call will be blocked or will cause a new occupation of the device.

Fig. 1 shows the conditions for a device according to case 2.

The probability of case 2, when the first arriving call occurs at the point of time x after the point of time T_0, will then be

$$-\left(1-e^{-\frac{x}{s}}\right)\varphi'_{r-1}(x)\,\varphi_r(t-x)\,dx$$

Case 2 will occur, however, for any value of $x < t$. The probability that case 2 will occur is then

$$-\int_{x=0}^{t}\left(1-e^{-\frac{x}{s}}\right)\varphi'_{r-1}(x)\,\varphi_r(t-x)\,dx$$

The probability $\varphi_r(t)$, that during the time t after a blocked call no other blocked call will occur, is equal to the sum of the probabilities of cases 1 and 2. We then get

$$\varphi_r(t)=\varphi_{r-1}(t)-\\ -\int_{x=0}^{t}\left(1-e^{-\frac{x}{s}}\right)\varphi'_{r-1}(x)\,\varphi_r(t-x)\,dx \qquad (13)$$

This integral equation is of basic importance for the study of the traffic conditions in a full availability group in a loss system. It contains in itself the *Erlang* formula but also offers the possibility of arriving at much more general conclusions on many different courses.

By similar reasoning we can now also deduce an integral equation for the distribution function $\omega_r(t)$ of the distances between the starting points of the occupation times of the rth device, that is, the distribution function for the distances between call arrivals for the traffic that is handled by the device. As has just been shown, this traffic has also limited after-effects. The function $\omega_r(t)$ expresses the probability that, during the time t after a call arrival leading to the occupation of the device, no new occupation occurs. In the same way as before for $\varphi_r(t)$ it can be shown that $\omega_r(t)$ also gives the probability for the case that no new occupation will occur during the time t after a blocked call. We then assume that at the point of time T_0 a call will arrive, causing an occupation of the device. The condition that during the subsequent time t the device will not again become busy will occur in two different mutually exclusive cases:

1 No new call will arrive during the time t. The probability of this is $\varphi_{r-1}(t)$.

2 At least one call to the device arrives during the time t but no call will cause a new occupation of the device. During the duration of the occupation starting at the point of time T_0, at least one call must arrive and be blocked. We assume that this will occur at the point of time $x < t$ after T_0. During the subsequent time $t-x$ the device should not be occupied again, the probability of this being,

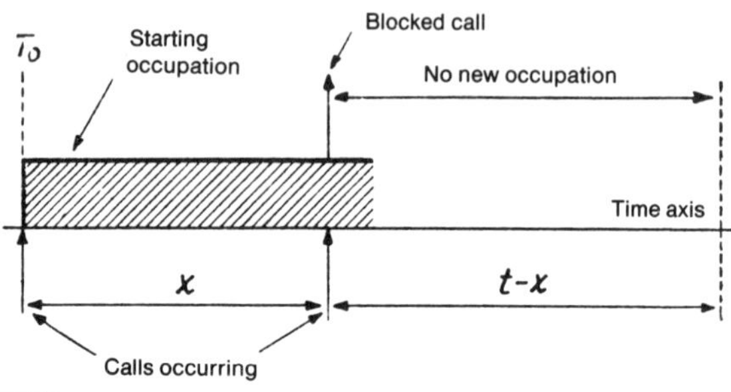

Fig. 2. Illustration of case 2 in the deduction of the integral equation (14).

$\omega_r(t-x)$ as was just mentioned. The probability of the whole process described is then obviously

$$-e^{-\frac{x}{s}}\varphi'_{r-1}(x)\,\omega_r(t-x)\,dx$$

The conditions in case 2 are shown in Fig. 2. Case 2 can now occur for any value of x between 0 and t so that the probability of case 2 can be obtained by integration between these limits of the probability deduced above. But $\omega_r(t)$ is equal to the sum of the probabilities of cases 1 and 2. Consequently the following equation is obtained:

$$\omega_r(t) = \varphi_{r-1}(t) - \\ -\int_{x=0}^{t} e^{-\frac{x}{s}}\varphi'_{r-1}(x)\,\omega_r(t-x)\,dx \qquad (14)$$

The mean value per unit time of the number of occupations of the rth device is c_r. Because of the general relation (6 b) in Chapter 2, it follows that

$$\frac{1}{c_r} = \int_0^\infty \omega_r(t)\,dt \qquad (15\text{ a})$$

For the mean value per unit time of the number of calls to the rth device that are blocked we get the relation in an analogous way

$$\frac{1}{y_r} = \int_0^\infty \varphi_r(t)\,dt \qquad (15\text{ b})$$

If we now integrate (13) we get

$$\int_{t=0}^{\infty}\varphi_r(t)\,dt = \int_{t=0}^{\infty}\varphi_{r-1}(t)\,dt - \\ -\int_{t=0}^{\infty}\int_{x=0}^{t}(1-e^{-\frac{x}{s}})\,\varphi'_{r-1}(x)\,\varphi_r(t-x)\,dx\,dt$$

As the integrands in the double integral are never negative, the order of the integrations can be reversed. We then get

$$\frac{1}{y_r} = \frac{1}{y_{r-1}} - \\ -\int_{x=0}^{\infty}\int_{z=0}^{\infty}(1-e^{-\frac{x}{s}})\,\varphi'_{r-1}(x)\,\varphi_r(z)\,dz\,dx$$

Here it is possible to integrate directly in respect of z, getting after some manipulation

$$y_r = -\,y_{r-1}\int_0^\infty e^{-\frac{x}{s}}\varphi'_{r-1}(x)\,dx \qquad (16)$$

By similar integration from (14) we get

$$c_r = y_{r-1} + y_{r-1}\int_0^\infty e^{-\frac{x}{s}}\varphi'_{r-1}(x)\,dx$$

This expression, together with (16), gives the self-evident relation $y_r + c_r = y_{r-1}$. In this way we can immediately compute the blocked and the handled traffic from the distribution function of the arriving traffic. The interesting relation (16) can also be deduced immediately by the following reasoning. Independently of whether a call is blocked or leads to the occupation of the device, the device should be busy immediately after the arrival of the call. The next arriving call will be blocked if this occupation is still in force at the arrival of the new call. The probability of this, as is readily realized, is:

$$-\int_0^\infty e^{-\frac{x}{s}}\varphi'_{r-1}(x)\,dx$$

Now any incoming call can be considered as such a next call. On the average there are y_{r-1} calls per time unit and the product of this mean number and the above integral must then give the mean number of blocked calls per time unit, i.e. is y_r. From that we obtain the relation (16).

For any device in a group with full availability in a loss system with sequential hunting, the integral relation has the form (13). If the group consist of n devices, then the blocked traffic for the whole group is identical with the blocked traffic for the nth device. To investigate the blocked traffic as a function of the incoming traffic, there exists a system of n integral equations. Before making a more systematic investigation of this system we shall illustrate one of its particular properties which is of great interest. If we differentiate (13) with respect to t, we get after some manipulation

$$\varphi_r'(t) = e^{-\frac{t}{s}} \varphi_{r-1}'(t) -$$
$$-\int_{x=0}^{t} (1 - e^{-\frac{x}{s}}) \varphi_{r-1}'(x) \varphi_r'(t - x) \, dx$$

For $t = 0$ we then get

$$\varphi_r'(0) = \varphi_{r-1}'(0)$$

and consequently

$$\varphi_n'(0) = \varphi_r'(0) = \varphi_0'(0) \qquad (17\text{ a})$$

The slope of the distribution at the point $t = 0$ is therefore equal for all blocked traffic and also equal to the slope of the distribution function for the arriving traffic. We now look at the case when the group is unlimited. Then, obviously, $y_r \to 0$ as $r \to \infty$, which will moreover be shown below on p. 24 in conjunction with the discussion of the set of formulas (24). If now $\varphi_0'(0) \neq 0$, which is trivial, the probability of call arrival intervals, with a limited value but differing from zero, does not converge towards zero because of (17 a), however great r may be. This means that with increasing r the blocked traffic converges towards clustering (regarding this concept see Chapter 2).

The second derivative at the point $t = 0$ also deserves attention. One readily realizes that

$$\varphi_r''(0) = \varphi_{r-1}''(0) - \frac{1}{s} \varphi_{r-1}'(0)$$

From this expression, because of (17 a), we obtain the general relation

$$\varphi_r''(0) = \varphi_0''(0) - \frac{r}{s} \varphi_0'(0) \qquad (17\text{ b})$$

If the incoming traffic is definite and has the distribution function $\varphi_0(t) = e^{-y_0 t}$ we get the formulas

$$\varphi_r'(0) = -y_0$$
$$\varphi_r''(0) = y_0^2 + \frac{r}{s} y_0$$

That the blocked traffic tends more and more towards clustering, the higher up in the group we go, is a not unexpected fact considering the hunting mechanism. On the occurrence of a blocked call, all devices of lower order are busy, and as long as this state continues, the intensity of the blocked calls is equal to the intensity of the call arrivals to the group (compare (17 a)). If a state of blocking is present, it has a self-sustaining tendency because of the arrivals of new calls. However, during the time interval when the devices of lower order are only partly busy, the probability of a rapid occupation of all devices is small.

The derivatives of the distribution function of the traffic handled in the zero point also deserve to be considered. By differentiating (14) we get

$$\omega_r'(0) = 0$$
$$\omega_r''(0) = \frac{1}{s} \varphi_0'(0)$$

If the incoming traffic is definite, the latter expression becomes

$$\omega_r''(0) = -\frac{y_0}{s}$$

That $\omega'_r(0)$ is always zero does not imply anything unexpected, because the occurrence of a new occupation presupposes two subsequent events, the ending of the first occupation and a new call arrival thereafter. That the second derivative always differs from zero shows that also the traffic carried tends towards clustering when $r \to \infty$, which can be explained in the same way as for the blocked traffic.

In order to study more closely, the solution to the set of integral equations that is obtained from (13) for $r = 1, 2, \dots n$, a series of functions is introduced, defined by

$$\psi_r(t) = \frac{1}{s}\int_{x=0}^{\infty} e^{-x\frac{t}{s}} \varphi_r(x)\, dx \qquad (18)$$

This gives an integral relation between $\psi_r(t)$ and $\varphi_r(t)$ the so-called *Laplace transform.* The function $\varphi_r(t)$ is never negative and $\varphi_r(t) \leqq \varphi_r(0)$ is always the case. As the integral (13) for $t = 0$ always equals $1/sy_r$ and is thus convergent, $\psi_r(t)$ is then always a limited non-negative function. If we differentiate (18) μ times we get

$$\psi_r^{(\mu)}(t) = \frac{1}{s}\left(\frac{-1}{s}\right)^{\mu}\int_{x=0}^{\infty} x^{\mu} e^{-x\frac{t}{s}} \varphi_r(x)\, dx$$

The μth derivative of $\psi_r(t)$ has then the sign $(-1)^{\mu}$. A function wih this property is called *completely monotonic.* This implies that the absolute values of the function itself as well as all its derivatives never increase.

Concerning the inversion of the Laplace transform a number of theorems are known (see the account in Chapter 10). Of those the following is interesting in this context. If a function $\psi_r(t)$ is given and furthermore we know that there exists a function $\varphi_r(t)$ that satisfies (18) and is limited and never negative within the interval of the integral, then this is the only limited non-negative function that can satisfy (18). For the determination of this function it is sufficient to know the values of $\psi_r(t)$ for $t = 1, 2, \dots \infty$.

The expression (13) can then be put in the form

$$\varphi_r(x) = \varphi_{r-1}(x) - \int_{z=0}^{x} (1 - e^{-\frac{z}{s}})\, \varphi'_{r-1}(z)\, \varphi_r(x - z)\, dz$$

If both sides are multiplied by

$$\frac{1}{s} e^{-x\frac{t}{s}}\, dx,$$

and we perform an integration between $x = 0$ and ∞ we get, considering the definition (18), the equation

$$\psi_r(t) = \psi_{r-1}(t) - \frac{1}{s}\int_{x=0}^{\infty}\int_{z=0}^{x} (1 - e^{-\frac{z}{s}})\, e^{-x\frac{t}{s}} \varphi'_{r-1}(z) \varphi_r(x - z)\, dz\, dx$$

By reversal of the order of the integrations in the double integral we get

$$\psi_r(t) = \psi_{r-1}(t) - \frac{1}{s}\int_{z=0}^{\infty} (1 - e^{-\frac{z}{s}})\, e^{-\frac{z}{s}t} \varphi'_{r-1}(z) \int_{x=z}^{\infty} e^{-\frac{x-z}{s}t} \varphi_r(x-z)\, dx\, dz$$

The integral over x can be carried out directly according to (18) and we get

$$\psi_r(t) = \psi_{r-1}(t) - \psi_r(t)\int_{z=0}^{\infty} (1 - e^{-\frac{z}{s}})\, e^{-\frac{z}{s}t} \varphi'_{r-1}(z)\, dz$$

But now, obviously,

$$\int_0^{\infty} e^{-\frac{z}{s}t} \varphi'_{r-1}(z)\, dz =$$

$$= \left[e^{-\frac{z}{s}t} \varphi_{r-1}(z) \right]_0^{\infty} + \frac{t}{s}\int_0^{\infty} e^{-\frac{z}{s}t} \varphi_{r-1}(z)\, dz =$$

$$= -1 + t\, \psi_{r-1}(t)$$

and in the same way

$$\int_0^{\infty} e^{-\frac{z}{s}(1+t)} \varphi'_{r-1}(z)\, dz = -1 + (1 + t)\, \psi_{r-1}(1 + t)$$

Finally we get

$$\psi_r(t) = \psi_{r-1}(t) - t\,\psi_r(t)\,\psi_{r-1}(t) + \\ + (1+t)\,\psi_r(t)\,\psi_{r-1}(1+t)$$

and from that

$$\psi_r(t) = \frac{\psi_{r-1}(t)}{1 + t\,\psi_{r-1}(t) - (1+t)\,\psi_{r-1}(1+t)} \tag{19}$$

By this recursive formula all $\psi_r(t)$ are unambigously determined by $\psi_0(t)$, which is obtained from the distribution function $\varphi_0(t)$ of the incoming traffic to the group. As all $\varphi_r(x)$ are now defined as probabilities that are limited non-negative functions it follows, from what has been said about the inversion of the *Laplace* transform, that if the system (13) has a solution with these properties, this solution is determined by (19).

We can by an analogous method determine the solution of the integral equation (14) for the distribution function of the traffic carried $\omega_r(t)$. The auxiliary function

$$\pi_r(t) = \frac{1}{s}\int_0^\infty e^{-x\frac{t}{s}}\,\omega_r(x)\,dx \tag{20}$$

is introduced. We then get in a similar way as before

$$\pi_r(t) = \psi_{r-1}(t) - \\ -\frac{1}{s}\int_{x=0}^{\infty}\int_{z=0}^{x} e^{-\frac{z}{s}}\,\varphi_{r-1}(z)\,e^{-x\frac{t}{s}}\,\omega_r(x-z)\,dz\,dx$$

After reversing the order of integration, both integrations can be carried out. After a minor manipulation we get

$$\pi_r(t) = \frac{1}{1+t}\cdot\frac{\psi_{r-1}(t)}{\psi_{r-1}(1+t)} \tag{21}$$

Thus the study of the distribution of the traffic carried by a device can be referred back in a simple way to the study of the distribution function for the incoming traffic to the device.

In order to investigate the distribution function $\varphi_n(x)$ of the blocked traffic of a group as a function of the distribution function $\varphi_0(x)$ of the incoming traffic to the group, we can now consider expressing $\psi_n(t)$ directly in terms of $\psi_0(t)$ using the recursive formula (19). It should then be observed that by multiplying both sides in (19) by t, a recursive formula with a simpler form is obtained for the function $t\,\psi_r(t)$. However, formally it is somewhat simpler to work with a series of constants defined by

$$\left.\begin{aligned} \beta_{r,0} &= \frac{1}{s}\int_0^\infty \varphi_r(x)\,dx \\ \beta_{r,\mu} &= \frac{\mu}{s}\int_0^\infty e^{-\mu\frac{x}{s}}\,\varphi_r(x)\,dx \end{aligned}\right\} \tag{22}$$

For $\mu > 0$ the constants are obtained by inserting $t = \mu$ in the function $t\,\psi_r(t)$. It has been remarked above that $\varphi_r(x)$ is unambiguously determined by the *Laplace* transform solely from the values of $\psi_r(t)$ for $t = 1, 2 \ldots$ It will moreover appear that the formulas obtained for the constants (22) differ only formally from the corresponding ones for the functions $\psi_r(t)$.

The expression (19) now gives the recursive formula for the constants (22)

$\underline{\mu = 0}$

$$\beta_{r,0} = \frac{\beta_{r-1,0}}{1 - \beta_{r-1,1}} \tag{23 a}$$

$\underline{\mu > 0}$

$$\beta_{r,\mu} = \frac{\beta_{r-1,\mu}}{1 + \beta_{r-1,\mu} - \beta_{r-1,\mu+1}} \tag{23 b}$$

The blocked traffic from the rth device $s\,y_r$ according to (15 b) will be equal to $1/\beta_{r,0}$. (23 a) can then also be written

$$y_r = y_{r-1}(1 - \beta_{r-1,1})$$

which evidently expresses the same as formula (16). The traffic volume $s\,y_0$ hunting for the group is equal to $1/\beta_{0,0}$.

From the form of the recursive formulas (23) it is evident that we can compute $\beta_{n,\mu}$ from the $n + 1$ quantities

$$\beta_{0,\mu},\ \beta_{0,\mu+1},\ \ldots\ \beta_{0,\mu+n}$$

It will be shown below that here the general formulas

$\mu = 0$

$$\beta_{n,0} = \sum_{\sigma=0}^{n-1} \binom{n-1}{\sigma} \frac{\beta_{0,0}\,\beta_{0,1}\cdots\beta_{0,\sigma}}{(1-\beta_{0,1})(1-\beta_{0,2})\cdots(1-\beta_{0,\sigma+1})} \qquad (24\text{ a})$$

$\mu > 0$

$$\beta_{n,\mu} = \frac{\displaystyle\sum_{\sigma=0}^{n-1} \binom{n-1}{\sigma} \frac{\beta_{0,\mu}\,\beta_{0,\mu+1}\cdots\beta_{0,\mu+\sigma}}{(1-\beta_{0,\mu+1})(1-\beta_{0,\mu+2})\cdots(1-\beta_{0,\mu+\sigma+1})}}{\displaystyle\sum_{\sigma=0}^{n} \binom{n}{\sigma} \frac{\beta_{0,\mu}\,\beta_{0,\mu+1}\cdots\beta_{0,\mu+\sigma-1}}{(1-\beta_{0,\mu+1})(1-\beta_{0,\mu+2})\cdots(1-\beta_{0,\mu+\sigma})}} \qquad (24\text{ b})$$

apply.

In (24 b) the first term in the denominator (for $\sigma = 0$) is set equal to 1. It may be remarked that, as is evident, all constants β are independent of the choice of the time unit and that also the functions ψ are independent of the unit for x. As the constants are always positive quantities we see from (23 a) that $\beta_{r,1}$ is always < 1. If (23 b) is written in the form

$$1-\beta_{r-1,\mu+1} = \beta_{r-1,\mu}\left(\frac{1}{\beta_{r,\mu}}-1\right)$$

it is clear that for $\beta_{r,\mu} < 1$, $\beta_{r-1,\mu+1}$ must also be < 1. By induction it then follows that $\beta_{r,\mu} < 1$ for every $\mu > 0$. All factors in the expressions (24) are thus positive.

The general validity of the formulas (24) can be proved by induction. As the computations are rather voluminous, only the general outline of the proof will be shown. For $n = 1$ the formulas (24) coincide with the formulas (23). If in (24 b) we set $\mu = 1$ and form $1-\beta_{n,1}$, we obtain an expression with a denominator which by means of the expression (24 a) can be written

$$\frac{1-\beta_{0,1}}{\beta_{0,0}}\beta_{n+1,0}$$

and with a numerator having the form

$$\sum_{\sigma=0}^{n} \binom{n}{\sigma} \frac{\beta_{0,1}\,\beta_{0,2}\cdots\beta_{0,\sigma}}{(1-\beta_{0,2})(1-\beta_{0,3})\cdots(1-\beta_{0,\sigma+1})} - \sum_{\sigma=0}^{n-1} \binom{n-1}{\sigma} \frac{\beta_{0,1}\,\beta_{0,2}\cdots\beta_{0,\sigma+1}}{(1-\beta_{0,2})(1-\beta_{0,3})\cdots(1-\beta_{0,\sigma+2})}$$

By means of the well known formula for the binomial coefficients

$$\binom{n}{\sigma} = \binom{n-1}{\sigma} + \binom{n-1}{\sigma-1}$$

this expression can be reduced to

$$\frac{1-\beta_{0,1}}{\beta_{0,0}}\beta_{n,0}$$

It is clear that thereby the expression

$$\frac{\beta_{n,0}}{1-\beta_{n,1}}$$

is transformed to the form $\beta_{n+1,0}$ according to (24 a). From that it follows that the formulas (24) satisfy the recursive formula (23 a). If by means of (24 b) we form the expression

$$1+\beta_{n,\mu}-\beta_{n,\mu+1}$$

for the general case $\mu > 0$, we can then, by a reduction similar to the above, put the expression $1-\beta_{n,\mu+1}$ into the form of a fraction in which the numerator is equal to the numerator of $\beta_{n,\mu}$ and the denominator is formed by multiplication of the denominator of $\beta_{n,\mu+1}$ by

$$\frac{\beta_{0,\mu}}{1-\beta_{0,\mu+1}}$$

If to this we add $\beta_{n,\mu}$ we obtain a sum of two fractions with the same numerator, namely the numerator $\beta_{n,\mu}$. It can be shown that the sum of the two denominators is equal to the denominator of $\beta_{n+1,\mu}$. After these reductions it is easy to see that the recursive formula (23 b) is satisfied.

The set of formulas (24) can be considered as a particularly comprehensive generalisation of *Erlang*'s loss formula (11). It makes it possible to compute the blocking conditions in all full availability groups in a loss system when the incoming traffic has limited after-effects, while the *Erlang* formula applies only to the special case where the incoming traffic is random. We thereby get the means of treating subgroups of larger groups separately.

In this connection it may be remarked that in (24) the coefficient

$$\binom{n-1}{\sigma}$$

of the terms on the right side of (24 a) causes $\beta_{n,0}$ to increase indefinitely as $n \to \infty$. Since now $sy_n = 1/\beta_{n,0}$ it follows that $y_n \to 0$ as $n \to \infty$.

By comparison of the analogously formed recursive formulas (19) and (23), it is realized that the formulas (24) that are derived from the latter should have similar equivalences as regards the functions $\psi_r(t)$. These are clearly obtained from (24) by substituting $\psi_r(t)$ for the constants $\beta_{r,0}$ and, when $\mu > 0$, $(t+\mu)\,\psi_r(t+\mu)$ for the constants $\beta_{r,\mu}$.

Instead of deriving the constants $\beta_{n,\mu}$ (for the traffic blocked from the group) from the constants $\beta_{0,\mu}$ (for the incoming traffic to the group) the formulas (24) could evidently be expressed in a formally more general way by deriving the constants for the blocked traffic for an arbitrarily chosen device within the group, for instance the rth, from the constants for the blocked traffic of an earlier device, the pth, $(p < r)$. In the formulas (24) we shall then have $\beta_{r,\mu}$ instead of $\beta_{n,\mu}$ and $\beta_{p,\mu}$ instead of $\beta_{0,\mu}$. In the binomial coefficients $r - p$ is then substituted for n.

The set of formulas worked out for the constants shows a remarkable invariance. The constants $\beta_{n,\mu}$ can be directly computed from (24). They can, however, also be derived in the way that we first compute the constants $\beta_{r,\mu}$ for the blocked traffic of the rth device $(r < n)$ from $\beta_{0,\mu}$ and then the constants $\beta_{n,\mu}$ from the quantities $\beta_{r,\mu}$. The pure algebraic proof of the fact that the formulas (24) really have these properties offers no basic difficulties. The computations become rather complicated, however, so they will be omitted here. It should be mentioned that we hereby get double sums that can be reduced by using the formula well-known from algebra[18]

$$\sum_{\mu=0}^{p} \binom{n-r-1}{\mu}\binom{r}{p-\mu} = \binom{n-1}{p}$$

Regarding the extent of validity for the formula system (24), it should be pointed out that the condition (introduced in this chapter) of sequential hunting within this group is not necessary. That is already evident from the fact that the time intervals when all devices in the group are occupied will occur regardless of the hunting method in the group, though of course on condition that the group works with full availability. The truth of this presumption will be proved strictly mathematically in Chapter 4.

As was done for the blocked traffic, it is likewise possible to introduce constants for the traffic carried by the devices. From the formula (21) it is clear how these constants can be computed in a simple manner from the constants for the incoming traffic. For other important distribution functions that can be derived from the call arrival distribution functions, e.g. the distribution of the next call arrival, we can also define constants that have a simple relation to the constants β.

Occasionally it may be convenient to introduce a new series of constants that simplifies the structure of the formulas. These constants are defined by

$$\gamma_{r,\mu} = \frac{\beta_{r,\mu}}{1-\beta_{r,\mu+1}} \tag{25 a}$$

From the recursive formulas (23) we then get

$\mu = 0$

$$\gamma_{r,0} = \beta_{r+1,0} \tag{25 b}$$

$\mu > 0$

$$\gamma_{r,\mu} = \frac{\beta_{r+1,\mu}}{1 - \beta_{r+1,\mu}} \tag{25 c}$$

Then the formulas (24) can be written as follows:

$\mu = 0$

$$\beta_{n,0} = \sum_{\sigma=0}^{n-1} \binom{n-1}{\sigma} \gamma_{0,0}\, \gamma_{0,1} \cdots \gamma_{0,\sigma} \tag{26 a}$$

$\mu = 0$

$$\beta_{n,\mu} = \frac{\sum_{\sigma=0}^{n-1} \binom{n-1}{\sigma} \gamma_{0,\mu}\, \gamma_{0,\mu+1} \cdots \gamma_{0,\mu+\sigma}}{\sum_{\sigma=0}^{n} \binom{n}{\sigma} \gamma_{0,\mu}\, \gamma_{0,\mu+1} \cdots \gamma_{0,\mu+\sigma-1}} \tag{26 b}$$

Finally the results will be shown for the important special case of purely random incoming traffic, so that $\varphi_0(t) = e^{-yt}$. We then get $\beta_{0,0} = 1/s\,y$ and for $\mu > 0$

$$\beta_{0,\mu} = \frac{\mu}{\mu + sy}$$

From (24) we then get after some transformations

$\mu = 0$

$$\beta_{n,0} = \frac{n!}{(sy)^{n+1}} \sum_{\sigma=0}^{n} \frac{(sy)^\sigma}{\sigma!} \tag{27 a}$$

$\mu > 0$

$$\beta_{n,\mu} = \frac{\sum_{\sigma=0}^{n} \frac{(n+\mu-\sigma)!}{\sigma!\,(n-\sigma)!} (sy)^\sigma}{(n+1) \sum_{\sigma=0}^{n+1} \frac{(n+\mu-\sigma)!}{\sigma!\,(n+1-\sigma)!} (sy)^\sigma} \tag{27 b}$$

The traffic volume that is blocked for the group we get from $1/\beta_{n,0}$. It is readily realized that (27 a) and the *Erlang* formula (11) agree.

It is now possible to investigate by a more direct method the properties of the functions $\varphi_r(t)$, for the case where the incoming traffic to the group constitutes random traffic. If the incoming traffic to the group has a definite call arrival process, then $\varphi_0(x) = e^{-yx}$ and hence $\psi_0(t) = 1/(t + s\,y)$. From the recursive formula (19) it then follows that $\psi_r(t)$ can always be expressed in the form of a ratio between two complete polynomials in t. Since $\psi_r(t) \to 0$ as $t \to \infty$, the degree of the numerator must be lower than that of the denominator. Even if we assume, somewhat more generally, that $\varphi_0(x)$ is the sum of a limited number of exponential expressions, we get $\psi_r(t)$ as a rational function of t.

We now assume that we can write $\psi_r(t)$ in the form

$$\psi_r(t) = \frac{A_r(t)}{B_r(t)}$$

where $A_r(t)$ and $B_r(t)$ are complete rational functions of t with the degrees a_r and b_r respectively. From the recursive formula (19) we then get

$$\frac{A_r(t)}{B_r(t)} = \\ = \frac{B_{r-1}(1+t)}{t\,B_{r-1}(1+t) + \frac{B_{r-1}(t)}{A_{r-1}(t)} \left\{ B_{r-1}(1+t) - (1+t) A_{r-1}(1+t) \right\}} \tag{28}$$

From this we might expect to find $a_r = a_{r-1} + b_{r-1}$ and $b_r = 2\,b_{r-1}$ (note that $a_{r-1} < b_{r-1}$). Generally, however, a considerable reduction of the degree takes place, which can be proved in the following manner.

Assume that we are concerned with a relation of the form

$$B_{r-1}(1+t) = K A_{r-1}(t) + (1+t)\,A_{r-1}(1+t) \tag{29}$$

where K is a complete, rational function of t of degree μ. Because of (29), (28) then passes into the form

$$\frac{A_r(t)}{B_r(t)} = \frac{B_{r-1}(1+t)}{t\,B_{r-t}(1+t) + K\,B_{r-1}(t)}$$

As this relation must constitute an identity, we get

$$A_r(t) = B_{r-1}(1+t) \tag{30 a}$$

$$B_r(t) = t\,B_{r-1}(1+t) + K\,B_{r-1}(t) \tag{30 b}$$

The relation (30 b) can also be written

$$B_r(1+t) = (1+t)\,B_{r-1}(2+t) + K\,B_{r-1}(1+t)$$

from which, by means of (30 a), we get

$$B_r(1+t) = (1+t)\,A_r(1+t) + K\,A_r(t) \tag{30 c}$$

This is the same relation as (29) but with index r instead of $r-1$. Hence it is clear that, if the relation (30 c) is valid for $r = 0$, i.e. also for $\psi_0(t)$, all relations (30) are valid for all values of r. If now for the incoming traffic $\varphi_0(x) = e^{-yx}$, then $A_0(t) = 1$ and $B_0(t) = t + s\,y$. For these values (30 c) is clearly satisfied and thus $K = s\,y$. Then for $\psi_r(t)$ the degree of the numerator will be r and the degree of the denominator $r+1$. As to the degrees, if $\mu > 0$, it holds generally that $a_r = b_{r-1}$ and $b_r = b_{r-1} + \mu$. In the cases of interest the incoming traffic to the group is either random in itself or constitutes the blocked traffic from another group, the incoming traffic to which is random. Also in the latter case, as will be readily understood, K will be a constant independent of t. It is therefore sufficient in the following account to look only at the case $\varphi_0(x) = e^{-yx}$ and thus $K = s\,y$. Following from (30 a) we then have

$$\psi_r(t) = \frac{B_{r-1}(1+t)}{B_r(t)} \tag{31}$$

$B_r(t)$ now gets the form

$$B_r(t) = \sum_{\sigma=0}^{r+1} \binom{r+1}{\sigma} t(1+t)(2+t)\ldots(r-\sigma+t)(sy)^{\sigma} \tag{32}$$

which can easily be shown by induction, as (32) satisfies the recursive formula (30 b) and for $r = 0$ reduces itself to $B_0 = t + s\,y$. The term in (32) for $\sigma = r+1$ will be equal to $(s\,y)^{r+1}$.

The roots that satisfy $B_r(t) = 0$ are of importance. (32) has the degree $r+1$ with respect to t, and all coefficients for the powers of t are positive. Accordingly no positive real roots exist. The $r+1$ roots may be denoted $-a_{r,0}$, $-a_{r,1}$, ... $-a_{r,r}$. Since, from (32), the coefficient for the highest power of t is unity, we get

$$B_r(t) = (t + a_{r,0})(t + a_{r,1})\ldots(t + a_{r,r})$$

We shall now show by induction that all roots $-a_{r,\sigma}$ are real, negative and satisfy the inequalities

$$a_{r,\sigma} > 1 + a_{r,\sigma-1} \tag{33 a}$$

We assume that the conditions are fulfilled for a certain value of r and look at the equation for the next higher value of r. From (30 b) we then obtain

$$B_{r+1}(t) =$$
$$= t(t + a_{r,0} + 1)(t + a_{r,1} + 1)\ldots(t + a_{r,r} + 1) +$$
$$+ sy(t + a_{r,0})(t + a_{r,1})\ldots(t + a_{r,r})$$

Due to the inequalities (33 a) we now find

$$B_{r+1}(0) = sy\,a_{r,0}\,a_{r,1}\ldots a_{r,r} > 0$$
$$B_{r+1}(-a_{r,0}) =$$
$$= -a_{r,0}(a_{r,1} + 1 - a_{r,0})\ldots(a_{r,r} + 1 - a_{r,0}) < 0$$

Thus $B_{r+1}(t) = 0$ has a root between 0 and $-a_{r,0}$.

Further we get

$$B_{r+1}(-a_{r,0} - 1) =$$
$$= -sy(a_{r,1} - a_{r,0} - 1)\ldots(a_{r,r} - a_{r,0} - 1) < 0$$
$$B_{r+1}(-a_{r,1}) =$$
$$= -a_{r,1}(-a_{r,1} + a_{r,0} + 1)(-a_{r,1} + a_{r,2} + 1)\ldots$$
$$\ldots(-a_{r,1} + a_{r,r} + 1) > 0$$

Thus $B_{r+1}(t) = 0$ has a root also between $-(a_{r,0} + 1)$ and $-a_{r,1}$. If we continue in this manner we realize that $B_{r+1}(t) = 0$ has a root between each $-(a_{r,\sigma} + 1)$ and $-a_{r,\sigma+1}$. Finally we get

$$B_{r+1}(-a_{r,r} - 1) =$$
$$= -sy(-a_{r,r} - 1 + a_{r,0}) \ldots (-a_{r,r} - 1 + a_{r,r-1})$$

with the sign $(-1)^{r+1}$. $B_{r+1}(-\infty)$ has the sign $(-1)^{r+2}$. Finally we thus get a root whose absolute value is greater than $a_{r,r} + 1$. Thereby we have determined all $r + 2$ roots of $B_{r+1}(t) = 0$ and have found the conditions

$$a_{r+1,0} < a_{r,0} \qquad (33\text{ b})$$

$$a_{r,\sigma} + 1 < a_{r+1,\sigma+1} < a_{r,\sigma+1} \qquad (33\text{ c})$$

$$a_{r,r} + 1 < a_{r+1,r+1} \qquad (33\text{ d})$$

From these conditions we get $a_{r+1,\sigma} > 1 + a_{r+1,\sigma-1}$. If we now take, by way of example, $a_{0,0} = sy$, it is thereby proved that all conditions (33) apply for all values of r.

As it has thereby been shown that $B_r(t) = 0$ has a total of $r + 1$ different real and negative roots, (31) can be written in the form

$$\psi_r(t) = \sum_{\sigma=0}^{r} \frac{C_{r,\sigma}}{t + a_{r,\sigma}} \qquad (34)$$

Then follows, according to (18),

$$\varphi_r(x) = \sum_{\sigma=0}^{r} C_{r,\sigma}\, e^{-a_{r,\sigma}\frac{x}{s}} \qquad (35)$$

whence we know that all $a_{r,\sigma}$ are positive and different.

The constants $C_{r,\sigma}$ can be defined by means of the usual rules for dividing up the rational functions into partial functions. We easily obtain the expression

$$C_{r,\sigma} =$$
$$= \frac{(-a_{r,\sigma}+a_{r-1,0}+1)\ldots(-a_{r,\sigma}+a_{r-1,r-1}+1)}{(-a_{r,\sigma}+a_{r,0})\ldots(-a_{r,\sigma}+a_{r,\sigma-1})(-a_{r,\sigma}+a_{r,\sigma+1})\ldots\ldots(-a_{r,\sigma}+a_{r,r})} \qquad (36)$$

The numerator and the denominator contain an equal number of factors. From the inequalities (33) it is clear that there are equally many factors with a negative sign in the numerator and the denominator. Thus the constants $C_{r,\sigma}$ are always positive. From this it follows that all the *distribution functions* $\varphi_r(x)$ are *completely monotonic.*

From the value of $\psi_r(0)$ we can obtain an expression for the blocked traffic volume containing the roots of the traffic which hunts for the device. We get

$$\psi_r(0) = \frac{1}{sy_r} =$$
$$= \frac{1}{(sy)^{r+1}}(a_{r-1,0}+1)(a_{r-1,1}+1)\ldots(a_{r-1,r-1}+1) \qquad (37)$$

As an example of the form of the distribution functions, we give here the explicit expression for the traffic that is blocked from the first device.

$$\varphi_1(x) = \left(\frac{1}{2} + \frac{1}{4\sqrt{sy+\frac{1}{4}}}\right) e^{-\left(sy+\frac{1}{2}-\sqrt{sy+\frac{1}{4}}\right)\frac{x}{s}} +$$
$$+ \left(\frac{1}{2} - \frac{1}{4\sqrt{sy+\frac{1}{4}}}\right) e^{-\left(sy+\frac{1}{2}+\sqrt{sy+\frac{1}{4}}\right)\frac{x}{s}} \qquad (38)$$

To get a notion of the numerical quantities, some information is given in the adjoining table and in Fig. 3 for a special case. For $sy = 2$ the distribution functions for $r = 0, 1, 2, 3$ have been computed. It should be mentioned that the search for the roots is in practice greatly facilitated by the conditions (33). In order to permit a direct comparison between the forms of the curves, the distribution functions have been reduced to the same mean value. Thus the areas under the respective curves are all alike. The curve for $r = 0$ gives the usual exponential function e^{-x}. The other curves are, in comparison with this function, flatter, which will be more pronounced the more devices the traffic has

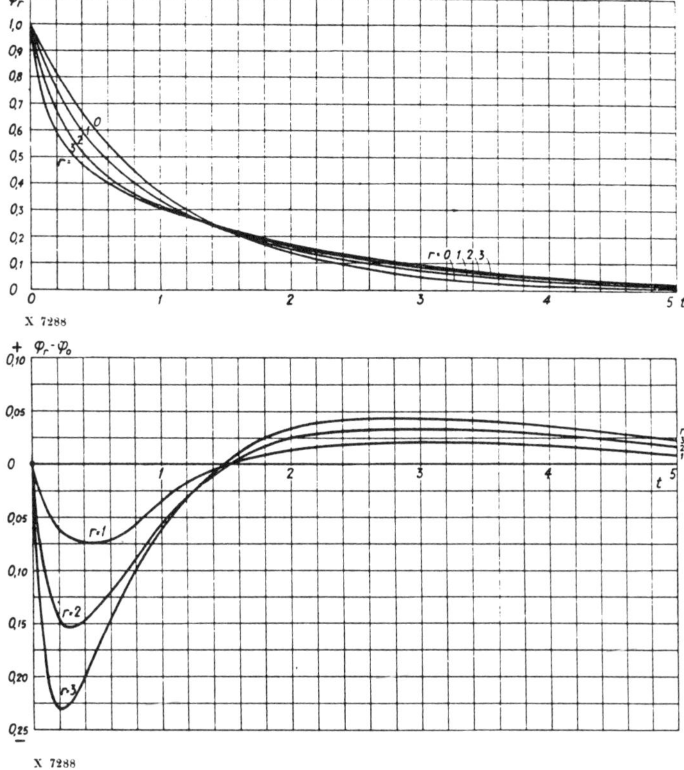

Fig. 3. The distribution functions $\varphi_r(t)$ for $r = 0, 1, 2$ and 3 and $y_0 = 2$. All curves are reduced to the same mean value = 1. The upper figure shows the distribution functions. In the lower figure $\varphi_r(t) - \varphi_0(t)$ is shown on a larger scale.

passed. The properties of blocked traffic, which differ from those of the definite process, are evidenced, for instance, in the form of an increased inability to utilize the devices and are apparent in the curves of the distribution functions by a higher degree of flatness in relation to the normal exponential function. It may then be of importance to have available a quantity which gives a measure of this flatness and thus also the deviation of the traffic from the pure random process. As a suitable such quantity the author has already in an earlier work[11] proposed the so-called *form factor* which is defined as the ratio between the second (quadratic) moment and the square of the mean value of the distribution function. For the distribution function for the interval in

r	$-a_{r,\sigma}$	$C_{r,\sigma}$	ε_r	y_r
0	2,0000	1,0000	2,0000	2,0000
1	1,0000 4,0000	0,6667 0,3333	2,4444	1,3333
2	0,5107 2,7108 5,7785	0,5769 0,2411 0,1820	2,8800	0,8000
3	0,2495 2,0000 4,3100 7,4406	0,5576 0,1818 0,1410 0,1196	3,1967	0,4211

time between successive calls $\varphi_r(t)$ the form factor will be

$$\varepsilon_r = \frac{\int\limits_0^\infty t^2 \varphi_r'(t)\, dt}{\left\{\int\limits_0^\infty t\, \varphi_r'(t)\, dt\right\}^2} \tag{39 a}$$

From the general formulas (7) and (18) we now get the following expression for the form factor

$$\varepsilon_r = -2 \frac{\psi_r'(0)}{\{\psi_r(0)\}^2} \tag{39 b}$$

For the pure exponential function $\varphi_0(t) = e^{-yt}$ we get $\varepsilon_0 = 2$. According to what is shown later in Chapter 6 this is the smallest value of the form factor for a completely monotonic function. The remaining ε_r are greater than 2.

From (31) we further get the formula

$$\frac{1}{2}\varepsilon_r = \frac{B_r'(0)}{B_{r-1}(1)} - \frac{B_r(0)\, B_{r-1}'(1)}{\{B_{r-1}(1)\}^2}$$

which after some transformations can be written

$$\frac{1}{2}\varepsilon_r = 1 + \frac{sy_r}{(sy)^r}\left\{B_{r-1}'(0) - \frac{y_r}{y} B_{r-1}'(1)\right\} \tag{40 a}$$

To deduce the derivatives we can use the development (32). The explicit formulas are easy to set up but do not seem to offer anything of interest. In numerical computations we usually have available (32) developed according to the powers of t.

Another form for ε_r that can easily be deduced is

$$\varepsilon_r = 2\, sy_r \left\{\sum_{\sigma=0}^{r} \frac{1}{a_{r,\sigma}} - \sum_{\sigma=0}^{r-1} \frac{1}{1 + a_{r-1,\sigma}}\right\} \tag{40 b}$$

In the above table the values of ε_r have been computed for the different distribution functions.

Chapter 4

General method for computing blocking

The introduction of general stochastic processes in theoretical investigations of telephone traffic problems has a double purpose. Firstly, the properties of traffic arriving in telephone plant can be better described than is possible with the definite process. Secondly, by means of the more general processes, certain problems can be treated which arise in conjunction with traffic handling and blocking in different groups of devices and which are very difficult to master with the earlier known procedures. The processes that are used for the treatment of these two questions are, however, very different in the two cases. In the next two chapters a definite process, modified in a comparatively simple manner, will be used to describe the properties of the arriving traffic. The results so obtained are closely connected with those we obtain when a pure definite process is used as starting point. For this reason all investigations of the conditions for arriving random traffic will still be very valuable. In this respect there already exist very many valuable results, most of which have been reached by comparatively simple mathematical means. There are, however, still many problems of this kind that have not found solutions suited for practical computations. It now appears that the use of more general stochastic processes offers certain possibilities for successful treatment of these problems. These are stochastic processes that arise from the transformation to which the random traffic is subjected within the groups and are quite different from the reshaped definite process just mentioned. From this standpoint the introduction of stochastic processes for the traffic appearing within the groups can be considered as an analytical aid. Sometimes, admittedly, it makes the mathematical treatment more complicated, but can give results that are easier to use for practical computation than the methods used hitherto. It is outside the scope of this work to complete the existing theories for conditions in groups with fully random arriving traffic; nevertheless it may be appropriate in connection with the presentation of the general stochastic properties of telephone traffic, to touch lightly upon the basis for the use of the above mentioned analytical aid. The investigations made in Chapter 3 can be considered as an introduction. In the present chapter some methods are given for the treatment of general processes. As an example of their use we take the computation of the blocking of an especially simple grading and show how much simpler this is than by the use of methods known hitherto.

The problem that will be discussed in this chapter can be described in its most general form as follows. Traffic hunts over a group of devices, the traffic being a stationary process of the general form that was studied in Chapter 2 and with known frequency functions. Certain rules are applied for the hunting within the group, so that the arriving calls can cause occupations of the devices. The hunting rules can be such that some of the calls are blocked from the group or cause occupations only after a certain time or with a certain probability. We can now select those calls that in some respect are exposed to the same treatment in the group (e.g. blocked from the group). These calls may then be considered to be traffic derived from the incoming traffic by selection. The task is to determine the properties of the traffic so derived.

This problem has been dealt with in Chapter 3 for the case of a full availability group in a loss system on the assumption that the traffic has a limited after-effect, random traffic being a special case. In this case the solutions can be represented in closed form. This is because, for traffic with limited after-effect, every call arrival is an equilibrium point. If on the contrary the arriving traffic lacks available equilibrium points, it is necessary in the treatment to look at the traffic process during an unlimited time backwards, which necessitates finding a limit value. The solutions will then usually appear as convergent series. Such a limit operation seems in principle to be applicable to every problem. However, it is of so complicated a nature that it could probably be used successfully only under more specific conditions than those presented in the general description of the problem above. Furthermore, as an examination of the main points of the treatment of this problem will in general necessarily be very abstract and difficult to survey, it seems best to represent the method by application to some specific cases.

As regards the occupation times of the devices, a presentation of the problem in its most general form should consider the possibility that these are dependent on the length of the previous occupation times and the positions of the previous call arrivals. No doubt there is such a correlation in reality. Every attempt at direct introduction of such a dependence seems, however, to entail such extraordinary complications that all possibility of attaining practical results is precluded. This does not, fortunately, exclude consideration of the correlation that exists in reality. Further on in this work an indirect way of achieving this will be indicated. For the present it should be assumed that, for the arriving traffic, the occupation times are independent of the previous course of events.

It is, however, appropriate to make one more restriction regarding the occupation times. It is well known that, if these are supposed to follow a special distribution function, the exponential, the treatment of all traffic problems will be extremely simplified. This is because, with this distribution, the probability of the remaining duration of the call is independent of the time it has already lasted. From the literature in this field it is clear, however, that successful treatment of cases in which the occupation times follow a more general distribution function is not impossible; but it seems best in this connection to take all measures that can simplify the mathematical presentation without impairing the general validity of the method. In this chapter it will be assumed that the probability of an occupation lasting at least the time t is expressed by $e^{-t/s}$ and that it is independent of the durations of the previous occupations as well as of the position of the previous call arrivals. It is easily seen that the constant s then expresses the mean occupation time. In order to simplify the method of notation we shall henceforward put $s = 1$. This will then signify that the mean occupation time will be chosen as the unit of time, and of course in no way impairs the general validity of the results.

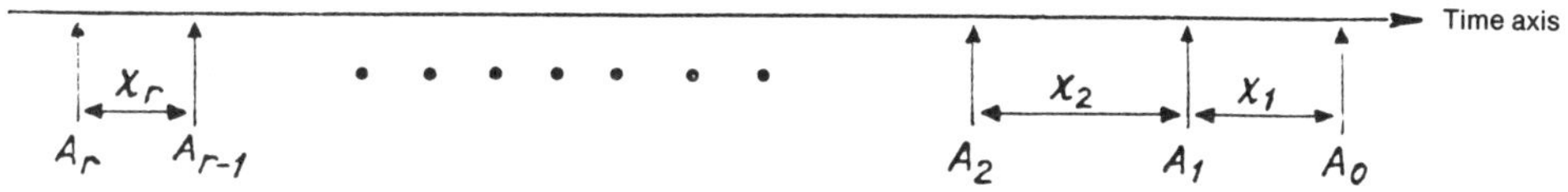

Fig. 4. Time diagram of the arrivals of $r + 1$ calls represented by $A_0, A_1, \dots A_r$.

A group of devices will be considered that, in respect of hunting rules, has such properties that it works as a full availability group and as a loss system. The group may consist of n devices and, furthermore, it is assumed that the traffic arriving in the group constitutes a general stochastic process, the properties of which are apparent from the expressions (1) – (9). Among the call arrivals appearing on the time axis an arbitrary call should be selected, that terminates a series of r inter-arrival times with the lengths x_1, $x_2, \dots x_r$ as is shown in Fig. 4. The probability that an arbitrarily chosen call arrival finishes such a series of inter-arrival times is then, according to (1 c),

$$F_r(x_r, x_{r-1}, \dots x_2, x_1)\, dx_r\, dx_{r-1} \dots dx_2\, dx_1$$

Now let $P_{\mu,\nu}$ be the probability that at the arrival of call A_ν according to Fig. 4, μ of the n devices are busy. When $\mu < n$ the call A_ν will occupy one of the $n - \mu$ free devices, so that the number of busy devices thereafter will be $\mu + 1$. If, on the other hand $\mu = n$, i.e. blocking prevails, the state of the group will not be changed by the arrival of call A_ν. The quantities $P_{\mu,\nu}$ are clearly functions of the previous inter-arrival times, thus from $x_{\nu+1}$ to x_r, as also of the state at the beginning of the course of events i.e. of $P_{\mu,r}$.

In order that there will be μ occupied devices at the arrival of call A_ν at least $\mu - 1$ devices will be occupied at the arrival of the immediately preceding call $A_{\nu+1}$. At the arrival of call $A_{\nu+1}$ σ devices will be busy, where $\mu - 1 \leqq \sigma < n$. The probability of this is $P_{\sigma,\nu+1}$. In order to have μ occupied devices at the arrival of the next call A_ν altogether $\sigma + 1 - \mu$ occupations must finish in the interval $x_{\nu+1}$. Now according to our assumption the probability that a certain occupation does not finish during this interval is $e^{-x_{\nu+1}}$, and the probability that it does finish $1 - e^{-x_{\nu+1}}$. Among the $\sigma + 1$ occupations prevailing immediately after the call arrival $A_{\nu+1}$, the μ occupations that exist at the next call arrival could now be selected in

$$\binom{\sigma+1}{\mu}$$

different ways. The probability that, at the arrival of the call $A_{\nu+1}$, a total of σ devices are busy and that further at the arrival of the next call A_ν totally μ devices are busy is then

$$\binom{\sigma+1}{\mu} e^{-\mu x_{\nu+1}} (1 - e^{-x_{\nu+1}})^{\sigma+1-\mu} P_{\sigma,\nu+1}$$

as it was assumed that the probabilities of the occupations of separate devices are independent of one another and of $P_{\sigma,\nu+1}$ and thus the different probabilities could be directly multiplied. When $\sigma = n$ we obtain a special expression as then the call $A_{\nu+1}$ is blocked. The probability that at the arrival of the call $A_{\nu+1}$ there are n occupied devices and at the arrival of the next call A_ν there are μ will then obviously be

$$\binom{n}{\mu} e^{-\mu x_{\nu+1}} (1 - e^{-x_{\nu+1}})^{n-\mu} P_{n,\nu+1}$$

The probability $P_{\mu,\nu}$ that the call arrival A_ν meets altogether μ busy devices is given as the sum of the above derived probabilities for $\sigma = \mu - 1, \mu, \mu + 1, \dots n$. After a simple transformation we get the relation

$$P_{\mu,\nu} = \binom{n}{\mu} e^{-\mu x_{\nu+1}} (1 - e^{-x_{\nu+1}})^{n-\mu} P_{n,\nu+1} + \sum_{\sigma=\mu}^{n} \binom{\sigma}{\mu} e^{-\mu x_{\nu+1}} (1 - e^{-x_{\nu+1}})^{\sigma-\mu} P_{\sigma-1,\nu+1} \quad (41)$$

For $\mu = 0$ the lower limit for the sum will be 1 instead of 0 as a state $P_{\sigma-1,\nu+1}$ for $\sigma = 0$ does not exist. Further the sum of probabilities of all occupation possibilities at every call arrival equals 1 so that the relation

$$\sum_{\mu=0}^{n} P_{\mu,\nu} = 1 \tag{41 a}$$

is always valid. We realize easily that the recursion formula (41) satisfies this property. By forming the sum of μ for both members from 0 through n we obtain after some transformations

$$\sum_{\mu=0}^{n} P_{\mu,\nu} = \sum_{\mu=0}^{n} P_{\mu,\nu+1}$$

By use of the recursion formula (41) we can clearly express the state quantities $P_{\mu,0}$ in terms of the corresponding quantities $P_{\mu,r}$. A similar operation can clearly also be undertaken when the occupation times do not follow the exponential law but have another distribution. Then, however, the ages of the present occupations must be included in the state quantities, which complicates the result considerably, whereas for the assumed distribution the probability that an occupation will last at least the time t is independent of its duration up till now, viz. simply e^{-t}.

By means of the expression of the probability derived earlier, that an arbitrarily chosen call is preceded by a sequence of inter-arrival times of the type just considered, one can easily realize that

$$P_{\mu,0} F_r(x_r, x_{r-1}, \ldots \ldots x_2, x_1)\, dx_r\, dx_{r-1} \ldots dx_2\, dx_1$$

expresses the probability that an arbitrarily chosen call is preceded by such a sequence and finds μ busy devices. If we now sum this probability for all possible values of the inter-arrival times $x_1, x_2, \ldots x_r$ then we get the probability that an arbitrarily chosen call, that is preceded by r calls with arbitrary inter-arrival times will find μ busy devices. This is, however, the same as an arbitrarily chosen call, any call. Thus

$$K_\mu = \int_{x_r=0}^{\infty} \int_{x_{r-1}=0}^{\infty} \cdots \int_{x_2=0}^{\infty} \int_{x_1=0}^{\infty} P_{\mu,0} F_r(x_r, x_{r-1}, \ldots \ldots x_2, x_1)\, dx_r\, dx_{r-1} \ldots dx_2\, dx_1 \tag{42}$$

expresses the probability that an arbitrarily chosen call, any call, will find totally μ devices busy. Now, however, we can, as just has been shown, express $P_{\mu,0}$ in the state quantities $P_{\mu,r}$ at the beginning of the procedure. The integrand of (42) becomes a function of the inter-arrival times $x_1, x_2, \ldots x_r$ and of the state quantities at the beginning of the course. It should be remarked that, after having made the integrations in the right-hand member, one always obtains the same K_μ independent of how many inter-arrival times the course considered contains.

In order to determine K_μ by means of (42), it seems to be necessary to know the state quantities $P_{\mu,r}$ (it being assumed, that the frequency function F_r is known). Now the determination of the quantities $P_{\mu,r}$ seems to be equally as difficult as the direct determination of $P_{\mu,0}$ or K_μ and therefore the formula (42) seems to offer only limited aid for the computation of K_μ. Now it holds for stochastic processes as a rule that the probability for a specific event is independent of the state in an unlimitedly distant earlier point in time. We thus have reason to believe, that if we let $r \to \infty$ in (42) we can make the integrations independent of the original quantities $P_{\mu,r}$. In order to evaluate such a limit condition we can proceed in the following way.

The quantities $P_{\mu,0}$ obtained in (42) will first be expressed by (41) with the state quantities prevailing for the call A_1. We then get an expression that contains $P_{\mu-1,1}$, $P_{\mu,1}$ etc. up to $P_{n,1}$. If these quantities are expressed in the state quantities valid for the call A_2 we get an expression that contains

$P_{\mu-2,2}$, $P_{\mu-1,2}$ up to $P_{n,2}$. If we continue in this manner we finally get an expression which contains the state quantities $P_{0,\mu}$, $P_{1,\mu}$, ... $P_{n,\mu}$. Hereby it is of course presumed that $r \geqq \mu$. As the integrand now contains all state quantities that are valid for the call A_μ, one of them can be eliminated by (41 a). In this way we obtain in the integrand a term that is independent of all P quantities. If now the other state quantities $P_{\varrho,\mu}$ are expressed in the quantities $P_{\varrho,\mu+1}$ by means of (41) then we obtain all the state quantities corresponding to call $A_{\mu+1}$. From these quantities we can now, by means of (41 a), derive a form independent of the P quantities. If we continue in this way, provided $r \geqq \mu$ for $P_{\mu,0}$, we arrive at an expression in the form

$$\begin{aligned} P_{\mu,0} = {} & C_1(x_1, x_2 \ldots x_\mu) \\ & + C_2(x_1, x_2 \ldots x_{\mu+1}) \\ & + \\ & \cdot \\ & \cdot \\ & \cdot \\ & + C_{r-\mu+1}(x_1, x_2 \ldots x_r) \\ & + Q \end{aligned}$$

where the C quantities are only functions of the corresponding inter-arrival times, while Q is a linear combination of the state quantities $P_{\varrho,r}$ and besides dependent on all inter-arrival times. Because of the form obtained for $P_{\mu,0}$, (42) will be divided up in terms which have the form

$$\int_{x_r=0}^{\infty} \int_{x_{r-1}=0}^{\infty} \cdots \int_{x_1=0}^{\infty} C_\nu(x_1, x_2 \ldots x_{\mu+\nu-1}) \cdot F_r(x_r, x_{r-1}, \ldots x_1)\, dx_r\, dx_{r-1} \ldots dx_1$$

Because of the relations (2) and (3) which are valid for the frequency functions, we can now directly integrate over the variables $x_{\mu+\nu}$, $x_{\mu+\nu+1}$, ... x_r and we get

$$\int_{x_{\mu+\nu-1}=0}^{\infty} \int_{x_{\mu+\nu-2}=0}^{\infty} \cdots \int_{x_1=0}^{\infty} C_\nu(x_1, x_2, \ldots x_{\mu+\nu-1}) \cdot F_{\mu+\nu-1}(x_{\mu+\nu-1}, x_{\mu+\nu-2}, \ldots x_1) \cdot dx_{\mu+\nu-1}\, dx_{\mu+\nu-2} \ldots dx_1 \tag{43}$$

These integrals can then be computed when the frequency functions are known. For K_μ according to (42) we then get a sum of such integrals for $\nu = 1, 2, \ldots (r - \mu + 1)$. In addition we get a residual term that consists of r integrals of Q multiplied by the frequency function F_r. We observe now that the integrand in (43) is independent of r and all inter-arrival times before the call $A_{\mu+\nu-1}$. The common form for the terms in the series that we have got for K_μ is then independent of the number of terms in the series. If we now let $r \to \infty$, we can expect that the residual term tends to the value 0, since K_μ according to the foregoing should be independent of the state at an infinitely distant earlier point of time. Further, the series that is obtained for K_μ must for physical reasons be convergent. Thus we have got an expression for the probability K_μ that μ devices are busy at the arrival of a call chosen at random, in the form of an infinite series. When using this method, one must always prove mathematically that the residual term tends to zero and that the series is convergent. As a rule this should not be very difficult; an example will be given below of the use of the method. However, it should not be impossible to prove this relation quite generally for an unspecified frequency function; one can here rely on the properties of the recursion (41). This, however, will not be investigated closer here.

In order to show the functioning of the method we will look at some special cases. We assume firstly that the group comprises only one device so that $n = 1$. From (41) we then get

$$P_{1,\nu} = e^{-x_{\nu+1}}(P_{1,\nu+1} + P_{0,\nu+1})$$

and hence because of (41 a)

$$P_{1,\nu} = e^{-x_{\nu+1}}$$

In this case the peculiar fact prevails that the state quantities are independent of the state at the previous call arrival. We need not carry

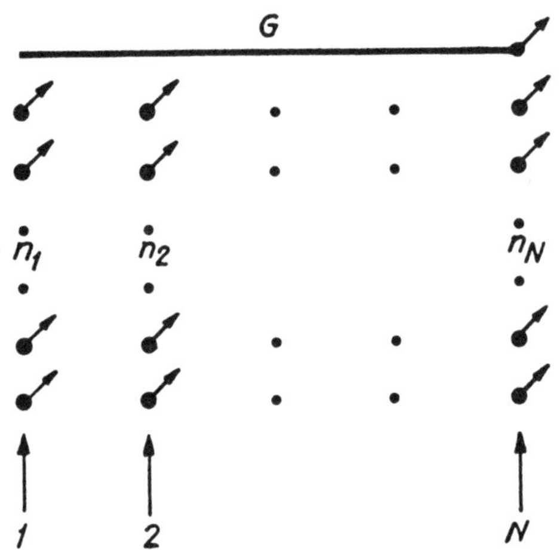

Fig. 5. Grading with N traffic flows, each of which hunts within its group of n_1, n_2 ... n_N devices respectively and the one common device G.

out any limit operation but by putting $r = 1$ get directly from (42)

$$K_1 = \int_0^\infty e^{-x} F_1(x)\, dx \tag{44}$$

K_1 expresses here the probability that the device is busy at an arbitrarily chosen call arrival, i.e. the blocking. For a simple device this is then independent of the higher density of the arriving traffic functions. Formula (44) is obviously a generalization of the formula (16), which is identical in principle, but was derived on the assumption that the arriving traffic has a limited after-effect. This formula is then valid also for arbitrary properties of the arriving traffic.

It may be of interest to show a simple but important application of formula (44). In order to compute the blocking at gradings, methods have been developed in the past which end up in linear equation systems.[7] These are, however, generally of such a high order that numerical computations will be extremely difficult. In Fig. 5 a special such grading is shown, the blocking of which up till now could only be computed numerically for the simplest cases. At this grading there are N arriving traffic flows, each of them hunting over a full availability group with n_1, n_2, ... n_N devices.

When all devices in one of these groups are busy, the arriving calls hunt also for the device G that is common for all groups. We say then that the group formed by all the devices is a grading. The traffic to the device G is formed by superposing the blocked traffic flows from the N full availability groups. If the N arriving traffic flows of the N groups are independent, then this is obviously also the case for the blocked traffic from the groups. The traffic hunting for G is formed by random superposing of the blocked traffic flows from the N groups. If the distribution function for the next call is respectively $\vartheta_1(t)$, $\vartheta_2(t)$, ... $\vartheta_N(t)$, then the distribution function for the next call from the traffic that hunts for G is expressed by $\Theta(t)$ according to (10). According to (8 b) the distribution function for the inter-arrival time is then

$$-\frac{1}{Y}\Theta'(t)$$

where Y is the average sum of the blocked calls from the N full availability groups per time unit. Further according to (5 b) the frequency function of the inter-arrival times for the traffic hunting for G is

$$F_1(t) = \frac{1}{Y}\,\Theta''(t)$$

From (44) we now get the expression for the probability that an arriving call to G is blocked at G.

$$\frac{1}{Y}\int_0^\infty e^{-t}\,\Theta''(t)\,dt$$

After having multiplied by the mean number Y of the calls that hunt for G per time unit, this expression gives the mean number of the calls to G that are blocked per time unit. To determine the blocking of the grading in Fig. 5 we have only to compute the integral

$$\int_0^\infty e^{-t}\,\Theta''(t)\,dt$$

By partial integration we then get

$$\Big[e^{-t}\,\Theta'(t)\Big]_0^\infty + \int_0^\infty e^{-t}\,\Theta'(t)\,dt$$

The expression within brackets is, however, equal to Y according to (8 b) and we get

$$Y + \int_0^\infty e^{-t}\,\Theta'(t)\,dt \qquad (45)$$

as an expression for the mean number of blocked calls to G per time unit. The integral in this expression is obviously negative and it gives, after change of sign the mean number of the calls attended to by G. If now the arriving traffic flows to the N full availability partial groups are fully random then one can determine, by means of the methods established in Chapter 3, the distribution functions that constitute $\Theta(t)$ according to (10). Then one can compute the integral in (45). To carry out this computation several methods can be developed; for instance one can use the roots of the polynomials (32) derived in Chapter 3. The details of the computation will not be discussed here. It should only be remarked that the results permit considerably simpler numerical computations than earlier known methods for determining the blocking in a grading of the type considered.

The traffic that is blocked from every partial full availability group in Fig. 5 has, according to Chapter 3, limited after-effect, if the arriving traffic is random. The traffic, formed by random superposing of such blocked traffic flows, that hunts for G, however, has not limited after-effects. Because of that, the method shown for the computation of the blocking, which is based upon the fact that the blocking is independent of the higher frequency functions, cannot simply be applied if the grading comprises more than one common device, as in such a case the blocking expression contains also the higher frequency functions.

A simple and at the same time typical example of the case of how the general method shown earlier functions for a fully available group is given for the number of devices $n = 2$. From the recursion formula (41) we get

$$P_{2,\nu} = e^{-2x_{\nu+1}}(P_{2,\nu+1} + P_{1,\nu+1})$$

$$P_{1,\nu} = 2\,e^{-x_{\nu+1}}(1 - e^{-x_{\nu+1}})(P_{2,\nu+1} + P_{1,\nu+1}) + e^{-x_{\nu+1}}P_{0,\nu+1}$$

$$P_{0,\nu} = (1 - e^{-x_{\nu+1}})^2(P_{2,\nu+1} + P_{1,\nu+1}) + (1 - e^{-x_{\nu+1}})P_{0,\nu+1}$$

and by means of (41 a)

$$P_{2,\nu} = e^{-2x_{\nu+1}} - e^{-2x_{\nu+1}}P_{0,\nu+1} \qquad (46\text{ a})$$

and

$$P_{0,\nu} = (1 - e^{-x_{\nu+1}})^2 + e^{-x_{\nu+1}}(1 - e^{-x_{\nu+1}})P_{0,\nu+1} \qquad (46\text{ b})$$

From (46 b) the following relation is obtained between $P_{0,0}$ and $P_{0,r}$, where for simplicity $a_\nu = e^{-x_\nu}$ and $b_\nu = 1 - e^{-x_\nu}$ are introduced

$$P_{0,0} = \sum_{\sigma=1}^{r} a_1 b_1\, a_2 b_2 \ldots a_{\sigma-1} b_{\sigma-1} b_\sigma^2 + a_1 b_1\, a_2 b_2 \ldots a_r b_r P_{0,r}$$

Now $a_\nu + b_\nu = 1$. Consequently $a_\nu b_\nu \leq \frac{1}{4}$.

The general term in the sum above is then

$$\leq \left(\frac{1}{4}\right)^{\sigma-1}.$$

The series that is obtained if $r \to \infty$ is thus convergent. Further, the residual term

$$\leq \left(\frac{1}{4}\right)^{r} P_{0,r},$$

and thus tends to zero as $r \to \infty$, since $P_{0,r}$ is a probablity and hence $\leqq 1$.

From (42) one now gets the probability that neither of the two devices is busy for an arbitrarily chosen call

$$K_0 = \sum_{\sigma=1}^{r} \int_{x_\sigma=0}^{\infty} \int_{x_{\sigma-1}=0}^{\infty} \ldots \int_{x_1=0}^{\infty} F_\sigma(x_\sigma, x_{\sigma-1} \ldots x_1) \cdot a_1 b_1\, a_2 b_2 \ldots a_{\sigma-1} b_{\sigma-1} b_\sigma^2\, dx_1\, dx_2 \ldots dx_\sigma + \int_{x_r=0}^{\infty} \int_{x_{r-1}=0}^{\infty} \ldots \int_{x_1=0}^{\infty} F_r(x_r, x_{r-1} \ldots x_1) \cdot a_1 b_1\, a_2 b_2 \ldots a_r b_r P_{0,r}\, dx_1\, dx_2 \ldots dx_r$$

Now from (2), (3) and (4) for every value of σ

$$\int_{x_\sigma=0}^{\infty} \int_{x_{\sigma-1}=0}^{\infty} \cdots \int_{x_1=0}^{\infty} F_\sigma (x_\sigma, x_{\sigma-1}, \ldots x_1) \cdot$$
$$\cdot\, dx_1\, dx_2 \ldots dx_\sigma = 1$$

It is clear that in the series for K_0 every term is

$$\leq \left(\frac{1}{4}\right)^{\sigma-1}$$

and that the residual term is

$$\leq \left(\frac{1}{4}\right)^{r}$$

If $r \to \infty$, one gets the following convergent series development for K_0

$$K_0 = \sum_{\sigma=1}^{\infty} \int_{x_\sigma=0}^{\infty} \int_{x_{\sigma-1}=0}^{\infty} \cdots \int_{x_1=0}^{\infty} F_\sigma (x_\sigma, x_{\sigma-1}, \ldots x_1) \cdot$$
$$\cdot\, a_1 b_1\, a_2 b_2 \ldots a_{\sigma-1} b_{\sigma-1}\, b_\sigma^2\, dx_1\, dx_2 \ldots dx_\sigma \tag{47 a}$$

From (46 a) one gets in the same way the probability K_2 that a call is blocked.

$$K_2 = \int_{x_1=0}^{\infty} a_1^2 F_1(x_1)\, dx_1 -$$
$$- \sum_{\sigma=2}^{\infty} \int_{x_\sigma=0}^{\infty} \int_{x_{\sigma-1}=0}^{\infty} \cdots \int_{x_1=0}^{\infty} F_\sigma (x_\sigma, x_{\sigma-1}, \ldots x_1) \cdot$$
$$\cdot\, a_1^2\, a_2 b_2\, a_3 b_3 \ldots a_{\sigma-1} b_{\sigma-1}\, b_\sigma^2\, dx_1\, dx_2 \ldots dx_\sigma \tag{47 b}$$

It is interesting to investigate the results given by these formulas for the simple cases where the arriving traffic is random or has limited after-effects. Suppose that the incoming traffic is random and that the mean number of calls per time unit is y. Then the frequency function is

$$F_\sigma (x_\sigma, x_{\sigma-1}, \ldots x_1) = y^\sigma e^{-y(x_1 + x_2 + \ldots + x_\sigma)}$$

The double integrals in (47) are then divided up in a number of simple integrals which are easy to compute. For instance after some transformation

$$K_2 = \frac{y}{2+y} - \frac{2}{2+y} \sum_{\sigma=1}^{\infty} \left\{ \frac{y}{(1+y)(2+y)} \right\}^\sigma$$

The geometrical series comprised in the expression above is rapidly convergent for normal values of y. By summation of the series one gets the closed expression

$$K_2 = \frac{y^2}{2 + 2y + y^2}$$

which as expected is identical to Erlang's formula (11) for $n = 2$, since $A = y$ when $s = 1$.

As a further example, the somewhat more general case is considered where the incoming traffic has limited after-effects. The distribution function of the inter-arrival times is given the notation $\varphi(t)$. Then the frequency function according to Chapter 2 is

$$F_\sigma (x_\sigma, x_{\sigma-1}, \ldots, x_1) =$$
$$= (-1)^\sigma \varphi'(x_\sigma)\, \varphi'(x_{\sigma-1}) \ldots \varphi'(x_1)$$

The expression (47 b) then simplifies to

$$K_2 = - \int_0^{\infty} e^{-2x} \varphi'(x)\, dx -$$
$$- \int_0^{\infty} e^{-2x} \varphi'(x)\, dx \int_0^{\infty} (1 - e^{-x})^2 \varphi'(x)\, dx \cdot$$
$$\cdot \sum_{\sigma=0}^{\infty} \left\{ - \int_0^{\infty} e^{-x} (1 - e^{-x})\, \varphi'(x)\, dx \right\}^\sigma$$

By analogy with the case of formula (22), a series of constants is now introduced

$$\beta_\sigma = \sigma \int_0^{\infty} e^{-\sigma x} \varphi(x)\, dx$$

By partial integration we then get

$$- \int_0^{\infty} e^{-2x} \varphi'(x)\, dx = 1 - \beta_2$$

$$- \int_0^{\infty} e^{-x} (1 - e^{-x})\, \varphi'(x)\, dx = \beta_2 - \beta_1$$

$$- \int_0^{\infty} (1 - e^{-x})^2 \varphi'(x)\, dx = 2\beta_1 - \beta_2$$

The above expression for K_2 then becomes

$$K_2 = 1 - \beta_2 - (1 - \beta_2)(2\beta_1 - \beta_2) \sum_{\sigma=0}^{\infty} (\beta_2 - \beta_1)^\sigma$$

The summation of the geometrical series then gives

$$K_2 = \frac{(1-\beta_1)\,(1-\beta_2)}{1+\beta_1-\beta_2}$$

and one can easily prove that this expression agrees with the results of formula (24 a).

Finally, it should be remarked that the general formula (47 b) for the blocking of a group of two devices offers a simple possibility for determining the higher limit of the blocking considered. As the series in (47 b) is always positive, the first term on the right side viz. the integral

$$\int_0^\infty e^{-2x} F_1(x)\,dx$$

must be greater than K_2. For traffic whose properties are arbitrary, one can thus determine an upper limit for the blocking for a group consisting of two devices, using only the frequency function of the inter-arrival times. Furthermore, the above integral is also equal to the integral that expresses the blocking in a group consisting of a single device, provided the arriving traffic has the same properties regarding call arrivals but for which the mean duration time is half as great, thereby implying half the traffic volume.

The examples discussed up to now have only been concerned with device groups with one or two devices. For greater device groups the computations will be more difficult because the recursion formula (41) becomes more complicated. However, for a general number of devices n one can also relatively easily treat the cases where the arriving traffic has limited after-effect.

Using the same expression for the frequency functions F_σ for a traffic with limited after-effect as before, (42) gets the form

$$K_\mu = \int\limits_{x_r=0}^{\infty} \int\limits_{x_{r-1}=0}^{\infty} \cdots \int\limits_{x_1=0}^{\infty} P_{\mu,0}\,(-1)^r \varphi'(x_r)\,\varphi'(x_{r-1}) \cdots \varphi'(x_1)\,dx_1\,dx_2 \ldots dx_r$$

As has been stated earlier, this is valid for every value of r, thus also for $r = 1$. In this case one gets

$$K_\mu = -\int\limits_{x_1=0}^{\infty} P_{\mu,0}\,\varphi'(x_1)\,dx_1$$

If now we use the recursion formula (41) to express $P_{\mu,0}$ in terms of $P_{\sigma,1}$, one gets from the above expression

$$K_\mu = -\binom{n}{\mu}\int_0^\infty e^{-\mu x_1}(1-e^{-x_1})^{n-\mu}\,P_{n,1}\,\varphi'(x_1)\,dx_1 - \sum_{\sigma=\mu}^{n}\binom{\sigma}{\mu}\int_0^\infty e^{-\mu x_1}(1-e^{-x_1})^{\sigma-\mu}P_{\sigma-1,1}\,\varphi'(x_1)\,dx_1$$

Now, however, $P_{\sigma,1}$ are the state probabilities for the call A_1 in a procedure according to Fig. 4, that only contains the calls A_0 and A_1. The quantities $P_{\sigma,1}$ are obviously independent of the size of the following inter-arrival time x_1. Thus the integrals in the expression above for K_μ can be computed. The different $P_{\sigma,1}$ appear then as factors. The following was then done. An arbitrarily chosen call A_0 has been considered and a preceding call A_1 with the inter-arrival time x_1. After that integration is performed for all possible values of x_1. Clearly then also the call A_1 can be considered as arbitrarily chosen. Then $P_{\sigma,1}$ is the probability that at an arbitrarily chosen call arrival, altogether σ devices are busy, but this is exactly the definition of K_σ according to (42). We thus have in this case $P_{\sigma,1} = K_\sigma$. The above expression for K_μ then reduces to the form

$$K_\mu = -\binom{n}{\mu}K_n\int_0^\infty e^{-\mu x}(1-e^{-x})^{n-\mu}\,\varphi'(x)\,dx - \sum_{\sigma=\mu}^{n}\binom{\sigma}{\mu}K_{\sigma-1}\int_0^\infty e^{-\mu x}(1-e^{-x})^{\sigma-\mu}\,\varphi'(x)\,dx$$

Thus we have got a homogeneous linear equation system with $n + 1$ equations for the $n + 1$ quantities K_μ, without the necessity of using any limit operations. The reason for this lies obviously in the fact that the frequency functions for traffic with limited after-effect have been expressed as products of a number of frequency functions of the interarrival times, and thus are connected with the condition that traffic with limited after-effects has an equilibrium point at each call arrival. In order to solve the set of equations, one more relation is needed, obtained from

$$\sum_{\mu=0}^{n} K_\sigma = 1$$

By means of the constants β_σ just introduced, the above equation for K_μ can, after the integrations have been performed, be written

$$K_\mu = \binom{n}{\mu} K_n \sum_{\varrho=0}^{n-\mu} (-1)^\varrho \binom{n-\mu}{\varrho} (1-\beta_{\mu+\varrho}) + \\ + \sum_{\sigma=\mu}^{n} \binom{\sigma}{\mu} K_{\sigma-1} \sum_{\varrho=0}^{\sigma-\mu} (-1)^\varrho \binom{\sigma-\mu}{\varrho} (1-\beta_{\mu+\varrho}) \tag{48}$$

As before for (41), it should be remarked that for $\mu = 0$ the lower limit for the sum is $\sigma = 1$.

The solving of the set of equations will not be treated here; only the method for setting up the equations is of interest in this connection. It should be remarked, however, that the solution for K_n, i.e. the blocking from the group, of course gives the same result as is obtained by the formula in Chapter 3 (24 a). The result obtained here is, however, somewhat more general, as assumptions concerning the hunting rules do not have to be introduced. The devices can thus be hunted for in any arbitrary order, the only condition being that no calls are blocked unless all devices are busy. The results in Chapter 3, however, was only deduced for sequential hunting.

The above examples of the treatment of the general problem that was put forward in the beginning of this chapter have only comprised the determination of the state in a group for an arbitrarily chosen call. Thereby a determination of the mean number of calls per time unit is arrived at for the derived, in this case the blocked, traffic. The general problem on p. 32 comprised also the question of the frequency functions of derived traffic. It is obvious that we can use methods analogous to those used above, although the computations will be considerably more complicated. Hence such examples will not be shown here. The method will, however, be rather simple in principle. We start from a course of events in the arriving traffic, the primary course, that can be as shown in Fig. 4. We assume that this course of events causes a certain other, secondary, course of events in the derived traffic. Depending upon the hunting rules and the definition of the derived traffic, certain calls in the primary course are fixed by the secondary course. The probability that a secondary course of events of the kind considered would appear, for instance, with reference to an arbitrarily chosen call, is obtained by the integration of the probability for the primary course of events for all possible values of the parameters in the primary course of events that are not fixed by the secondary course of events. As in the above examples, a number of quantities thereby appear which usually cannot be determined by direct methods. They describe the state of the group at the beginning of the primary course of events. In order to avoid these quantities, one performs a limit operation of the same kind as in the examples by displacing the start of the primary course of events backwards to an indefinitely distant point of time. The success of this operation is based upon the fact that the probability of the secondary course of events should be independent of the state of the group in an unlimitedly distant earlier point of time.

It should be remarked that the possibility of carrying through operations of this kind is closely connected to the assumption of stationariness as this concept was defined in Chapter 2.

It is not intended here to develop further the methods that have been hinted at for the theoretical treatment of general traffic problems; the rest of this work will be devoted to problems that have more direct ties with practical applications. It has, however, been considered suitable to present first a general theorem which often seems to be of value in continued development of the methods that have been shown in the foregoing.

Expressions have been derived in Chapter 3 for the blocking of a fully available group in a loss system when the arriving traffic has a limited after-effect. Hereby an assumption of sequential hunting was introduced. The same results have been obtained above without the introduction of this assumption. The same result is obtained in the treatment of some other traffic problems and it seems probable that the blocking in a fully available group is always independent of the hunting method in the group. However, no proof of this theorem has ever been put forward, although it is rather easy to prove that not only the blocking but also the location of blocked calls on the time axis, and thereby all the statistical properties of the blocked traffic, are fully independent of the hunting order in a fully available group in a loss system. This fact is important, as one can proceed in steps of one device at a time to derive the properties of blocked traffic, as was the case with the investigations in Chapter 3. This means that one assumes sequental hunting. The mathematical treatment can then be simpler than if one must treat the whole group at the same time, which is necessary if the hunting rules are not known. However, the theorem considered shows that the properties of the blocked traffic must be the same with both methods. One can then apply the simplest.

In order to prove the theorem, a fully available group in a busy system is considered with n devices for which stationary traffic with arbitrary statistical properties is hunting. One compares the traffic properties within the group for two different hunting rules A and B. In both cases, however, the condition applies that calls arriving when all devices are busy are totally blocked from the group and that calls arriving when devices are free will cause occupations within the group. These assumptions just imply that we consider a full availability group working as a loss system. Assume now that there is a point of time at which no devices are busy either for the hunting rules A or B. Let such a point of time be termed a zero point. If we now consider the traffic conditions after such a zero point, we readily realize that, as the starting conditions in both cases are equal, always equally many devices will be busy at the same time in both cases and the occupations start and stop at the same points of time although, because of the different hunting rules, they could concern different devices in the two cases. This applies also to the state in the group when all devices are busy and the calls arriving at that time are blocked in both cases. After a zero point the blocked traffic will be identical for the two hunting rules.

The problem has now been reduced to the question of whether such zero points exist. If we can prove that an arbitrarily chosen point of time has always been preceded by such a zero point, it follows that the blocked traffic is fully independent of the hunting rules.

Now it is obvious that, if the probability is greater than zero, that during a finite but otherwise arbitrarily chosen time T, at least one state should appear in the group where no devices are busy, then the appearance of at least one such state is guaranteed, i.e. the probability tends to 1 the further back in time we look. We then have to prove that the probability is greater than zero that during such a time T a state will occur where no devices are busy, both in case A and in case B.

In order to show this, a group with an unlimited number of devices is considered, over which the same traffic as above hunts. In this case no blocked calls occur and it is obvious that in such an unlimited group the probability that no devices are busy is fully independent of the hunting rules. But at a point of time when no devices are busy in the unlimited group, no devices in the original group of n devices could be busy either, both in case A and in case B. Such a point of time is thus a zero point.

The problem is now reduced to showing that the probability is greater than zero that, during an arbitrarily chosen time span T, a state would appear in a group of unlimited size where no devices are busy. This should be true under very general assumptions for the properties of the traffic, although it is of course possible to define traffic that does not fulfil the condition. It is here sufficient to prove the condition for one assumption that is always fulfilled in investigations of telephone traffic problems, i.e. that all arriving traffic is always generated only from random traffic flows. By this is meant that all calls at any point of time have belonged to random traffic flows from which they are selected in some way. (It is then not necessary that the whole traffic considered comes from one and the same random traffic.) From this it obviously does not follow that the arriving traffic must be random. Through the selection, which may be arbitrary, the random properties could have got lost. Assume now that the selection has taken place from a total of r random traffic flows. If we further imagine that the r traffic flows each hunt for one group with an unlimited number of devices, then it is clear that at a point of time when all r groups are free, the traffic considered must also have a zero point. If random traffic hunts for a group with an unlimited number of devices, then, as is well known, the probability is greater than zero that, during a time span T, a state will occur, where no devices are busy. As now the r traffic flows are random and thus independent of each other, the probability that on a certain occasion all devices in all r groups are free is the product of the probabilities that the devices in each of the groups are free. This product will be greater than zero, as all factors are greater than zero. The occurrence of zero points is then guaranteed. In a fully available group in a loss system the blocked traffic is thus independent of the hunting rules, under the assumption that the arriving traffic is derived from purely random traffic flows.

This argument cannot be applied unchanged to delay systems, since for these the comparison between a limited group and an unlimited group is not possible.

INTENSITY VARIATIONS AS A STARTING POINT FOR METHODS OF TREATING TELEPHONE TRAFFIC PROBLEMS

Chapter 5

Slow intensity variations

All theoretical investigations which have been made up to now on problems associated with traffic handling in telephone plant have, deliberately or not, been based on the assumption that the traffic directly originated by the subscribers has in some sense had a pure random character. The results, some of which have been touched upon and supplemented in the foregoing, have not always been fully satisfactory in practical applications. It seems to be established that the reason for this may be found in the fact that the properties of the telephone traffic in real life cannot be described with sufficient accuracy by assumed random schemes. It does not seem possible to solve the problems of traffic that arise in the planning of telephone plant on the basis of practical experience or measurements. Thus reliance on results arrived at by theoretical investigations will to a considerable extent be unavoidable. It seems therefore highly desirable that the theoretical investigations are based upon assumptions that describe the real conditions better than those used at present. The main purpose of this work is to put forward some proposals in this context. The theories that have been presented in the foregoing regarding general stochastic processes applicable to telephone traffic are then considered primarily as an introduction and a base for the continued, more special investigations.

Concerning the design of a probability scheme that better accords with the properties of real telephone traffic than the random traffic scheme, it is not sufficient to have as aim that this accordance should be as good as possible. It is equally important that the scheme has a sufficiently simple structure so that it can be used in theoretical treatment of traffic problems without severe complications. It would be specially advantageous if it were possible to design a scheme that could make use of the abundant results that have been obtained earlier on the basis of the scheme for random traffic. A wish of a more special kind is also that the new scheme should comprise one or more parameters that, by their variation, could lead to a consideration of the special character of different kinds of traffic. A self-evident requirement is, finally, that the agreement between the properties of the scheme and real telephone traffic can be checked by measurements without great difficulty.

In order to get guidance for the design of a probability scheme that covers the properties of real telephone traffic in the best way, we can partly fall back on ideas based upon experience regarding the casual relation between different calls. The introduction of the scheme for random traffic has thus been based on the conception that the wish of a subscriber to make a call at a certain point of time is not influenced by the number and the occurrence of other calls in the vicinity of this point of time. This motivation has some justification. We can, however, imagine that calls arriving close to each other in time and originating from different subscribers could have a causal dependence between them. Generally, however, such calls are very seldom included in the traffic arriving to the same group. More doubtful, on the contrary, is whether we can assume that calls from one and the same subscriber are independent of each other in a statistical sense. When a subscriber gets a wrong connection or the wanted number is busy, or when a subscriber is blocked, there often arises a reason to repeat the call after a short while. Furthermore, we can often observe that a call will cause new calls in a short time. Measurement of the positions of a subscriber's calls would probably show a clustering that is not consistent with a random distribution. The influence of this fact on the properties of the total traffic of a great number of subscribers can, however, only be small. This line of thought should then make it probable that the traffic from not too small numbers of subscribers would have a random character. A weakness of this line of thought, however, is that it does not take into account the possibility of less direct dependence between the positions of the calls. That such dependences exist can be seen from the fact that each traffic has strong intensity variations over the day which undoubtedly are not of a pure random nature. The direct dependences can surely be derived from man's prevailingly common customs regarding work times, sleep, mealtimes, etc. They thus depend on the course of the common life in society. The conditions can also be described as follows. At every point of time there exists for every subscriber a certain probability of his originating a call. This probability can then to a certain extent be considered independent of the calls of other subscribers in the vicinity of the point of time considered. It changes, however, with absolute time, partly at random, partly in the same way as for most subscribers.

By this line of thought we arrive at the following tentative probability scheme for description of the properties of telephone traffic. During very short time intervals the laws for random traffic are valid; the probability of call arrivals varies, however, with absolute time. The mathematical expressions for the properties of random traffic comprise one parameter for the call arrivals, viz. the mean number of call arrivals per time unit y, and really this *call intensity* (call arrival intensity) should obviously be assumed to vary with absolute time. This probability scheme can be described mathematically as follows. The probability that a call arrival occurs during an unlimitedly short interval dT following immediately after a point of time T is $y(T)\,dT$ and is independent of the position of the other earlier call arrivals. The function $y(T)$ is then a known function of absolute time. It should be observed that the probability of a call arrival is obviously not independent of the points of time of earlier call arrivals when the values of $y(T)$ are considered to be unknown, because in this case the probability of a certain value of this quantity will be influenced by the number of earlier call arrivals.

If a group of devices is hunted for by a traffic with the said properties, and further we know an initial state within the group, then no fundamental difficulties exist for computation of the traffic conditions within this group (more details are given in Chapter 7). It is even unnecessary to assume that the traffic is stationary. If we want to use the results of such a computation in practice, however, we must establish a suitable form of the function $y(T)$ by measurement. Now it seems fundamentally impossible to determine the value of the call arrival intensity y at a certain point of time. The only thing that can be done is to investigate the number of call arrivals and their positions in the vicinity of this point of time. From these quantities we can obviously draw conclusions regarding the probability that the call arrival intensity has certain values. The real value at the point of time in question cannot, however, be determined. To avoid this difficulty, it seems to be necessary to introduce the assumption that the variations of the call arrival intensity are so slow that we can assume that the traffic conditions at every point of time in a statistical sense are the same as for a completely random traffic in which the call arrival intensity is equal to the call arrival intensity at the point of time in question. If the call arrival intensity changes, then transient conditions develop in which the traffic conditions adapt themselves to the new traffic level. From a theoretical viewpoint such a transient condition ends only after an unlimited time. However, most physical transient conditions can with good accuracy be considered to end within a comparatively short, limited time. Whether this way of looking at things is permissible in this case cannot be determined without closer investigation. In Chapter 7 a theoretical investigation is made which shows that the results that are obtained without considering the influence of the transient condition are in a certain sense valid also if the transient condition is considered. The decision whether the assumption made above is justified or not in the treatment of normal traffic problems (that the variations of the call arrival intensity are very slow) can only be made on the basis of measurements. These measurements should show how the conditions which are theoretically arrived at from the probability scheme coincide with reality. The measurements which are reported at the end of this work give strong support to the possibility of successful work with a probability scheme for the traffic based upon the assumptions given above.

We can now pose the question as to how the assumption about the slow character of the variation of the call arrival intensity solves the related difficulty in determination of the intensity function. The answer is the following: by disregarding the influence of the transient conditions on the variation of intensity, the derived traffic conditions will be independent of the time sequence of the intensity changes. In order to determine the traffic conditions we need not know the function $y(T)$ in detail, but only a function that is an integration of $y(T)$. This function gives the intensity distribution. Contrary to $y(T)$, this function can, however, in principle be determined uniquely by measurements. The subject will be further discussed in subsequent parts of this work.

Let us denote the function that is to replace the complicated and indeterminable function $y(T)$ by *the distribution function $G(x)$ for the call arrival intensity*. It is defined as the probability that, at a point of time chosen at random for a certain traffic, the call arrival intensity is $< x$. (In this connection it is found appropriate to define the distribution function as increasing and not, as in most earlier parts of this work, as decreasing.) We can also say that $G(x)$ indicates the part of the total time with a call arrival intensity $< x$.

As $G(x)$ is independent of the point of time at which the traffic conditions are observed, we must assume that the mean value of the call arrival intensity does not change with the absolute time. This can be expressed in a more accurate way as follows. We assume that the traffic process goes on during an unlimited time before and after any arbitrary point of time. We look at a time span T that starts at an arbitrarily chosen point of time. The ratio between that part of the time span T, when the call arrival intensity is $< x$, and T is in this time span $G_0(x)$. As $T \to \infty$, $G_0(x)$ must converge in probability towards $G(x)$, regardless of the choice of the starting point of time. The same condition must apply for any time span T positioned before an arbitrarily chosen point of time. We easily find that these assumptions imply that the process is stationary in the sense defined in Chapter 2. Due to the conditions of slow variations in the call arrival intensity the assumption of this stationary property does not imply any further limitation in respect to the applicability of these findings in practice.

The traffic process that is defined by the probability scheme given above will be described as a *random traffic modified by slow variations of call arrival intensity* or, shorter, *traffic with slow variations*. In those cases when the impact of the transient phenomena at a change of the call arrival intensity cannot be neglected, the traffic process should be described as *traffic with fast variations*.

As regards variations of the call arrival intensity of real traffic, a certain physical continuity should be present, so that the variations during very small time intervals are all very small. Because of that we can assume that any distribution function $G(x)$ for the intensity that should represent the intensity distribution of real traffic should be continuous. Because of the theoretical treatment it will be advantageous, however, not to limit ourselves to continuous functions, but to permit $G(x)$ to have a limited number of jumps. As $G(x)$ is defined as an increasing distribution function, it starts at 0 for $x = 0$ although there is nothing to prevent a jump already at the point $x = 0$. $G(x)$ never decreases with increasing x and approaches the value one as $x \to \infty$. As an example of an important discontinuous distribution we can look at the case where the call arrival intensity does not change with time but always has the value y, i.e. the case of normal random traffic. The intensity distribution function $G(x)$ is then 0 for $x < y$ and for $x > y$ it is 1. It follows that $G(x)$ has a jump of size 1 at the point $x = y$. With such an intensity distribution, which in what follows constitutes a very important special case, we will talk of *zero variation*.

In the domains where $G(x)$ has a limited derivative $g(x)$, the jump within an unlimited small interval dx is a differential quantity $dG(x) = g(x)\,dx$. In the points where $G(x)$ has a jump which differs from zero this should also be designated as $dG(x)$. As the sum of the probabilities of all possibilities is equal to 1, then

$$\int_0^\infty dG(x) = 1 \tag{49 a}$$

If the derivative $g(x)$ is limited and defined in all points, then (49 a) can be written as an ordinary integral

$$\int_0^\infty g(x)\,dx = 1$$

If on the contrary $G(x)$ has jumps in a limited number of points, we can interpret (49 a) as a *Stieltjes* integral. The *Stieltjes* integral concept can, as is well known, be interpreted in the following simple way. In the domains where the derivative exists, the integration is carried out in the usual way and to this the jumps in $G(x)$ are added. The simple expression (49 a) then substitutes the sum of a limited number of terms that correspond to the jumps together with the sum of the integrals within the continuous domains between the jumps. As the *Stieltjes* integral concept is often used in this work, the following further remark will be made. We can interpret the more general integral

$$\int_0^\infty F(x)\,dG(x)$$

as the sum of the usual integrals in the domains where the derivative of $G(x)$ exists and the sum of $F(x)$ multiplied by the value of $G(x)$ in points with jumps greater than 0.

Because of the definition of the probability scheme for traffic with slow variations, whether it is a quantity not equal to 0 or a differential, $dG(x)$ gives the probability that, at an arbitrarily chosen point of time, we find a call arrival intensity between x and $x + dx$ or, expressed more simply, that it equals x. We can also say that $dG(x)$ expresses the part of the total time when the arrival intensity has the value x. Then it is at once clear that

$$y = \int_0^\infty x \, dG(x) \tag{49 b}$$

expresses the mean of the call arrival intensity for processes with slow variations. It may be remarked that this expression is also valid for processes with fast variations.

When it comes to applying principles for traffic with slow variations of the call arrival intensity to the traffic conditions in groups of devices we must take into account that the traffic contains two elements, i.e. not only the call arrivals but also the call durations. The latter can also change their character with absolute time. It seems to be justified, therefore, to try to take this into account by introducing a corresponding variation concept also for the call durations. Thereby the difficulty arises that not only the mean call duration s but also its distribution function may vary with time. Now investigations of the conditions of random traffic have shown that they are in general independent of or very little dependent on the form of the distribution function of the call durations.[9] It is then sufficient to take into account only the variations of the mean of the call duration s. There is no doubt that such variations are present to a certain extent in real telephone traffic. Their magnitude and impact on the traffic conditions, however, seem to be insignificant in comparison with the impact of the variations of the call arrival intensity. As a first approximation the mean call duration s can therefore be considered to be constant. At the end of this chapter this question will be more closely considered and it will be evident that the treatment will be much more complicated if the variations in s are taken into account.

In traffic research up to now, expressions have been presented on the assumption of random traffic for a great number of mean values of different kinds that are of great importance for estimation of the traffic conditions in telephone plant. These mean values can be referred to two different types designated *time mean values* and *call mean values*. A time mean value is the fraction of the total traffic time in which a certain state of traffic in the considered telephone plant exists. An example of this is the risk time or *time congestion* that designates the mean value of the fraction of the traffic time when all devices in a group available for a certain traffic are busy. A call mean value is the mean of the number of calls that have a certain property or are treated in a certain way in the telephone plant considered. This number will then be referred to a certain time span, e.g. to each time unit, or to a different quantity of calls. As an example the *call congestion* can be mentioned that gives the relation of the mean number of calls in a group blocked in a certain way to the mean number of calls that hunt for the group in the same time span and that could be blocked in the same way. Another example is the distribution function of the inter-arrival times that designates the mean value of the fraction of the total number of calls for which the time distance to the subsequent call arrival is greater than a certain time.

We will now let $F(x)$ represent a time mean value that is valid for random traffic with the call intensity x. Then $F(x)$ gives the mean value of the fraction of the total time when a certain state, to be represented by A, occurs in a telephone plant. We assume now that the traffic is not random but has slow variations with the properties described above. Over a long time span T the call arrival intensity should be x during the time $T\,dG_0(x)$. During this time, then, A will occur on an average in the time $TF(x)\,dG_0(x)$. Consequently A will occur within the total time T on an average during the time

$$T\int_0^\infty F(x)\,dG_0(x)$$

The ratio between this time and T, viz.

$$\int_0^\infty F(x)\,dG_0(x)$$

then expresses the mean value of the fraction of the time T during which A occurs. When now $T \to \infty$, then $G_0(x) \to G(x)$ according to the definition of stationariness given earlier in this chapter. Consequently

$$\int_0^\infty F(x)\,dG(x) \tag{50 a}$$

expresses the corresponding mean value for traffic with slow variations that for random traffic is expressed by $F(x)$.

For call mean values the conditions are somewhat different. Let $f(x)$ denote the mean value of the relative number of calls of a certain kind which is denoted by A, and let this number refer to the total number of calls of random traffic whose call arrival intensity is x. If we consider, as above, a time span T, for traffic with slow variations, then during the part of this time span $T\,dG_0(x)$ in which the call intensity is x, on average $T\,x\,f(x)\,dG_0(x)$ calls of type A will occur. Hence during the total time T on an average

$$T\int_0^\infty x\,f(x)\,dG_0(x)$$

calls of type A will occur. If now, according to (49 b), we denote by y the total number of calls in the traffic with slow variations, then during the time T on an average $y\,T$ such calls will occur. The mean value of the fraction of all calls of type A during the time T is then

$$\frac{1}{y}\int_0^\infty x\,f(x)\,dG_0(x)$$

For $T \to \infty$ this expression will pass into

$$\frac{1}{y}\int_0^\infty x\,f(x)\,dG(x) \tag{50 b}$$

The expression (50 b) then gives the same call mean value for traffic with slow variations as will be expressed by $f(x)$ for random traffic.

By means of the expressions (50 a) and (50 b), all the results obtained for random traffic can be applied to traffic with slow variations. (In some special cases other expressions may be necessary, but these are generally of minor interest. They have been passed over here, as they can, when needed, easily be derived.) From the analogy between random traffic and traffic with slow variations we should not be led to assume that most conditions of random traffic directly correspond to those of traffic with slow variations, which in fact is a rather general process with unlimited after-effect. It seems to be appropriate here to enter in some detail into some important applications of formula (50) that will clearly illustrate the different conditions of random traffic and traffic with slow variations.

For a group of n devices with full availability in a loss system the mean fraction of the total time when all devices are busy is $E_{1,n}(A)$ according to formula (11). This formula only applies to random traffic whose call arrival intensity is y and whose mean call duration is s, so that $A = sy$ is the traffic quantity. $E_{1,n}(A)$ is then a mean value of time that expresses the time congestion. The corresponding time congestion in a traffic with slow variations will then according to (50 a) be

$$\int_0^\infty E_{1,n}(sx)\,dG(x) \tag{51 a}$$

For random traffic $E_{1,n}(A)$ also expresses the call congestion, i.e. the relative number of blocked calls. From formula (50 b) it now follows that for traffic with slow varations this will be

$$\frac{1}{y}\int_0^\infty x\, E_{1,n}\,(sx)\; dG\,(x) \qquad (51\text{ b})$$

However, that is not the same expression as for time congestion. Call congestion and time congestion are always different, except for zero variation, since this means that the traffic is entirely random and the two expressions coincide. There are many examples of similar cases in which quantities that coincide for random traffic have completely different values for traffic with slow variations.

A very important quantity for the use of the theory for slow variations is the distribution function for the inter-arrival times. For random traffic with call intensity x, e^{-xt} expresses the mean value of the proportion of call arrivals for which the time distance to the subsequent call arrival is at least t. It is apparently a question of a call mean value and, according to (50 b), the inter-arrival time distribution function at traffic with slow variations will be

$$\varphi\,(t) = \frac{1}{y}\int_0^\infty x e^{-xt}\, dG\,(x) \qquad (52)$$

that is, a decreasing distribution function that for $t = 0$ has the value 1, which is obtained from (49 b). That the condition (6 b) is fulfilled also follows from (49 a).

The distribution introduced in Chapter 2 for the next call arrival $\vartheta\,(t)$ is now obtained by carrying out the integration in (8 a). We get

$$\vartheta\,(t) = \int_0^\infty e^{-xt}\, dG\,(x) \qquad (53)$$

This expression can also be easily obtained if we observe that $dG\,(x)$ gives the probability that a point of time chosen arbitrarily will be within a domain on the time axis where the call arrival intensity is x.

The distribution functions for inter-arrival time and for the next call arrival are thus not the same for traffic with slow variations (except by zero variation). This is a further example of quantities that coincide for random traffic but not for traffic with slow variations.

Finally an expression will be given for one more traffic quantity that has a great importance for the treatment of measurements, that will be discussed later on. In a group with so great a number of devices that blocking can be disregarded, the mean value for random traffic of the fraction of the total time when totally p but not more devices are busy is expressed by[8]

$$\frac{(sy)^p}{p!}\, e^{-sy}$$

This is a time mean value. Thus we get, according to (50 a), the following corresponding expression at slow variations:

$$\int_0^\infty \frac{(sx)^p}{p!}\, e^{-sx}\, dG\,(x) \qquad (54)$$

The distribution hereby expressed has been investigated lately (1940) by *O. Lundberg*[6] in connection with the treatment of problems within the statistics of illness and accidents. It appears also in this case as the result of a kind of slow variation, though in *Lundberg*'s work it is a question not of a time variation but of a variation of the homogeneity of the statistical data. The expressions of the general type (50) introduced above are to be denoted *integrated traffic functions*. Before their properties are investigated in detail in the next chapter, we will consider the question earlier touched upon, viz. how to present variations in the mean value of the call duration s. In principle this can be treated in the same way as for the variations of the intensity of the call arrivals. We assume that the variations in both y and s are so slow that we can apply the results from random traffic, as both quantities have almost constant values. We need then only introduce a distribution function of two variables $G_{sy}(x_s, x_y)$ that are defined so that $dG_{sy}(x_s, x_y)$ gives the probability of finding at an arbitrarily chosen point of time a mean duration of calls x_s and a call arrival intensity x_y. A mean traffic value, for example a time mean value, that has the value $F(x_s, x_y)$ for random traffic has, for slow variations in both s and y, the value

$$\int_{x_s=0,\, x_y=0}^{\infty,\,\infty} F(x_s, x_y)\, d\, G(x_s, x_y)$$

In order to apply such formulas the function $G_{sy}(x_s, x_y)$ must be available for determination or at least for estimation by measurements. This seems to imply serious difficulties. For the call intensity distribution function introduced earlier we can, on the contrary, find several principles for a determination by measurements, as will be shown later. We easily obtain the relation

$$\int_{x_s=0}^{\infty} dG_{sy}(x_s, x_y) = G(x_y)$$

but by this evidently $G_{sy}(x_s, x_y)$ is not determined. A probability scheme that is based upon slow variations in s as well as in y does not seem to fulfill the desire expressed above, viz. that it could be used in a simple way for computations in practice. This fact need probably not imply any severe disadvantages. As has just been mentioned above, the variations of the mean occupation time s should be small by comparison with the variations of the call arrival intensity. Still more important is that most mean traffic values are functions only of the product $s\,y$ and not of the quantities s and y separately. It is therefore generally sufficient to look at the variations in the *traffic intensity* A and its distribution function can be measured in a similar way as for the call arrival intensity.

It is thus appropriate to introduce a distribution function $G_A(x_A)$ defined such that $dG_A(x_A)$ is the probability that the product $s\,y$ has the value x_A at an arbitrarily chosen point of time. A remark should be inserted here about the notations used. The traffic quantity A is usually defined as the sum of the lengths of occupations during a certain time divided by the time in question. We can then also speak of the instantaneous traffic quantity, which is the number of occupations at a given moment. However, in the case of the quantity x_A introduced above it is not a question of momentary variations in the traffic quantity but of the slow variations in the product $s\,y$, where s and y are the defined mean values at random traffic. It is therefore appropriate, to avoid misunderstandings, to speak of variations in traffic intensity instead of in traffic quantity. We now easily find that, through the use of the aforementioned distribution function $G_A(x_A)$ we obtain the same formulas as were derived earlier in this chapter for $G(x)$ by putting $s = 1$. We can also appropriately use the formula without any changes by deleting the index A, as in every separate case it is a matter of finding out whether the distribution function $G(x)$ considered can be classified as belonging to the traffic intensity or to the call arrival intensity. In accordance herewith we assume in all later derivations that the mean occupation time is not subject to any intensity variations. The possible variations of s in real traffic will show up when we determine the distribution function $G(x)$ by measurements, by our getting somewhat different results depending upon the quantities to which the measurements refer.

Chapter 6

Properties of integrated traffic functions

In this chapter we will treat a series of general properties of integrated traffic functions that appear in traffic with slow variations. For this purpose a general expression of the form

$$F_s = \int_a^b F(x)\, d\, G(x) \tag{55}$$

is considered.

It is assumed that the function $dG(x)$ within the integration interval is never negative. Further, the integral (55) is assumed to be limited. On the other hand no condition of the form (49 a) needs to be introduced in advance. The function $F(x)$, which will be designated a *kernel function*, can arbitrarily vary and thus also assume negative values. However, it is always assumed that its first and second derivatives are limited in the domain a to b. The integration limits can be positive as well as negative, though the latter will never appear in the applications in further treatments.

To investigate the quantity F_s that results from the summation (55), a function

$$H(\xi) = \int_a^b F(m + (x - m)\,\xi)\, d\, G(x) \tag{56}$$

is considered. The condition $a \leqq m \leqq b$ for the quantity m applies (we can always assume $a < b$). With the condition $0 \leqq \xi \leqq 1$ the variable

$$m + (x - m)\,\xi$$

is clearly always within the interval a to b and hence the integral (56) is completely defined for all values of ξ between 0 and 1. Further

$$H(0) = F(m) \int_a^b d\, G(x) \tag{56 a}$$

and

$$H(1) = F_s \tag{56 b}$$

Assume now that the ν first derivatives of $F(x)$ exist and are limited in the interval a to b, then

$$H^{(\nu)}(\xi) = \int_a^b (x - m)^\nu\, F^{(\nu)}(m + (x - m)\,\xi)\, d\, G(x)$$

is also limited. If now $H(\xi)$ for $\xi = 1$ is expanded by *Maclaurin*'s series we get

$$H(1) = \sum_{\sigma=0}^{\nu-1} \frac{F^{(\sigma)}(m)}{\sigma\,!} \int_a^b (x - m)^\sigma\, d\, G(x) +$$

$$+ \frac{1}{\nu\,!} \int_a^b (x - m)^\nu\, F^{(\nu)}(m + (x - m)\,\xi)\, d\, G(x)$$

where ξ in the residual term lies in the interval 0 to 1. If the notation

$$m_0 = \int_a^b d\, G(x) \tag{57 a}$$

is introduced, m can always be chosen so that

$$\int_a^b x\,dG(x) = m \int_a^b dG(x) = m\,m_0 \qquad (57\text{ b})$$

where m satisfies the condition $a \leqq m \leqq b$ which was set up above. The above development for $H(1)$, which is equal to F_s, then gives

$$F_s = m_0 F(m) + \sum_{\sigma=2}^{\nu-1} \frac{F^{(\sigma)}(m)}{\sigma!} \int_a^b (x-m)^\sigma\,dG(x) + $$

$$+ \frac{1}{\nu!} \int_a^b (x-m)^\nu F^{(\nu)}(m + (x-m)\xi)\,dG(x) \qquad (58)$$

A special case of this formula, which is important for further applications, occurs for $\nu = 2$ and $m_0 = 1$. This is the case when $G(x)$ is a distribution function, whose whole variation takes place in the interval a to b. We then get

$$F_s = F(m) + \frac{1}{2} \int_a^b (x-m)^2 F''(m + (x-m)\xi)\,dG(x) \qquad (59)$$

In the integrand on the right-hand side the quadratic expression is always positive except for $x = m$. It is assumed that $G(x)$ has no zero variation, so that $x = m$ is not the only place where $G(x)$ varies. If now $F''(x)$ differs from zero in the entire interval a to b and always has the same sign, then clearly the integral in (59) will differ from zero and have the same sign as $F''(x)$. The same is also true when $F''(m) = 0$. From (59) we get the following theorem:

If $G(x)$ has no zero variation and $F''(x) > 0$ and < 0 respectively for all x in the domain where $G(x)$ varies, then $F_s > F(m)$ and $< F(m)$ respectively, m being the mean value of the variable x. The proposition is true also if $F''(m) = 0$. For zero variation $F_s = F(m)$.

It should be remarked that, if $F(m) = 0$, then the theorem is true also if $m_0 \neq 1$, i.e. also if the interval a to b does not comprise the whole domain for the variation of $G(x)$.

As a first application of this theorem the moments of the distribution $G(x)$ will be considered, where in accordance with the present applications we assume that the variation of $G(x)$ takes place wholly within the domain 0 to ∞. The nth moment is defined, as is well known, by

$$m_n = \int_0^\infty x^n\,dG(x) \qquad (60\text{ a})$$

so that m_1 is equal to the earlier introduced mean value m and $m_0 = 1$. If now x^n is considered as a kernel function, then its second derivative will be $n(n-1)x^{n-2}$, which is positive for all positive x and $n > 1$. From this proposition we then immediately get

$$m_n > m_1^{\,n}$$

This inequality can, however, be made considerably sharper. We introduce $z = x^{n-1}$. (60 a) will then pass over to

$$m_n = \int_0^\infty z^{\frac{n}{n-1}}\,dG(z^{\frac{1}{n-1}})$$

Further

$$m_{n-1} = \int_0^\infty z\,dG(z^{\frac{1}{n-1}})$$

We notice in the first place that, while $G(x)$ never decreases with increasing x, also

$$G(z^{\frac{1}{n-1}})$$

never decreases with increasing z, always under the assumption that $n > 1$. If we now consider

$$z^{\frac{n}{n-1}}$$

as a kernel function, then the variable z has the mean value m_{n-1} according to the relation above. The inequality just derived then immediately gives

$$m_n > m_{n-1}^{\frac{n}{n-1}}$$

and hence

$$m_1 < m_2^{\frac{1}{2}} < m_3^{\frac{1}{3}} < \ldots < m_n^{\frac{1}{n}} \qquad (60\text{ b})$$

where for zero variation equality signs should be substituted for all $<$ signs.

The relation (60 b) is well known and can be found in most textbooks on probability. It is of interest, as was already hinted at in Chapter 2, that (60 b) can be made still sharper if the distribution function is completely monotonic. The distribution function is then suitably defined as decreasing. Each such completely monotonic distribution function can then be written in the form

$$G(x) = \int_0^\infty e^{-x\eta}\, dg(\eta)$$

where $g(\eta)$ is a function which never decreases with increasing η with the limit values $g(0) = 0$ and $g(\infty) = 1$. The nth moment of the distribution is defined by

$$M_n = -\int_0^\infty x^n\, dG(x) = \int_0^\infty \int_0^\infty x^n e^{-x\eta}\, \eta\, dg(\eta)\, dx \tag{61 a}$$

If the order of integration is changed and if the integration with respect to x is carried out, we get

$$\frac{M_n}{n!} = \int_0^\infty \frac{1}{\eta^n}\, dg(\eta)$$

We introduce here a new variable

$$z = \frac{1}{\eta^{n-1}}$$

with which for $n > 1$ the limits will be interchanged as $z = 0$ for $\eta = \infty$ and $z = \infty$ for $\eta = 0$. We then get

$$\frac{M_n}{n!} = \int_\infty^0 z^{\frac{n}{n-1}}\, dg\left(z^{-\frac{1}{n-1}}\right)$$

Now, however,

$$g\left(z^{-\frac{1}{n-1}}\right)$$

is a function decreasing with z, for which reason we suitably introduce the function increasing with z

$$g_1(z) = 1 - g\left(z^{-\frac{1}{n-1}}\right)$$

whereby the limits of the integral will be interchanged, and we obtain

$$\frac{M_n}{n!} = \int_0^\infty z^{\frac{n}{n-1}}\, dg_1(z)$$

In the same way we get

$$\frac{M_{n-1}}{(n-1)!} = \int_0^\infty z\, dg_1(z)$$

Now according to the assumptions

$$\int_0^\infty dg(\eta) = 1$$

from which follows

$$\int_0^\infty dg_1(z) = 1$$

We can then use the inequality expressed in (59) and consider

$$z^{\frac{n}{n-1}}$$

as a kernel function of the variable z, which according to the above relation has the mean value $M_{n-1}/(n-1)!$. We then get the inequality

$$\frac{M_n}{n!} > \left\{\frac{M_{n-1}}{(n-1)!}\right\}^{\frac{n}{n-1}}$$

and from that

$$\frac{M_1}{1!} < \left(\frac{M_2}{2!}\right)^{\frac{1}{2}} < \left(\frac{M_3}{3!}\right)^{\frac{1}{3}} < \ldots < \left(\frac{M_n}{n!}\right)^{\frac{1}{n}} \tag{61 b}$$

which implies a significant sharpening of (60 b). For zero variation the equality sign will be substituted for the sign $<$.

The call congestion according to (51 b) will be chosen as the next example. If we here put $s = 1$, in accordance with the remarks at the end of the last chapter, then the kernel function will be

$$\frac{x}{y} E_{1,n}(x)$$

In the appendix it will be shown that the second derivative of this function is always positive. The expression (51 b) is then greater than the one which is obtained if in the kernel function we put x equal to the average y, i.e. greater than $E_{1,n}(y)$. Thereby it is proved that the call congestion by traffic with slow variations is always greater then by pure random traffic. An exception – besides the one always applicable for zero variation – arises for $y = 0$, when both quantities become zero. Correspondingly, totally trivial exceptions from the inequality proposition naturally always arise when the average of the variable equals zero. Regarding the time congestion according to (51 a), no corresponding relation applies to blocking for random traffic. The second derivative of the function $E_{1,n}(x)$ is as a matter of fact positive for smaller x-values and negative for larger. From that it can be seen that, for small averages of x, the time congestion for slow variations must be greater than for random traffic, whilst the reverse condition must arise at greater averages of x.

Furthermore, the properties of the distribution function (52) for the distances between call arrivals will be investigated by means of the inequality theorem. This distribution function is e^{-yt} for random traffic. For zero variation, which will not be considered any further, (52) will pass into this form. The two curves (52) and e^{-yt} have now each the value 1 for $t = 0$, which is then an intersection point for the curves. To show that, besides this, there is always one more but only one more intersection point, the derivative of $\varphi(t)$ in the zero point will be considered. From (52) we get primarily

$$\varphi'(t) = -\frac{1}{y}\int_0^\infty x^2 e^{-xt}\, dG(x) \qquad (62)$$

Accordingly it follows that

$$\varphi'(0) = -\frac{1}{y}\int_0^\infty x^2\, dG(x) \qquad (62\text{ a})$$

When x^2 is considered as a kernel function of the variable x that has the mean value y, then its second derivative is positive and $= 2$, the inequality theorem giving

$$-\varphi'(0) > y$$

Now, the derivative of e^{-yt} in the zero point is $-y$. Therefrom it follows that $\varphi(t)$ decreases more sharply immediately behind $t = 0$ than the curve e^{-yt}. For small positive t-values, $\varphi(t)$ must then lie under the curve e^{-yt}. For large t-values on the other hand, $\varphi(t)$ lies over e^{-yt}, which can be proved in the following way.

As $G(x)$ has no zero variation, we can always find a positive quantity $a < y$ with the property that within the interval 0 to a, the variations of $G(x)$ are greater than 0. Then it follows that

$$0 < \frac{1}{y}\int_0^a x\, e^{-xt}\, dG(x) < \varphi(t)$$

The latter inequality follows from the fact that the contributions which with certainty differ from zero for $x > y$ are not contained in the integral. Now for all x within the interval 0 to a

$$e^{-xt} \geq e^{-at}$$

from which follows that

$$\frac{1}{y}\int_0^a x\, e^{-xt}\, dG(x) \geq \frac{e^{-at}}{y}\int_0^a x\, dG(x) = Ke^{-at}$$

where $K > 0$ with certainty. From the next to last inequality we get

$$0 < Ke^{-at} < \varphi(t)$$

Since $K > 0$ and $a < y$, t can always be chosen so large that

$$Ke^{-at} > e^{-yt}$$

viz.

$$K > e^{-(y-a)t}$$

It follows that it is always possible to find a t-value so large that $\varphi(t) > e^{-yt}$. Since, as has just been shown, $\varphi(t) < e^{-yt}$ for small positive t-values, the two curves must have an intersection point between these two t-values. Call the abscissa of the intersection t_0. Then we shall prove that there cannot be any further intersection point either for positive or negative t. Let us assume that t_1 gives an intersection point and that t_1 differs from 0 and t_0. The following two relations would then be satisfied:

$$\int_0^\infty e^{-xt_0}\frac{x}{y}\,dG(x) = e^{-yt_0}$$

$$\int_0^\infty e^{-xt_1}\frac{x}{y}\,dG(x) = e^{-yt_1}$$

If we now introduce a new variable $z = e^{-xt_0}$ and further

$$dG_1(z) = -\frac{x}{y}\,dG(x)$$

the two relations will pass over to

$$\int_0^1 z\,dG_1(z) = e^{-yt_0}$$

$$\int_0^1 z^{\frac{t_1}{t_0}}\,dG_1(z) = e^{-yt_1}$$

In order to apply the inequality theorem in the integration interval 0 to 1, the relation

$$\int_0^1 dG_1(z) = \int_0^\infty \frac{x}{y}\,dG(x) = 1$$

must be satisfied; this is true, however, for the mean expression (49 b). In the two formulas above we can use the expression

$$z^{\frac{t_1}{t_0}}$$

as a kernel function of the variable z which has the mean value e^{-yt_0}. Now the second derivative of the kernel function

$$\frac{d^2}{dz^2}\left(z^{\frac{t_1}{t_0}}\right) = \frac{t_1(t_1-t_0)}{t_0^2}\,z^{\frac{t_1}{t_0}-2}$$

but this is always positive, or always negative as, according to the assumption, t_1 is different from 0 and t_0. The inequality theorem then gives

$$\int_0^1 z^{\frac{t_1}{t_0}}\,dG_1(z) \neq \left(e^{-yt_0}\right)^{\frac{t_1}{t_0}} = e^{-yt_1}$$

which contradicts the fact that t_1 is an intersection point between $\varphi(t)$ and e^{-yt}. There is then no further intersection between $\varphi(t)$ and e^{-yt} than the one at $t = 0$ and the one at $t = t_0$. The only exception occurs for zero variation, in which case the curves coincide.

For our further investigations it is interesting to study closely the *difference curve for the inter-arrival times*

$$D(t) = \varphi(t) - e^{-yt}$$

From the conditions just shown for the points of intersection we already get a good picture of the general course of this function. Fig. 6 a shows a difference curve that is derived from an intensity function $G(x)$ with jumps different from zero in three points and otherwise without variation. Fig. 6 b shows the course of $G(x)$. In the points $x = 0.4, 0.6, 0.8$ the jumps in $G(x)$ are 0.625, 0.250, and 0.125 respectively. The distribution function from (52) has the form

$$\varphi(t) = 0.5\,e^{-0.4t} + 0.3\,e^{-0.6t} + 0.2\,e^{-0.8t}$$

The mean value of the call arrival intensity is $y = 0.5$ so that the equation for the difference curve is

$$D(t) = \varphi(t) - e^{-0,5t}$$

The derivative of the difference curve for $t = 0$ is -0.04. We observe that the value of the ordinate is everywhere very small. This condition, which is characteristic for the difference curve for the inter-arrival times and also applies to very strong intensity variations, is of importance for the determination of the intensity function by measurements.

For a closer study of the general course of the difference curve we can use the expansion (58). In the expression (52) for the distribution function

$$\varphi(t) = \frac{1}{y}\int_0^\infty xe^{-xt}\,dG(x)$$

the part

$$\frac{x}{y}e^{-xt}$$

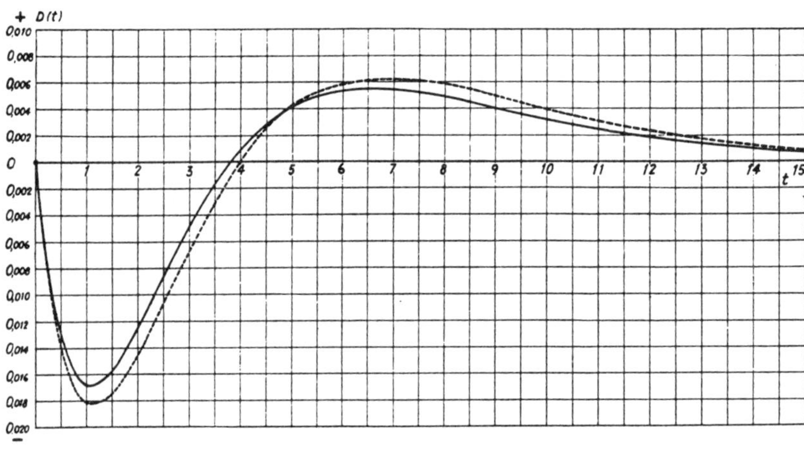

6 a

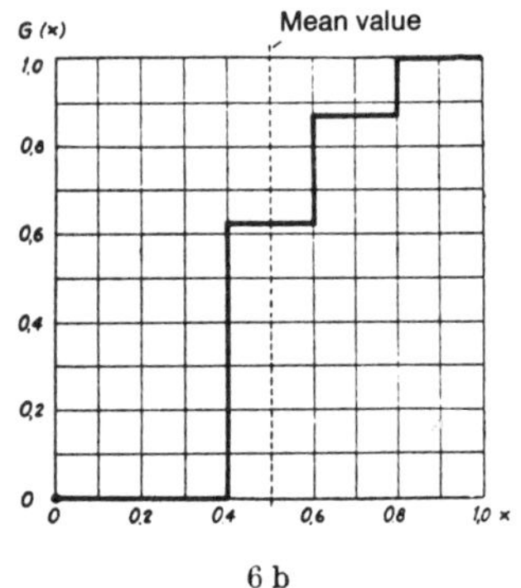

6 b

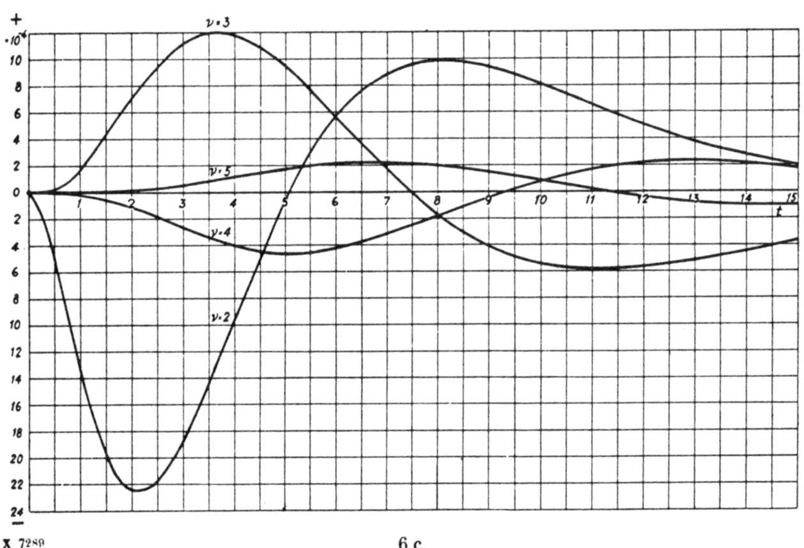

6 c

Fig. 6. Example of a difference curve and its approximate values. The distribution function is constructed from an intensity distribution according to Fig. 6 b with the mean value $y = 0.5$. Fig. 6 a shows the difference curve $D(t)$ with a full line and the approximate curve $D_2(t)$ with a dotted line. Fig. 6 c shows on a larger scale $D_\nu(t) - D(t)$ for $\nu = 2, 3, 4$, and 5.

is then considered as a kernel function for the variable x that has the mean value y according to (49 b). Now, the νth derivative of the kernel function with respect to x is

$$\frac{(-1)^\nu}{y} t^{\nu-1}(tx - \nu)\, e^{-xt}$$

which can easily be proved by induction. When we then, for the νth moment of the distribution function $G(x)$ in respect of the axis $x = y$, introduce the designation

$$(-y)^\nu m_{y,\nu} = \int_0^\infty (x-y)^\nu \, d\,G(x) \qquad (63)$$

we obtain according to the expansion (58)

$$D(t) = e^{-yt} \sum_{\sigma=2}^{\nu} \frac{(yt)^{\sigma-1}}{\sigma!}(yt-\sigma)\, m_{y,\sigma} + R_{\nu+1} \qquad (64)$$

where $R_{\nu+1}$ is a residual term, the form of which can be obtained from (58).

If we investigate the expression

$$D_\nu(t) = e^{-yt} \sum_{\sigma=2}^{\nu} \frac{(yt)^{\sigma-1}}{\sigma!}(yt-\sigma)\, m_{y,\sigma} \qquad (65)$$

we find that it will usually very rapidly converge towards $D(t)$. The very first term

$$D_2(t) = \frac{1}{2} yt(yt - 2) e^{-yt} m_{y,2}$$

in which $m_{y,2}$ is always positive gives a good picture of $D(t)$. Accordingly $D_2(t)$ starts for $t = 0$ with the value zero, then has negative values and a minimum, an intersection point with the t-axis, a positive maximum and for large t-values tends asymptotically towards the t-axis. The intersection with the t-axis occurs at $yt = 2$. From the derivative

$$D_2'(t) = -\frac{y}{2}\{(yt)^2 - 4yt + 2\} e^{-yt} m_{y,2} \quad (66)$$

we obtain, according to the definition of $m_{y,2}$ in (63)

$$D_2'(0) = -y\, m_{y,2} = -\frac{1}{y}\int_0^\infty x^2\, dG(x) + y \quad (66\text{ a})$$

From (62 a), however, it is clear that this is the same expression as we obtain for the derivative of $D(t)$ for $t = 0$. Both curves, then, start with the same slope. From (66) we obtain further the following values for the zero points of the derivatives

$$\left.\begin{matrix} yt_1 \\ yt_2 \end{matrix}\right\} = 2 \pm \sqrt{2} = \left\{\begin{matrix} 0.58579 \\ 3.41421 \end{matrix}\right. \quad (66\text{ b})$$

where t_1 gives the minimum and t_2 the maximum. The value of $D_2(t)$ in these points is

$$\begin{aligned} D_2(t_1) &= -0.23058\, m_{y,2}; \\ D_2(t_2) &= +0.07944\, m_{y,2} \end{aligned} \quad (66\text{ c})$$

The values obtained for $D_2(t)$ of position and magnitude of the extreme points as well as for the intersection with the t-axis are generally in very good accordance with the corresponding values for $D(t)$, although systematic displacements occur for certain types of intensity functions; more will be said about this later. In order to get a more accurate representation of $D(t)$ we can also investigate the higher $D_\nu(t)$. Then it is necessary, however, in order to get a meaningful discussion, to know the relations between the different $m_{y,\nu}$. Such an investigation will be carried out later.

In Fig. 6 a some curves are drawn in order to show the accuracy of the approximation by $D_\nu(t)$ of $D(t)$ for the distribution function in question. In this case

$$\begin{aligned} m_{y,2} &= +0{,}08 \\ m_{y,3} &= -0{,}024 \\ m_{y,4} &= +0{,}0176 \\ m_{y,5} &= -0{,}0096 \end{aligned}$$

For $D_2(t)$ the intersection with the t-axis is at $t = 4$ whilst for $D(t)$ it is at $t = 3.80$. For $D_2(t)$, furthermore, the minimum is at $t = 1.17$ and with the magnitude 0.0184, whilst the corresponding values of $D(t)$ are 1.00 and 0.0169. The maximum for $D_2(t)$ falls at the point $t = 6.83$ with magnitude 0.00635. The corresponding values for $D(t)$ are 6.70 and 0.00552. In Fig. 6 c the curves $D_\nu(t) - D(t)$ for $\nu = 2$, 3, 4 and 5 are all drawn in 10-times ordinate scale. From this we can see the extraordinarily good representation of $D(t)$ that can be obtained from an expansion (64) that contains only a small number of terms.

In Chapter 3 we have given a quantity, the *form factor* ε, to judge the general form of a distribution function and its consequent properties. The form factor is defined as the second moment of the distribution function divided by the squared mean value. For a pure exponential function the form factor is 2 and for all other completely monotonic functions > 2, which also follows from (61 b). For the distribution function $\varphi(t)$ for the inter-arrival times the form factor according to (52) is

$$\varepsilon = -y^2 \int_0^\infty t^2 \varphi'(t)\, dt = 2y \int_0^\infty \frac{dG(x)}{x} \quad (67)$$

An approximate value for the form factor can be found by using the relation

$$\varphi(t) \approx e^{-yt} + D_2(t)$$

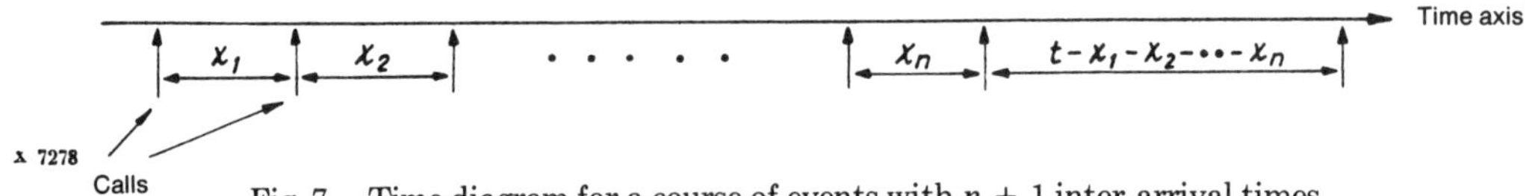

Fig. 7. Time diagram for a course of events with $n + 1$ inter-arrival times.

From (65) we then get

$$\varepsilon \approx 2 + 2\, m_{y,\,2} \tag{68 a}$$

If the series in the expansion (64) is convergent and the residual term tends towards zero, we get from the definition of $D(t)$ the exact expression

$$\varepsilon = 2 + 2\sum_{\sigma=0}^{\infty} m_{y,\,\sigma} \tag{68 b}$$

According to (63) and (67) this has the same significance as

$$\int_0^{\infty} \left\{ \frac{y}{x} - \sum_{\sigma=2}^{\infty} \left(\frac{y-x}{y} \right)^{\sigma} \right\} d\,G(x) = 1$$

This is a rather interesting expression. If the variation of $G(x)$ falls only in the range zero to $2y$ of x, the geometrical series in the integrand is always convergent; in this case the relation is trivial. However, it is true also for $x \geqq 2y$ where the series is divergent. The form factor for the distribution function chosen as example in Fig. 6 is 2.135. This value shows that $\varphi(t)$ comes very close to a pure exponential distribution. That is also clear from the relatively small values that $D(t)$ adopts. From the approximate formula (68 a) we get in this case the form factor 2.160.

Because of the importance of the distribution function of the inter-arrival times for the determination of the variations of the intensity by measurements, the properties of this function have here been investigated in some detail. In such measurements, however, an inconvenience appears owing to the fact that the distribution function $\varphi(t)$ only insignificantly differs from the pure exponential function. A relatively voluminous measurement material is then often necessary. It is clear, however, that this inconvenience can be avoided by regarding the inter-arrival times not in relation to the immediately subsequent calls but instead to calls farther away. Unfortunately this method has shortcomings in other respects and because of that it has not been used. It is, however, conceivable that future measurements can be based on such methods, so that it seems appropriate to touch upon the properties of the relevant functions.

The probability that the distance between a call chosen at random and the $(n+1)^{\text{th}}$ subsequent call arrival, is at least t is $\varphi_n(t)$. The above considered distribution function for the inter-arrival times will then be given the designation $\varphi_0(t)$. In order to determine $\varphi_n(t)$ for slow intensity variations we must first determine the corresponding quantity for purely random traffic. The simplest way is first to determine the frequency function, i.e. the probability that the distance between a call arrival and the next $(n+1)^{\text{th}}$ call arrival falls between t and $t + dt$. Fig. 7 shows such a course of events. The distances between the subsequent call arrival are respectively x_1, $x_2, \ldots x_n$ and $t - x_1 - x_2, \ldots -x_n$. If the calls arrive at random with an average y per time unit, then the probability that after an arbitrarily chosen call arrival such a course of events will occur is

$$y e^{-y x_1}\, dx_1\, y e^{-y x_2}\, dx_2 \,\ldots$$
$$\ldots\, y e^{-y x_n}\, dx_n\, y e^{-y(t - x_1 - \cdots - x_n)}\, dt$$

or

$$y^{n+1} e^{-yt}\, dx_1\, dx_2 \ldots dx_n\, dt$$

In order to get the probability that the time is t between the first and the last call arrival, this expression should be integrated n times, viz.

x_n from 0 to t
x_{n-1} from 0 to $t - x_n$
. . . .
x_1 from 0 to $t - x_n - x_{n-1} \dots - x_2$

The multiple integral so obtained is easy to work out and we get from the above expression

$$y \frac{(yt)^n}{n!} e^{-yt} dt$$

This is the frequency function looked for, i.e. the probability that the course of events lasts exactly the time t. The distribution function $\varphi_n(t)$, i.e. the probability that the course of events lasts at least the time t, we then get by integrating this expression in respect of t, between the limits t and ∞. We then get

$$\varphi_n(t) = \left(1 + yt + \frac{(yt)^2}{2!} + \dots \frac{(yt)^n}{n!}\right) e^{-yt}$$

As is the case for all decreasing distribution functions, this expression equals 1 for $t = 0$ and tends towards 0 as $t \to \infty$. The mean of the distribution we get from

$$-\int_0^\infty t\,\varphi'_n(t)\,dt = \frac{n+1}{y}$$

which is self-evident, as the course of events comprises $n + 1$ inter-arrival times.

The resulting expression for $\varphi_n(t)$ is now true for random traffic. In order to get the corresponding distribution function for traffic with slow intensity variations we have to observe that it expresses a call average. With the aid of formula (50 b) we then get

$$\varphi_n(t) = \frac{1}{y}\int_0^\infty x\left(1 + xt + \frac{(xt)^2}{2!} + \dots + \right.$$
$$\left. + \frac{(xt)^n}{n!}\right) e^{-xt}\, d\,G(x) \qquad (69)$$

as an expression of the distribution function sought for a course of events of $n + 1$ inter-arrival times in the traffic with slow intensity variations. Hereby, y is the mean value of the variable x and we find as before, that $(n + 1) / y$ is the mean of the distribution $\varphi_n(t)$. For $n = 0$ (69), as would be expected, goes over into (52).

If in (69) we now consider

$$\frac{x}{y}\left(1 + xt + \frac{(xt)^2}{2!} + \dots + \frac{(xt)^n}{n!}\right) e^{-xt}$$

as a kernel function, the second derivative in respect of x will be

$$\frac{1}{y}\, t\,(xt - n - 2)\,\frac{(xt)^n}{n!}\, e^{-xt}$$

In order to get an idea of the properties of (69) we can use the expansion (58) and include only the first term for $\sigma = 2$. As an approximate expression for the difference between $\varphi_n(t)$ for slow intensity variations according to (69) and for pure random traffic, i.e. at zero variation, we then obtain

$$\frac{1}{2}(yt - n - 2)\,\frac{(yt)^{n+1}}{n!}\, e^{-yt}\, m_{y,2} \qquad (70)$$

From this we see that $D_2(t)$ according to (65) is a special case for $n = 0$.

The difference curve (70) starts with the value 0 for $t = 0$ and has a further intersection point with the t-axis at $yt = n + 2$. Between $t = 0$ and this t-value (70) is negative and for greater t-values positive. At $t = 0$ the curve is a tangent to the t-axis in an n-fold point. For large t-values the curve goes asymptotically towards the t-axis. The derivative in t of (70) is

$$-\frac{y}{2}\left\{(yt)^2 - 2(n+2)\,yt + (n+1)\,(n+2)\right\}\frac{(yt)^n}{n!}\, e^{-yt} m_{y,2}$$

This has two zero points, viz. for

$$yt = n + 2 \pm \sqrt{n+2}$$

The lower t-value corresponds to a minimum of (70) and the higher to a maximum. Of special interest is now that these extreme values of (70) in their absolute values become considerably greater than the corresponding values of the difference curve $D(t)$ of the distance between the call arrivals and also increase with increasing n. Thus already for $n = 2$ the minimum is more than double and the maximum more than three times the values for $n = 0$. For $n = 7$ the minimum is about 10 times and the maximum 20 times greater than for $n = 0$.

Thus when measuring courses of events of $n + 1$ inter-arrival times where $n > 0$, we obtain greater deviations from the conditions prevailing with random traffic than on measurements from simple inter-arrival times.

Of the integrated traffic functions presented in Chapter 5 the distribution of next call arrival (53) will also be briefly touched upon, although it does not seem to have any great importance for the carrying out of measurements. In (53) we can consider e^{-xt} as kernel function of the variable x with the mean value y. The second derivative of the kernel function with respect to x is now always positive (with the exception of $x = 0$). From the inequality theorem derived from (59) it then follows that (53) is greater than e^{-yt}. This is due to the fact that, although e^{-yt} is the function that we get from (53) at zero variation, $\vartheta(t)$ generally has not the same mean value as the distribution e^{-yt}. The mean value of $\vartheta(t)$ will according to (53) be

$$-\int_0^\infty t\,\vartheta'(t)\,dt = \int_0^\infty \frac{dG(x)}{x}$$

With the help of (67) this can be written as

$$\frac{1}{y}\,\frac{\varepsilon}{2}$$

Now, except at zero variation, as shown earlier, ε is always > 2, from which it follows that the mean value of the distribution $\vartheta(t)$ is greater then $1/y$, i.e. greater than the mean value of the distribution e^{-yt}. Applying the inequality theorem to (53), we find that contrary to the cases that have been investigated earlier, there is not obtained any comparison earlier, no comparison is obtained with an exponential function having the same mean value.

Finally, one more integrated function should be investigated which is of greatest importance for later considerations, viz. the mean value expressed by (54). It is appropriate to choose $s = 1$ in accordance with the remarks at the end of the preceeding chapter. Then from (54)

$$P_n = \int_0^\infty \frac{x^n}{n!}\,e^{-x}\,dG(x) \tag{71}$$

which expresses the mean value of the part of the total time during which n devices, but not more, are busy at the same time in a full availability group. The traffic arriving at the group has the mean value y. The expression is true only on the assumption that the group has such a great number of devices that the blocking can be neglected.

In (71) we consider

$$\frac{x^n}{n!}\,e^{-x}$$

as a kernel function. Its second derivative is then

$$\left\{x^2 - 2\,n\,x + n\,(n-1)\right\}\frac{x^{n-2}}{n!}\,e^{-x}$$

In order to get an idea of the course of (71) we use the expansion (58) where only the first term for $\sigma = 2$ is included in the series. We then get from (58)

$$P_n \approx \frac{y^n}{n!}\,e^{-y} + \frac{1}{2}\left\{y^2 - 2ny + n\,(n-1)\right\}\frac{y^n}{n!}\,e^{-y}\,m_{y,2} \tag{71 a}$$

or, if we introduce $P_{n,0} = y^n\,e^{-y}/n!$ to designate the value of P_n at zero variation,

$$\frac{P_n}{P_{n,0}} \approx 1 + \frac{1}{2}\left\{n^2 - (2\,y+1)\,n + y^2\right\} m_{y,2} \tag{71 a}$$

The formula also applies to the special cases $n = 0$ and 1.

If we investigate the squared expression on the right side of (71 a)

$$n^2 - (2\,y+1)\,n + y^2$$

as a function of n, we will find that it is positive for small values of n and decreases for increasing n. For the integer value of n that is next to and higher than

$$y+\frac{1}{2}-\sqrt{y+\frac{1}{4}}$$

the expression becomes negative. At an n-value $y + 1/2$ the expression will be a minimum. Then it increases without limit for increasing n and becomes positive at the integer value next to and higher than

$$y+\frac{1}{2}+\sqrt{y+\frac{1}{4}}$$

The integrated traffic function (71) can then, depending on the value of n, become either greater or smaller than the corresponding value for pure random traffic. Of special interest is that for values of n large in relation to y the relation (71 a) can be considerably greater than 1. This signifies that groups that are dimensioned for normal blocking values at pure random traffic can have much higher blocking if they are exposed to traffic with intensity variations. If we assume, e.g. that $y = 20$ and $n = 35$, then at an intensity variation as shown in Fig. 6 with $m_{y,2} = 0.08$ the right member in (71 a) will be 8.6. In such a case a considerable increase of the blocking from the 35th device on takes place.

Chapter 7

Rapid intensity variations

Among the conditions that were introduced in the theories for slow intensity variations in Chapter 5, the very assumption of slowness seems to be the most critical with regard to the possibilities of applying the results to real telephone traffic. This assumption seems to entail that we can apply the results only to traffic whose intensity variations take place so slowly that we can at least approximately count upon the presence of an equilibrium at every moment. However, it seems to be rather probable or at least possible, that real telephone traffic to a considerable extent also shows rather rapid intensity variations, with the result that the equilibrium condition cannot always be considered to be fulfilled with sufficient accuracy. Fortunately, however, this condition does not essentially influence the applicability of the results based only on slow intensity variations. In order to demonstrate this, the conditions in conjunction with rapid intensity variations will be considered in this chapter, viz. cases in which equilibrium is not a priori assumed. As it is not possible to study in detail all those traffic problems that have previously been treated in the literature on the assumption of equilibrium, it has proved necessary to limit the account to an investigation of the most important traffic quantitites. The results, however, seem to permit rather general conclusions about the connection between the traffic conditions during rapid and slow intensity variations.

In this connection it should be remarked that the results in this chapter also seem to be of value as a starting point for investigations in other domains than those pertinent to the present instance. As examples of such domains may be mentioned the after-effect of blocking and the waiting-time conditions resulting from occasional overloads in delay systems.

We now consider traffic whose call intensity $y(T)$ is a function of the absolute time T. The probability of a call arrival within the time interval T to $T + dT$ is then $y(T)\,dT$ but is otherwise independent of the position of earlier call arrivals. This traffic is assumed to hunt in a group with an unlimited number of devices, so that no consideration need be paid to the influence of blocking. As in the treatment in Chapter 3, we assume that the holding times follow an exponential distribution so that

$$e^{-\frac{t}{s}}$$

expresses the probability that a holding time will last at least the time t, the mean holding time being s. This assumption is introduced only to simplify the mathematical treatment and does not seem to be necessary for the validity of the results.

The probability that at the point of time T altogether n (but no more) of the devices in the group are busy, i.e. that the state n is present, will now be designated $f_n(T)$. If $y(T)$ is constant with time and equal to y_0 and the process has already gone on for an unlimited time, then f_n, as is well known, takes the *Poisson* expression[8]

$$f_n = \frac{(sy_0)^n}{n!} e^{-sy_0} \tag{72}$$

In order to derive $f_n(T)$ when $y(T)$ is variable, we look at the probability $f_n(T + dT)$ that at the time $T + dT$ the state n applies. This will be the case if any of the following three mutually exclusive events has occurred:

1 At the point of time T the state n applied and during the time dT there was no new call arrival and none of the ongoing calls ended.

2 At the point of time T the state $n + 1$ applied. During the time dT there was no new call arrival but one of $n + 1$ ongoing calls ended.

3 At the point of time T the state $n - 1$ applied. During the time dT there was one new call arrival but none of the $n - 1$ ongoing calls ended.

If $n = 0$ there is evidently no case 3. Further it should be remarked that the state n, naturally, can arise from a state $n \pm r$ present at the time T with $r > 1$. In such a case, however, at least two events must occur with probabilities each proportional to dT (as will be shown below). The probabilities for such courses of events will then be proportional to higher powers of dT and these disappear when $dT \to 0$. Thus, from the outset we can disregard the possibility of such changes of state.

Now the probability of event 1 above is the product of the probabilities of the following three events which are independent of each other:

- At the point of time T the state n is in force. The probability for that is $f_n(T)$.
- During the short interval dT there is no call arrival. The probability of this is $1 - y(T)\,dT$.
- During the short interval dT none of the ongoing calls ends. The probability of this is, as is easily seen,

$$1 - \frac{n}{s}\,dT.$$

The probability of the whole course of events is then

$$(1 - y(T)\,dT)\left(1 - \frac{n}{s}\,dT\right) f_n(T)$$

In the same way we obtain the probability of event 2

$$(1 - y(T)\,dT)\frac{n+1}{s}\,dT\,f_{n+1}(T)$$

and the probability of event 3

$$y(T)\,dT \cdot \left(1 - \frac{n-1}{s}\,dT\right) f_{n-1}(T)$$

The sum of the probabilities of the three possible events must then be equal to $f_n(T + dT)$. We thus get the following relation

$$f_n(T + dT) = (1 - y(T)\,dT)\left(1 - \frac{n}{s}\,dT\right) f_n(T) + (1 - y(T)\,dT)\frac{n+1}{s}\,dT\,f_{n+1}(T) + y(T)\,dT\left(1 - \frac{n-1}{s}\,dT\right) f_{n-1}(T)$$

If here $dT \to 0$ we get the following relation

$$f_n'(T) + \left(y(T) + \frac{n}{s}\right) f_n(T) = \frac{n+1}{s} f_{n+1}(T) + y(T) f_{n-1}(T) \tag{73 a}$$

For $n = 0$ there is no case 3 and we get the following special relation

$$f_0'(T) + y(T) f_0(T) = \frac{1}{s} f_1(T) \tag{73 b}$$

As the sum of the probabilities of all possible states at every point of time must equal 1, we further get the relation

$$\sum_{n=0}^{\infty} f_n(T) = 1 \tag{73 c}$$

In order to treat the resulting system of linear differential equations, a generating function is introduced, defined by

$$F(T, x) = \sum_{n=0}^{\infty} f_n(T)\, x^n \tag{74}$$

From (73 c) it follows that the series is absolutely convergent for $|x| \leqq 1$. When now (73 a) is multiplied by x^n and the resulting relations for $n = 0, 1, 2,$ and so on are added, we get

$$\sum_{n=0}^{\infty} f_n'(T)\, x^n + y(T) \sum_{n=0}^{\infty} f_n(T)\, x^n +$$

$$+ \frac{1}{s} \sum_{n=1}^{\infty} n\, f_n(T)\, x^n = \frac{1}{s} \sum_{n=0}^{\infty} (n+1)\, f_{n+1}(T)\, x^n +$$

$$+ y(T) \sum_{n=1}^{\infty} f_{n-1}(T)\, x^n$$

By means of the expression (74) this can be rewritten in the form

$$\frac{\partial F(T,x)}{\partial T} + y(T)\, F(T,x) + \frac{1}{s} \sum_{n=1}^{\infty} n f_n(T)\, x^n =$$

$$= \frac{1}{s} \sum_{n=0}^{\infty} (n+1)\, f_{n+1}(T)\, x^n + y(T)\, x\, F(T,x)$$

From (74) we easily get

$$\sum_{n=1}^{\infty} n\, f_n(T)\, x^n = x\, \frac{\partial F(T,x)}{\partial x}$$

and

$$\sum_{n=0}^{\infty} (n+1)\, f_{n+1}(T)\, x^n = \frac{\partial F(T,x)}{\partial x}$$

The relation above is then simplified to:

$$\frac{\partial F(T,x)}{\partial T} + \frac{1}{s}(x-1)\, \frac{\partial F(T,x)}{\partial x} =$$

$$= y(T)\,(x-1)\, F(T,x) \tag{75}$$

This is a linear partial differential equation of the first order. Its general solution is obtained, as is well known,[14] in the form of an arbitrary function of two mutually independent integrals to the system

$$dT = s\, \frac{dx}{x-1} = \frac{dF}{y(T)\,(x-1)\, F}$$

From the first relation we get the integral

$$\int dT = s \int \frac{dx}{x-1}$$

or

$$(x-1)\, e^{-\frac{T}{s}} = c_1 \tag{76}$$

where c_1, is an integration constant. From the relation

$$\frac{dF}{F} = y(T)\,(x-1)\, dT$$

we further get the integral

$$\ln F = \int y(T)\,(x-1)\, dT$$

In order to work out the integral on the right-hand side we first eliminate $x - 1$ by means of (76). We then get

$$\ln F = c_1 \int y(T)\, e^{\frac{T}{s}}\, dT$$

If we here eliminate the integration constant c_1, again by means of (76), we get after some transformations

$$F\, e^{-(x-1)\, e^{-\frac{T}{s}} \int y(T)\, e^{\frac{T}{s}}\, dT} = \text{const.}$$

The general solution of (75) we then get by equating the left-hand side of this expression to an arbitrary function R of the left-hand side of (76). We then get

$$F(T,x) = e^{(x-1)\, e^{-\frac{T}{s}} \int y(T)\, e^{\frac{T}{s}}\, dT}\, R\left\{(x-1)\, e^{-\frac{T}{s}}\right\}$$

The integral in the exponent is an undetermined integral. Nothing prevents us, however, from putting it in determined form, as the integration constant can be included in R. We then get the expression

$$F(T,x) = R\left\{(x-1)\, e^{-\frac{T}{s}}\right\} e^{(x-1)\, e^{-\frac{T}{s}} \int_0^T y(z)\, e^{\frac{z}{s}}\, dz} \tag{77}$$

For the undetermined function R we then get a condition by putting $x = 1$. From (73 c) we get $F(T, 1) = 1$ and then (77) gives $R(0) = 1$. In order to determine R completely, however, a knowledge of the initial state is necessary, for instance for $T = 0$. We then have

$$F(0, x) = R(x-1) \tag{78}$$

We shall firstly assume that the call intensity $y(T)$ is constant over time and equal to y_0. From (77), after having carried out the integration, we then get

$$F(T, x) = R\left\{(x-1)\, e^{-\frac{T}{s}}\right\} e^{sy_0 (x-1)(1-e^{-\frac{T}{s}})}$$

If we here let $T \to \infty$ (noting that always $R(0) = 1$), we get

$$\lim_{T \to \infty} F(T, x) = e^{sy_0 (x-1)} \tag{79}$$

If we expand the right-hand side in increasing powers of x, we see that the coefficient of x^n is precisely f_n according to (72). At a constant call intensity the traffic conditions within the group, *regardless of the initial state*, with increasing time tend to the equilibrium according to the *Poisson* distribution.

If now at a point of time T the generating function has the form

$$F(T, x) = e^{A(x-1)}$$

we find from a comparison with (79) that the traffic conditions are the same as in the state of equilibrium and at a constant traffic intensity A. We shall then name this quantity A *the equilibrium intensity of the traffic* or more shortly, *equilibrium traffic*. As a rule it is a function of T but not necessarily equal to the traffic intensity $s\,y(T)$ at the time – in this chapter designated *real traffic*. The equilibrium traffic can then be considered as a fictitious traffic. Through its introduction the traffic conditions can be compared with those prevailing at equilibrium. We shall now assume that at an arbitrarily variable call intensity the initial state is determined by the right-hand side of (79), so that the equilibrium traffic is sy_0 at the point $T = 0$. From (78) we then get

$$R(x-1) = e^{sy_0 (x-1)}$$

and from that

$$R\left((x-1)\, e^{-\frac{T}{s}}\right) = e^{sy_0 (x-1)\, e^{-\frac{T}{s}}}$$

It then follows from (77)

$$F(T, x) = e^{(x-1)\left\{e^{-\frac{T}{s}} \int_0^T y(z)\, e^{\frac{z}{s}}\, dz + sy_0\, e^{-\frac{T}{s}}\right\}}$$

The equilibrium traffic at the point of time T is thus in this case

$$\int_0^T y(z)\, e^{-\frac{T-z}{s}}\, dz + sy_0\, e^{-\frac{T}{s}} \tag{80}$$

The term

$$sy_0\, e^{-\frac{T}{s}}$$

will be named *the equilibrium traffic at the point of time zero attenuated to the point of time T*. Further the expression

$$\int_0^T y(z)\, e^{-\frac{T-z}{s}}\, dz$$

will be named the *traffic smoothed during the time interval* 0 *to* T.

Instead of the state of departure at the point of time $T = 0$ we can choose any point of time T_0. The expression (80) can then be interpreted in the following way:

The equilibrium traffic at a point of time T equals the sum of the equilibrium traffic at the point of time T_0 attenuated to the time T and the traffic smoothed during the time interval T_0 to T.

This important result should be named *the addition theorem of the equilibrium traffic*. It is proved above only for a group which is so great that the blocking has no influence. However, it seems to have a considerably wider ap-

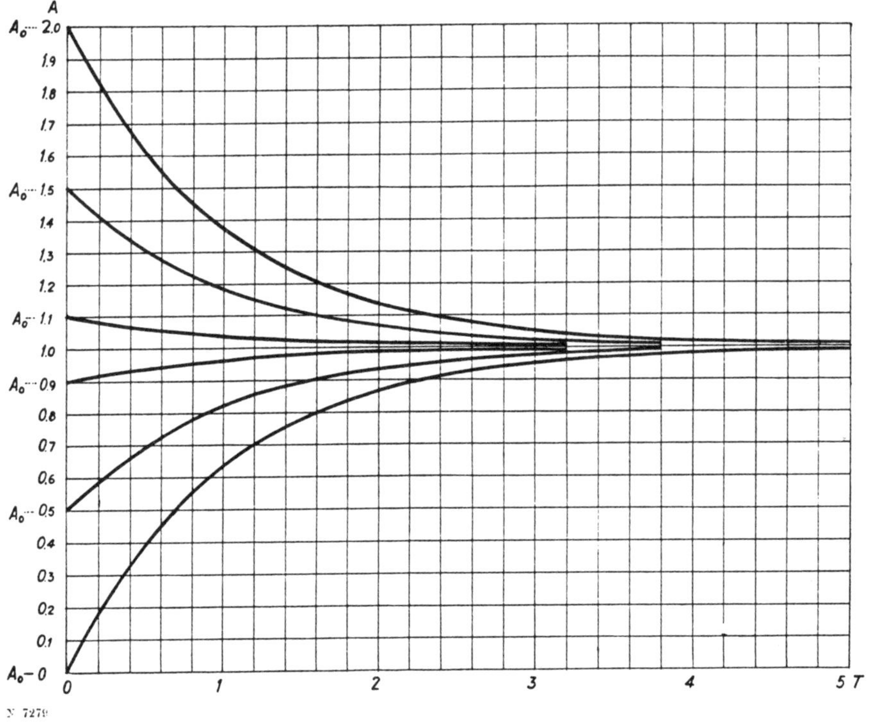

Fig. 8. The equilibrium traffic A as a function of the time T at the equilibrium traffic A_0 as point of departure. From the point of time $T = 0$ the real traffic is 1. The time is measured with the mean holding time s used as unit.

plicability and is in any case valid also for all limited groups of a loss system, though the proof becomes more complicated.

The conditions during discontinuous variations of the call intensity will be investigated as an application of the addition theorem. Let the equilibrium traffic at the point of time T_0 be sy_0. During the time T_0 to T_1 the call intensity is constant and equal to y_1. During the time T_1 to T_2 the call intensity is constant and equals y_2, and so on, so that finally during the time T_{r-1} to T_r the call intensity is constant and equal to y_r. The expression for the equilibrium traffic at the point of time T_r will then according to (80) be

$$sy_r - \sum_{\nu=0}^{r-1} s\,(y_{\nu+1} - y_\nu)\, e^{-\frac{T_r - T_\nu}{s}} \tag{81}$$

If $s\,(y_{\nu+1} - y_\nu)$ is designated as a positive-going traffic discontinuity we then get from (81) the following result:

During variation of the call intensity by discontinuities the equilibrium traffic at a certain point of time is equal to the traffic occurring at the same time reduced by all previous discontinuities of traffic attenuated to the point of time in question.

If we imagine that an arbitrary variation of the call intensity has developed exclusively by discontinuities, which may be of infinitesimal magnitudes, the stated theorem gives directly a formulation of the general expression (80) for the equilibrium traffic.

The attenuation to equilibrium is clearly dependent only upon the value of the mean

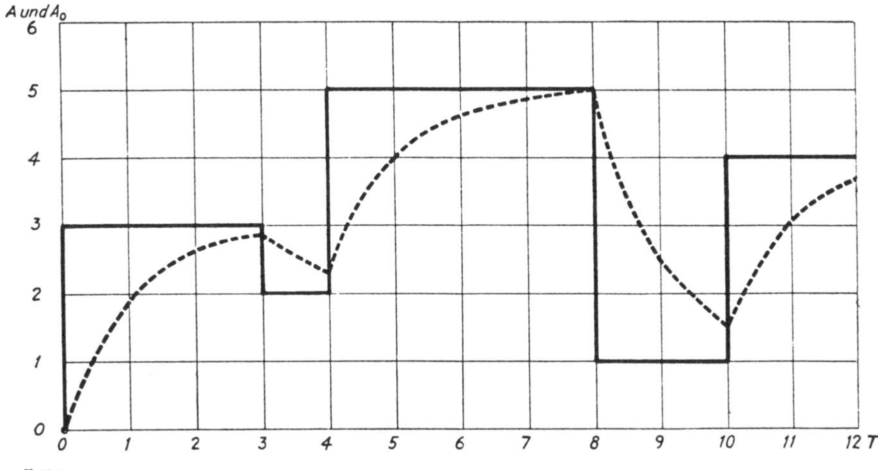

Fig. 9. The equilibrium traffic A as a function of the time T with variation of the real traffic A_0 in steps. Starting state for T_0 is $A = A_0 = 0$.

- - - - - A ——— A_0

holding time s. The smaller s is in relation to the speed of the variations of the call intensity, the closer does the equilibrium traffic approach the value for the actual traffic, and the traffic conditions can be computed from the theory of slow variations with so much greater accuracy. An idea of the speed of the attenuation can be obtained from Fig. 8 which shows the variation with T of the equilibrium traffic after a discontinuity at the point of time 0 from the equilibrium traffic A_0 to the traffic value 1. The time T is measured in terms of the mean holding time s as unit. As we see in Fig. 8, the equilibrium traffic very rapidly approaches the actual traffic. After 3 to 4 mean holding times, the difference even for large discontinuities is without significance. This shows that in practice under normal traffic conditions the transient approach to the new condition after a sudden traffic variation can be considered that in practice under normal traffic conditions the transient approach to the new condition after a sudden traffic variation can be considered to be terminated within 10 minutes. In groups with very short holding times, for example registers with 10 s mean holding time, equilibrium can be reached after only half a minute.

For the graphic representation of the course of the equilibrium traffic with discontinuous intensity variations we need only start from an increasing and a decreasing curve of the same kind as in Fig. 8. The later variation of the equilibrium traffic is determined by the relation between the equilibrium traffic and the real traffic. Where the latter shows discontinuities we have to move to another part of the auxiliary curve. The curve of the equilibrium traffic has kinks at these points. Fig. 9 shows a curve that has been drawn in this way for an equilibrium traffic with a traffic intensity varying in steps.

For determination of the indefinite function R in the solution (77) we considered initial conditions where the probability distribution corresponds to the one prevailing at equilibrium and a certain traffic. We can now also regard initial conditions where a certain state prevails in the group, for instance when μ of the devices in the group are busy at the same time. Such cases have less interest for the present comparison between slow and rapid variations. But as they could have an importance for the solution of problems in other fields, they will be considered here briefly.

When neither more nor less than μ of the devices in the group are busy at the time 0, this signifies that the probability of the state μ at the same point of time is 1 and the probability of any other state is 0. Then $F(0, x)$ and according to (78), also the function $R(x-1)$ therewith equals x^{μ}. We then have

$$R\left\{(x-1)e^{-\frac{T}{s}}\right\} = \left\{1+(x-1)e^{-\frac{T}{s}}\right\}^{\mu}$$

and the solution (77) gives

$$F(T,x) = \left\{1+(x-1)e^{-\frac{T}{s}}\right\}^{\mu} e^{(x-1)e^{-\frac{T}{s}}\int_0^T y(z)e^{\frac{z}{s}}dz} \qquad (82\text{ a})$$

For $\mu = 0$ this becomes the same solution as for an equilibrium traffic with the value 0 as starting state, which is quite natural. When the call intensity is constant in time and equal to y we get from (82 a)

$$F(T,x) = \left\{1+(x-1)e^{-\frac{T}{s}}\right\}^{\mu} e^{sy(x-1)\left(1-e^{-\frac{T}{s}}\right)} \qquad (82\text{ b})$$

When we here want to determine the coefficients for x^{ν} we generally get a rather complicated sum. Apart from $\nu = 0$ a special simplification for $\nu = \mu$ occurs, as we get

$$f_{\mu}(T) = e^{-sy\left(1-e^{-\frac{T}{s}}\right)-\mu\frac{T}{s}} \cdot$$

$$\cdot \sum_{\sigma=0}^{\mu} \frac{1}{\sigma!}\binom{\mu}{\sigma}\left\{sy\, e^{\frac{T}{s}}\left(1-e^{-\frac{T}{s}}\right)^2\right\}^{\sigma} \qquad (82\text{ c})$$

In order to get a more detailed comparison between the traffic conditions for rapid and slow intensity variations it seems appropriate to look at periodical variations of $y(T)$. We assume firstly that the call intensity has a constant and a sine wave component, so that

$$y(T) = y_0 + y_1 \cos \omega T \qquad (83)$$

After having made the integration in (77) the solution takes the form

$$F(T,x) = R\left\{(x-1)e^{-\frac{T}{s}}\right\} e^{(x-1)sy_0\left(1-e^{-\frac{T}{s}}\right)} \cdot$$

$$\cdot\, e^{(x-1)\frac{sy_1}{1+s^2\omega^2}\left\{\cos\omega T + s\omega\sin\omega T - e^{-\frac{T}{s}}\right\}}$$

If we here let $T \to \infty$ then always, as has been remarked earlier, $R \to 1$ independently of the starting state and we get in the limit

$$F(T,x) = e^{(x-1)s\left\{y_0+\frac{y_1}{1+s^2\omega^2}(\cos\omega T + s\omega\sin\omega T)\right\}}$$

When the transient period is over, the traffic conditions at a point of time T are the same as for a constant call intensity with the value

$$y_0 + \frac{y_1}{1+s^2\omega^2}(\cos\omega T + s\omega \sin \omega T)$$

The mean holding time s multiplied by this expression in this case thus gives the equilibrium traffic. This can be put into a simpler form by the introduction of an auxiliary angle α, defined by

$$\left.\begin{aligned}\sqrt{1+s^2\omega^2}\,\sin\alpha &= s\omega\\ \sqrt{1+s^2\omega^2}\,\cos\alpha &= 1\end{aligned}\right\} \qquad (84)$$

The equilibrium traffic then takes the form

$$sy_0 + \frac{sy_1}{\sqrt{1+s^2\omega^2}}\cos(\omega T - \alpha) \qquad (85\text{ a})$$

and the generating function is

$$F(T,x) = e^{(x-1)\left\{sy_0+\frac{sy_1}{\sqrt{1+s^2\omega^2}}\cos(\omega T-\alpha)\right\}} \qquad (85\text{ b})$$

For a sine wave variation of the traffic intensity, all equilibrium traffic thus appears at the limit of $T \to \infty$ to vary with the same periodicity and to have the same constant term as the real traffic. The equilibrium traffic has a certain phase shift backwards in time and the amplitude of the oscillations is attenuated in relation to the real traffic. However, it is quite clear that the equilibrium traffic and the real traffic have the same mean value. The phase shift as well as the attenuation are determined by the value of $s\omega$, that is, 2π times the number of periods per mean holding time. When $s\omega \to 0$ the equilibrium traffic will equal the real traffic. If on the other hand $s\omega \to \infty$ then the equilibrium traffic in (85 a) approaches sy_0 and the traffic conditions remain as for pure random traffic without any intensity variation. Very rapid periodical intensity variations thus have no influence on the traffic conditions within a group.

Of interest is now the mean value during a period of the solution expressed by (85 b). For integration over a complete period we can disregard the phase shift and the wanted mean value becomes

$$\frac{\omega}{2\pi} e^{sy_0(x-1)} \int_0^{\frac{2\pi}{\omega}} e^{\frac{sy_1}{\sqrt{1+s^2\omega^2}}(x-1)\cos\omega T} dT \quad (86\text{ a})$$

If $\omega \to 0$, this expression will take the form

$$\frac{\omega}{2\pi} e^{sy_0(x-1)} \int_0^{\frac{2\pi}{\omega}} e^{sy_1(x-1)\cos\omega T} dT \quad (86\text{ b})$$

which evidently is the mean value that we obtain from the theory of slow variations for call intensity variations according to (83). It is, of course, possible to write (86 b) in the form of an integral of type (50 a). In this case, however, when the call intensity is known directly as a function of time, it is a roundabout way first to compute a distribution function G for the call intensity.

We meet no difficulty in extending the results for the call intensity to cases with an arbitrary number of oscillation terms. Assume that a call intensity has the form

$$y(T) = y_0 + \sum_{\nu=1}^{r} y_\nu \cos(\omega_\nu T + \gamma_\nu)$$

Then we find an equilibrium traffic of the form

$$sy_0 + \sum_{\nu=1}^{r} \frac{sy_\nu}{\sqrt{1+s^2\omega_\nu^2}} \cos(\omega_\nu T + \gamma_\nu - \alpha_\nu)$$

where the phase shifts α_ν are determined by expressions similar to (84). As the oscillations generally need not have any common period, to obtain the mean value of the generating function, we must consider the integral

$$\frac{1}{T_0}\int_0^{T_0} F(T, x)\, dT$$

and determine its limit for $T_0 \to \infty$. In this case we cannot generally disregard the phase shifts.

The results for the periodical intensity variations can be summarized as follows. If we wish to use the methods that have been presented in the theory for slow intensity variations we can take into account the fact that the variations in reality are not idealistically slow by using a modified traffic, the equilibrium traffic. In this the oscillations will show some phase displacements and, in comparison to the oscillations of the real traffic, are in a certain manner attenuated. The values obtained in this way are always between those found for the real traffic without modification and those obtained for pure random traffic with the mean intensity of the real traffic. When the oscillation periods are very small in comparison to the mean holding time, the traffic can be treated as pure random traffic. When, on the other hand, the oscillation periods are large in comparison to the mean holding time, the modifications will be of minor importance. The intensity distribution of the real traffic can be used with good approximation. The attenuation factor

$$1/\sqrt{1+s^2\omega^2}$$

will then be close to 1 and the phase displacement measured in time units close to the mean holding time s. As a numerical example we can consider a period equal to 100 times the mean holding time. The attenuation factor is then 0.998 and the phase displacement 0.0616 radians.

The investigations up to now have comprised the states in a group of such a size that the influence of blocking can be disregarded. If consideration is paid to the blocking, the mathematical treatment will be more complicated but the results do not seem to deviate in any essential respect. With regard to other traffic quantities than the states, however, the conditions may be different. An example of these quantities of special interest is the distribution function for inter-arrival times. We shall therefore investigate to what extent this is influenced by the speed of the intensity variations.

The occupations of the devices in a group constitute a physical state within the group with a certain inertia, apparent from the fact that a change in the call intensity only gradually causes a change in the equilibrium of the states. No counterpart to this exists for the inter-arrival times. Here equilibrium prevails at any point of time between the arrivals of the calls and the magnitude of the instantenous call intensity. In spite of that, as will be shown below, there is a certain influence of the speed of the variations which is due to the fact that the intensity varies between successive calls.

If the call intensity only varies stepwise, it is clear that the inter-arrival times that occur entirely during intervals when the intensity has one and the same constant value are not influenced by the size of the steps. Only the inter-arrival times within which intensity steps appear will have a different distribution of their lengths compared to the conditions at constant call intensity. If the steps appear rather seldom, this has little significance and we can apply the rules for slow variations. If on the other hand the intensity varies continuously, then all inter-arrival times will be exposed to the influence of the variation of intensity. A closer investigation is needed in order to estimate the amount of this influence.

The call intensity as a function of the absolute time T will be represented as above by $y(T)$. $\Theta(T, t)$ is the probability that no call arrives during the interval t after the point of time T. The probability that no call arrives during the short interval dt following the time t is now $1 - y(T + t)\, dt$. From this it is clear that

$$\Theta(T, t)\{1 - y(T + t)\, dt\}$$

expresses the probability that no call arrives either during the time t or during the interval dt. But this probability is also expressed by $\Theta(T, t + dt)$ and we get the relation

$$\Theta(T, t + dt) = \Theta(T, t)\{1 - y(T + t)\, dt\}$$

whence at the limit for $dt \to 0$ we get

$$\frac{d}{dt}\Theta(T, t) = -y(T + t)\,\Theta(T, t)$$

If we integrate this expression, taking into account the boundary condition $\Theta(T, 0) = 1$, we get

$$\Theta(T, t) = e^{-\int_0^t y(T+z)\,dz} \tag{87 a}$$

If we consider the call intensity to be constant during the interval t, we further get

$$\Theta_0(T, t) = e^{-y(T)t} \tag{87 b}$$

which thus corresponds to the expression that applies on the assumption of slow intensity variations.

If we now integrate $\Theta(T, t)\, dT$ within a certain range, for instance 0 to T_1, and divide by the length of the integration range – in this case with T_1 – we get, as is clear, the probability that no call arrives within the time t after an arbitrarily chosen point of time within the integration range. Thus

$$\vartheta(t) = \frac{1}{T_1}\int_0^{T_1} e^{-\int_0^t y(T+z)\,dz}\, dT \tag{88 a}$$

expresses the distribution of the next call arrival. The corresponding expression with the assumption of slow intensity variations becomes according to (87 b)

$$\vartheta_0(t) = \frac{1}{T_1}\int_0^{T_1} e^{-y(T)t}\, dT \tag{88 b}$$

The expressions (88 a) and (88 b) apply within the interval considered 0 to T_1. In order to get the general distribution for the next call we must generally look at the limit value of the expressions as $T_1 \to \infty$.

The expressions (88 a) and (88 b) now permit a comparison of the distributions for rapid and slow intensity variations. The expression for the distribution of the next call arrival applicable to slow variations also applies for rapid ones if we introduce a fictitious call intensity of the form

$$\frac{1}{t}\int_0^t y(T+z)\,dz \qquad (89)$$

(compare with the fictitious equilibrium traffic introduced earlier). In the present case, the comparison is more difficult as t appears in (89). Further, it is only the distribution of the next call arrival that permits a direct comparison. The distribution of the call inter-arrival times becomes, according to formula (8 b) in Chapter 2 for rapid variations,

$$\frac{1}{\int_0^{T_1} y(T)\,dT}\int_0^{T_1} e^{-\int_0^t y(T+z)\,dz}\, y(T+t)\,dT \qquad (90\text{ a})$$

and for slow variations

$$\frac{1}{\int_0^{T_1} y(T)\,dT}\int_0^{T_1} e^{-y(T)\,t}\, y(T)\,dT \qquad (90\text{ b})$$

These expressions can generally not be transformed to one another by introducing a fictitious call intensity.

We shall now consider the case that the call intensity is a periodic function of the form (83). The fictitious call intensity (89) is then

$$\frac{1}{t}\left\{y_0\,t + \frac{y_1}{\omega}(\sin\omega(T+t) - \sin\omega T)\right\}$$

and after some transformations

$$y_0 + y_1\frac{\sin\frac{\omega t}{2}}{\frac{\omega t}{2}}\cos\omega\left(T+\frac{t}{2}\right) \qquad (91)$$

For determination of the distribution function for the next call (88 a) it is now sufficient to carry through the integration over one period. Further we observe that the phase displacement in (91) does not influence the value of the integral; it can then be disregarded. We then get

$$\vartheta(t) = \frac{\omega}{2\pi}e^{-y_0 t}\int_0^{\frac{2\pi}{\omega}} e^{-\frac{2y_1}{\omega}\sin\frac{\omega t}{2}\cos\omega T}\,dT \qquad (92\text{ a})$$

Compare this with the expression for slow variations obtained from (88 b)

$$\vartheta_0(t) = \frac{\omega}{2\pi}e^{-y_0 t}\int_0^{\frac{2\pi}{\omega}} e^{-y_1 t\cos\omega T}\,dT \qquad (92\text{ b})$$

The difference between the distributions for rapid and slow variations thus depends only on the ratio between the amplitudes of the oscillations of the fictitious and real call intensity. This ratio is

$$\frac{\sin\frac{\omega t}{2}}{\frac{\omega t}{2}} \qquad (93)$$

and is numerically always smaller than 1, except for $t = 0$. When $\omega \to \infty$, (93) approaches 0 and the distribution function (92 a) approaches $e^{-y_0 t}$. For very rapid intensity variations the call distribution will be the same as for pure random traffic without any intensity variations at all. In normal cases ωt is indeed very small and thus (93), as known, is very close to 1. For traffic of normal order of magnitude the mean inter-arrival time for instance, and thus the mean value of t, is equal to 6 s. For an intensity variation with a period of 2 hours the value of (93) at $t = 6$ s equals 0.99999828. In such a case the difference between (92 a) and (92 b) must be quite insignificant. The corresponding inter-arrival time distribution functions that are obtained as derivations of these expressions must then also very nearly coincide.

We see then that, regarding the call distribution, we can generally use the formulas for slow variations with good accuracy also for relatively rapid intensity variations. The errors that appear depend upon the ratio between the length of the period of the intensity variations and the mean value of the inter-arrival times. For the states within a group treated earlier the corresponding errors depend, on the other hand, upon the ratio between the period length and the mean holding time. As the latter is usually many times greater than the mean inter-arrival time, it may be expected that the call conditions would generally be considerably less influenced by the possible speed of the intensity variations than is the case for the occupations within the group.

It is easy to generalize the findings that have been made for an intensity variation of the form (83) to the case where an arbitrary number of oscillations with different periods and phase displacements are present. We then get, as is easily understood, a fictitious call intensity that is built up in the same way as (91) by giving every member an attenuation factor and a phase displacement forward of $t/2$. The phase displacement can also in this case be disregarded in the construction of the integral for the call distribution.

The most important result of the investigations in this chapter is the proof of the existence of an equilibrium traffic. This means that whatever intensity variations the traffic has – rapid or slow, continuous or in steps – the traffic conditions can always be calculated from the results of the theory for ideally slow variations if modified traffic is used – the equilibrium traffic. The intensity variations thus appear as ideally slow, regardless of their real speed.

It must be possible to determine the equilibrium traffic in question and its intensity distribution in order that these theoretical results shall have a value for practical use. Later on investigation will be made of the possibilities which exist to determine the properties of real traffic from measurements. Here the following general viewpoints will be put forward. If we try to estimate the intensity distribution of the traffic from the duration of the states for a certain traffic flow by means of the formulas for slow variations, we obviously get a result that refers to the intensity variations of the equilibrium traffic and not of the real traffic (except of course for cases where the equilibrium traffic and the real traffic are identical). Now, however, it is exactly the intensity variations of the equilibrium traffic and not of the real that we want to determine in order to be able to use the formulas for slow variations. We are in this way totally independent of the real intensity variations of the real traffic in that we work with the variations of the equilibrium traffic both in measurements and computations. We can express the matter to the effect that, in all measurements that can be made to determine intensity distributions, we automatically obtain a levelling out of the variations in such a way that the formulas for slow variations can be applied.

Considering that all applications in practice must be founded on measurements of the intensity distribution, the discussion will then have the result that *the assumption of the slowness of the intensity variations does not imply any limitation regarding the applicability*. Thus, if the other assumptions that have been discussed in Chapter 5 regarding the theory for the intensity variations can be considered to be fulfilled, we can apply the formulas for slow variations regardless of the speed of the variations. A necessary prerequisite is, however, that the intensity distribution will be determined in such a way that it applies to the equilibrium traffic. That is the case in all empirical determinations.

The theoretical investigations of rapid variations in this chapter have shown that it is possible to define an equilibrium traffic for the states as well as for the inter-arrival times. These two equilibrium traffic flows are, however, not identical. Strictly speaking, we have for the inter-arrival times a certain equilibrium traffic of the form (89) for every value t of the inter-arrival times. The difference for different t-values will generally be very small and can be disregarded. It is also possible that for other traffic quantities, not investigated here, other kinds of equilibrium traffic flows appear. It would, of course, be very inconvenient if, in order to use the formulas for slow variations, we were forced to introduce several different equilibrium traffic flows which also have to be determined by measurements. Generally, however, it seems to be sufficient to know the equilibrium traffic that applies to the states and, because of that, for most quantities that are important in dimensioning. On closer analysis of the traffic properties the equilibrium traffic flows of the inter-arrival times are also of interest since they can be expected, as has been shown earlier, to coincide very closely with the real traffic. For the equilibrium traffic of the states, on the contrary, the more rapid intensity variations are much more attenuated. We can then get an idea of the speed of the variations in real traffic by comparing the equilibrium traffic flows of the states and of the inter-arrival times. It must be observed however, as was already shown in Chapter 5, that the intensity distributions of the inter-arrival times and the states can also differ because the mean of the holding times is subject to a variation over time.

Chapter 8

Superposition

Anyone engaged in the dimensioning of telephone plant will very often be faced with questions that relate to the superposing of different traffic flows. If all traffic flows could be regarded as pure random processes, such questions would be very simple to treat. However, every individual traffic flow shows more or less strong intensity variations and we may ask how we can determine the change of the variations that occurs on superposing. In the usual rather rough manner of treating traffic questions the problem can be formulated as follows: We know the traffic during the busy hour for each individual of a number of traffic flows; if now these traffic flows are superposed, which will then be the traffic flow of the resulting traffic during the busy hour? Usually we cannot apply any normal summation, as the maxima of the partial traffic flows do not coincide in time. To treat such problems, certain traffic reductions, so called group reductions, are often introduced, which are determined by experience. The justification till now for the magnitude of the group reductions seems to be unconvincing. The results of systematic treatment of the attempts made so far cannot be considered satisfactory. They often do not satisfy even the simplest formal requirements, e.g. that the result of superposing should be unambiguous and thereby not dependent on the order of the superposing of the different traffic flows. We could perhaps venture the opinion that the development of a solution of the superposing questions is at present one of the most important problems of traffic research.

The question of the properties of superposed traffic has its importance not only in the dimensioning of groups, whose traffic flows are constituted by superposing of known traffic flows; it is also of basic importance for the entire traffic research, since the properties of telephone traffic cannot be described by the magnitude of the traffic alone but we must also know other quantities, for instance the distribution of the intensity variations. Then the question arises how these other quantities can be determined in practice. Even if we may assume that the quantities could be determined by measurements at each time for each traffic flow, the question is not solved; for the dimensioning of planned telephone plant, namely, only data with approximate values of the traffic flows and the general character of the subscriber population are available. More detailed information on all properties of the traffic flows, for instance the intensity distribution, can hardly be asked for and in certain cases, when dealing with telephone traffic in telephone plant not yet installed, cannot be asked for at all. We must then fall back on the only information at our disposal apart from the magnitude of the traffic, and that is the general character of the subscriber population, which in principle should allow prediction of the traffic properties. This requires, however, not only an extensive experience of the character of telephone traffic in different types of communities such as cities, suburbs and rural areas, but also a detailed knowledge of how the properties derived from the general character vary with different magnitudes of the traffic flow. The assembly and maintenance of such a great empirical material as would be necessary for this purpose must be ruled out as impossible in practice. It is in this respect that we can get valuable help from the study of traffic superposing conditions.

We shall now consider a rather big telephone plant in a city. It will contain numerous groups of cooperating devices for instance preselectors, registers, group selectors, final selectors, trunk circuits to other plant, perhaps also toll lines, etc. In each such group there is traffic whose properties are dependent on the habits of the subscribers from whom and to whom the calls go. In a group of preselectors, all calls from a limited subscriber group, for instance about 500, are handled. The traffic properties are exclusively determined by the character of the subscribers in the group. In group selectors on the contrary, a certain part of the calls often comes from a considerably greater number of subscribers. The closer a group selector stage is to the final selectors, the greater is the influence on the traffic properties of the habits of the called subscribers and the smaller the influence of the callers. Now in groups where the calls originate from a very great number of subscribers and at the same time are through-connected to a very great number of subscribers, the traffic can be considered to be selected to a large extent by chance from the total traffic to or from the subscribers in question. We may therefore expect that the traffic flows in different such groups have similar properties, as they have arisen by random selection from a common large traffic. Also for groups where the number of calling or called subscribers is somewhat more limited, a similar condition can be expected since the subscriber group in each such traffic flow can be considered to a certain extent to constitute a random sample from a much larger subscriber population. In such cases, however, quite naturally, remarkable differences may also exist as, for instance, in many cases the business traffic, in others the residential traffic, is the greater part. However, the geographical distribution of business and residential telephones can be considered rather arbitrary, at least if we are concerned with not too large a homogeneous area. Furthermore, the selection of the subscriber numbers that belong to one and the same traffic within a limited area is often to a certain extent random.

A line of thought of this kind makes it probable that different telephone traffic flows in many cases have a common character, the origin of which can be explained by the random selection of the different traffic flows from a large common traffic. Also for small rural offices where the population conditions have as a consequence a more uneven distribution, we can expect somewhat similar conditions. We can therefore hope that, at least for the majority of the traffic flows, it will be possible to establish not too great a number of types which must have such properties that all traffic flows of the same type can be derived by random selection from a common traffic. The empirical material that is required to establish the existence of such types need not be very great.

It is obvious that, if all traffic flows belonging to the same type arise by random selection from a common large traffic, we must get a resulting traffic of this type when superposing traffic flows of the same type. Therefore the theory for the description of the traffic types in question will be tied to the theory of superposing. Further elaborations will show that these ties are rather close. In the treatment of superposition questions we shall arrive at a concept *traffic classes* that can be considered as a generalization of the abovementioned type concept. The introduction of traffic classes proves to be absolutely necessary to get any useful results from the superposition theory. We cannot expect that superposing of traffic flows that lack any common character will give results that permit fruitful treatment.

The problems treated earlier in this work were concerned without exception with the conditions for one individual traffic flow and the frequency functions introduced in Chapter 2 of the form (1) are in principle sufficient to determine all probabilities that occur. In a comparison between several traffic flows, however, a knowledge of the probabilities of different simultaneous courses of events within the traffic flows considered is necessary, and under such conditions the frequency functions do not give any explanation. A traffic process is thus not completely defined by the frequency functions of its own course of events only. If a process is to be completely defined, it should be possible to determine the probability of each course in the process, if at the same time we assume arbitrary courses in any other arbitrary process, also including a fixed time scale. Random traffic is an example of a completely defined process where the calls arrive fully independently of other conditions. The processes derived in Chapter 3 with limited after-effects, where the calls occur fully independently of other occurences, is another example. A further example is random traffic modified by intensity variations, where the value of the intensity is given as a function of time. If, on the other hand, only the distribution function of the intensity is known, the traffic is not completely defined.

Next it should be pointed out that the validity of the concept of stationariness that was introduced in Chapter 2 can without difficulty be extended to several simultaneous traffic flows. We must then imagine that these will continue from an arbitrary point of time during an infinite time forward and backward. We regard a point of time T and investigate the probability for one of the traffic flows that a randomly chosen call among the N calls immediately after this point of time has a certain position in relation to a combination of different courses of events in the different traffic flows, which in turn have fixed positions in relation to each other. It might then happen that this probabaility tends towards a certain limit if $N \to \infty$. Consider then the probability that, starting at the point of time T, a randomly chosen call among the N nearest arriving earlier calls has a certain position in relation to a similar combination of courses of events. If this probability as $N \to \infty$ converges to the same limit as previously and this occurs for every combination of the different traffic flows, the traffic flows considered are mutually stationary. We assume that the traffic flows treated below satisfy this condition.

Already in Chapters 2 and 4 the results of certain special superpositions were considered. The traffic superpositions which will be treated now have a more general character. The treatment will, however, be strictly limited to random traffic flows modified by slow intensity variations, as there are possibilities that all fresh traffic flows, i.e. traffic flows that are not distorted in the previous selector stages, could be represented in this form. A superposition theory based on more general assumptions would be very complicated and would probably not permit any practical applications.

We look now at two traffic flows 1 and 2 having call intensity distribution functions $G_1(x)$ and $G_2(x)$ and mean call intensities y_1 and y_2. Let the derivatives of the intensity distribution functions be $g_1(x)$ and $g_2(x)$ and let these be taken as being generally limited. This assumption is not necessary, but the presentation will be easier and the following applications will not be subject to any limitation because of that.

If the intensity variations were independent of each other, then the probability that traffic flow 1 at an arbitrary point of time has the intensity x would always be $g_1(x)\,dx$ regardless of the intensity prevailing in traffic flow 2 at the same time. However, we have to assume in the general case that a certain correlation exists between the two simultaneous intensity values of the traffic flows concerned. In order to describe this correlation we introduce

$$F(x_1, x_2)\,dx_1\,dx_2$$

as an expression of the probability that at a randomly chosen point of time we have a call intensity x_1 in traffic flow 1 and a call intensity x_2 in traffic flow 2. By integration of the expression above for all possible values of x_2 we obviously obtain the probability that traffic flow 1 at a randomly chosen point of time has the intensity x_1 regardless of the simultaneously prevailing intensity in traffic flow 2. From this we get the relation

$$g_1(x_1) = \int_{x_2=0}^{\infty} F(x_1, x_2)\,dx_2$$

and in the same way

$$g_2(x_2) = \int_{x_1=0}^{\infty} F(x_1, x_2)\,dx_1$$

These relations are obviously not sufficient to determine the function $F(x_1, x_2)$.

If we now superpose the traffic flows 1 and 2, then the call intensity of the resulting traffic is at every point of time equal to the sum of the simultaneous call intensities of traffic flows 1 and 2. The probability that the resulting traffic at a randomly chosen point of time has the intensity z we get by summation of all probabilities for the case that traffic flow 1 has the intensity x and traffic flow 2 the intensity $z - x$, with x taking all values between 0 and z. We then find that

$$\int_0^z F(x, z - x)\,dx$$

is the frequency function of the call intensity of the resulting traffic. Naturally this can also be written in the form

$$\int_0^z F(z - x, x)\,dx$$

From the superposing function $F(x_1, x_2)$ the intensity distribution of the resulting traffic is then fully determined. We can now easily prove that the mean number of calls in the resulting traffic equals the sum of the mean number of calls in traffic flows 1 and 2. The latter according to (49 b) are defined by

$$y_1 = \int_0^{\infty} x g_1(x)\,dx$$

and

$$y_2 = \int_0^{\infty} x g_2(x)\,dx$$

The mean number of calls in the resulting traffic is found in the same way from the expressions derived above for the frequency function of the resulting traffic z, namely

$$\int_{z=0}^{\infty} \int_{x=0}^{z} z F(x, z - x)\,dx\,dz$$

or after having changed the integration order

$$\int_{x=0}^{\infty} \int_{z=x}^{\infty} z F(x, z - x)\,dz\,dx$$

This can also be written

$$\int_{x=0}^{\infty} \int_{\xi=0}^{\infty} x F(x, \xi)\,d\xi\,dx + \int_{x=0}^{\infty} \int_{\xi=0}^{\infty} \xi F(x, \xi)\,d\xi\,dx$$

by introducing the new variable $\xi = z - x$. The first double integral is equal to y_1, which is easily obtained from the condition presented earlier for the function F as well as from the definition of y_1. If we then change the integration order in the second double integral, we find in a similar way that this is equal to y_2. The mean number of calls in the resulting traffic is thus equal to $y_1 + y_2$. It follows of course that, by superposing an arbitrary number of traffic flows, the mean number of calls in the resulting traffic flow is obtained by adding the means of the superposed traffic flows.

In order to consider superposition under more general conditions, we assume that for each value of y there is one and for the time being only one traffic flow with the mean number of calls y. The probability that the n traffic flows that have the mean numbers of calls $y_1, y_2, \ldots y_n$ should have the respective call intensities $x_1, x_2, \ldots x_n$ at a randomly chosen point of time is denoted

$$F_n\begin{pmatrix} x_1, x_2, \ldots x_n \\ y_1, y_2, \ldots y_n \end{pmatrix} dx_1\, dx_2 \ldots dx_n \qquad (94)$$

This function is defined by the intensity variations of the traffic flows considered for every positive integer value of n and for every value (of course non-negative ones) of the variables x and y. The only exception is that the y-parameters must have unequal values. For equal values the function is not yet defined. For $n = 1$ the superposition function F_n will reduce itself to the frequency function of a single traffic flow, to be represented by $g_y(x)$.

We can now imagine that for every y-value an unlimited number of other traffic flows with the same y-value will be added to the traffic group considered. These should satisfy the condition that the superposition function of an arbitrary number of the original or added traffic flows of the group always has the same value as F_n according to (94). Further it is assumed that the superposition functions that we get by superposing traffic flows with equal or different y-values become invariant when the superposed traffic flows are replaced by other traffic flows with the same y-values. The expression (94) is then defined unambiguously for all, including equal, values of the y-parameter in the expression. If these conditions are satisfied then all the traffic flows considered are said to form a *complete traffic group*.

A complete traffic group then comprises an unlimited number of traffic flows for each value y of the mean number of calls. When superposing traffic flows from a complete traffic group, any traffic flow can be replaced by any other with the same y-value and belonging to the group without change of the intensity distribution of the resulting traffic.

We note that the traffic flows in a complete traffic group need not in themselves be completely defined. Two traffic flows with the same y-value that belong to the same complete traffic group can, therefore, in certain respects have different properties. They can, e.g. on superposing with a traffic not belonging to the group, give different resulting processes.

We can also imagine that one does not start from the traffic flows in defining a complete traffic group but from a given series of functions of the type F_n. When these functions are unambiguously determined for all values of the variables x and y and for every positive integer value of n, then clearly a complete traffic group is defined with these F_n as superposition functions. We note that the functions F_n must satisfy certain conditions. Firstly, they must always be symmetrical in respect of the pair of variables $x_\nu y_\nu$, so that the pair in expression (94) can be arbitrarily changed without a change in the value of (94). Further, if we integrate (94) for all values of one of the variables, e.g. x_1 we obviously get the probability that at a randomly chosen point of time $n - 1$ traffic flows with the mean values of calls $y_2, y_3 \dots y_n$ show the respective intensities $x_2, x_3 \dots x_n$. We then get the equation

$$\int_{x_1=0}^{\infty} F_n \begin{pmatrix} x_1, x_2, \dots x_n \\ y_1, y_2, \dots y_n \end{pmatrix} dx_1 = F_{n-1} \begin{pmatrix} x_2, x_3, \dots x_n \\ y_2, y_3, \dots y_n \end{pmatrix} \tag{95}$$

From a certain F_n, thus, all superposition functions with lower n-index are completely determined. For $n = 2$ we get the relation

$$\int_{x_1=0}^{\infty} F_2 \begin{pmatrix} x_1, x_2 \\ y_1, y_2 \end{pmatrix} dx_1 = g_{y_2}(x_2) \tag{95 a}$$

and further

$$\int_0^{\infty} g_y(x)\, dx = 1 \tag{95 b}$$

Finally the mean number of calls y is determined by

$$y = \int_0^{\infty} x g_y(x)\, dx \tag{95 c}$$

The superposition functions of every complete traffic group must satisfy all these conditions, as also the symmetry condition mentioned above.

The definition of complete traffic groups now offers the possibility of deciding whether a certain traffic A belongs to a certain complete traffic group or not. The necessary and sufficient condition for A to belong to the traffic group in question is, clearly, that the superposition functions that apply between A and an arbitrary number of traffic flows within the complete traffic group are identical to the corresponding superposition functions within the traffic group, whereby A is replaced by a traffic flow with the same mean number of calls and belonging to the traffic group.

It may now happen that, on superposing of two traffic flows belonging to the same complete traffic group, we get a resulting traffic also belonging to the same traffic group. It now seems suitable to introduce the following definition.

If on superposing two arbitrary traffic flows belonging to the same complete traffic group the resulting traffic always belongs to the traffic group in question, we say that this group constitutes a traffic class.

When a traffic group constitutes a class, we can clearly superpose an arbitrary number of traffic flows from the group and the resulting traffic will always belong to the group.

We then consider further n traffic flows for which (94) gives the probability that the call intensities at a randomly chosen point of time have the values $x_1, x_2, \dots x_n$. When we now superpose two of the traffic flows, e.g. with the mean number of calls y_1 and y_2, then the resulting traffic has the mean value of $y_1 + y_2$ calls and at the point of time in question the call intensity $x_1 + x_2$. The probability that at a randomly chosen point of time the intensity of this resulting traffic is z and the intensities of the other traffic flows are still $x_3, x_4, \dots x_n$ is then obtained by integrating (94) for all values of x_1 and x_2 that satisfy $x_1 + x_2 = z$. When the traffic flows considered belong to the same traffic class, however, we can express the result with the aid of the superposition function for this class. We then get the relation

$$\int_{x=0}^{z} F_n \begin{pmatrix} x, z-x, x_3 \dots x_n \\ y_1, \ y_2, \ \ y_3 \dots y_n \end{pmatrix} dx = \\ = F_{n-1} \begin{pmatrix} z, \quad x_3 \dots x_n \\ y_1 + y_2, y_3, \dots y_n \end{pmatrix} \tag{96}$$

This is clearly a necessary and sufficient condition for a complete traffic group to constitute a traffic class. The relation must, clearly, be satisfied for all possible values of the parameters belonging to it.

The line of thought that has been followed at the beginning of this chapter showed that in many cases of traffic flows in a telephone plant we can expect properties that arise from random selection from a large common traffic. We shall then investigate more closely what results we can get from a probability scheme based upon such a procedure. We consider a large traffic, the base traffic, with the mean call intensity Y. The base traffic should have been constituted by superposing a large number N of partial traffic flows with the mean call intensities $y_1, y_2, \ldots y_N$. At each point of time the call intensity of the base traffic is equal to the sum of the simultaneous call intensities of the partial traffic flows. If these are $x_1, x_2, \ldots x_N$ then at each point of time

$$\xi Y = x_1 + x_2 + \cdots + x_N$$

where ξY represents the call intensity of the base traffic. By superposition we now form a selected traffic of n partial traffic flows where the n traffic flows at each point of time are randomly chosen among the N partial traffic flows present. The call intensity of the selected traffic is then at each point of time equal to the sum of n among $x_1, x_2, \ldots x_N$ randomly selected quantities. It is then quite clear that, at a point of time at which the call intensity of the base traffic is ξY, the probable intensity of the selected traffic is

$$\frac{n}{N} \xi Y$$

Another way of expressing this is that during the total time in which the intensity of the base traffic is ξY, the mean intensity of the selected traffic is

$$\frac{n}{N} \xi Y$$

Further it is clear that the mean value of the selected traffic becomes

$$\frac{n}{N} Y$$

during the total time.

If we then assume that the total number of traffic flows N is very great in relation to the selected number n, then the composition of the collective traffic formed by the partial traffic flows will be insignificantly changed by removing the n traffic flows. However, at each point of time we can choose at random n_1 among the remaining $N - n$ partial traffic flows, to form a new selected traffic. If we assume N to be unlimitedly great at a given call intensity of the base traffic, i.e. at a given value of ξ, the probability of a certain call intensity of the new selected traffic will be independent of the simultaneous call intensity of the first selected traffic.

From the base traffic we can in this way form unlimited numbers of selected traffic flows whose intensity variations depend upon the value of ξ but in all other respects are independent of each other. In general we cannot in this way get traffic flows with arbitrary mean call intensities, as these are totally determined by the number of the chosen partial traffic flows and the mean of the call intensities of all partial traffic flows. But in order to be able to form a complete traffic group from the traffic flows this forming must be possible for any arbitrary mean call intensity. For this purpose we let the mean value of the mean call intensities of the partial traffic flows tend towards 0, while simultaneously the number of selected partial traffic flows for each traffic flow increases so that the sum tends to the wanted mean call intensity. It should be remarked that the mean call intensities of all partial traffic flows need not tend towards zero, but only the mean value of the mean call intensities of all partial traffic flows.

By means of this principle of random selection we can then form a complete traffic group from a base traffic. Herewith the base traffic is considered as infinitely great and composed of partial traffic flows with, on average, indefinitely small mean call intensities. Now we can easily see that the group so formed constitutes a traffic class. If we form two traffic flows and assume them to be superposed, we execute the same procedure as if we directly formed a traffic whose mean call intensity is equal to the sum of the mean call intensities of the superposed traffic flows.

We now write $H(\xi)$ for the probability that the call intensity of the base traffic at a randomly chosen point of time is $\leqq \xi Y$. Then evidently

$$\int_0^\infty dH(\xi) = 1 \tag{97 a}$$

The mean call intensity Y of the base traffic we get from

$$Y = \int_0^\infty \xi Y \, dH(\xi)$$

from which it follows that

$$\int_0^\infty \xi dH(\xi) = 1 \tag{97 b}$$

We consider further a traffic flow with call intensity y that is chosen from the base traffic by the random procedure. We write $f_y(x, \xi)\,dx$ for the probability that the traffic flow considered has the call intensity x at a randomly chosen point of time, at which the base traffic simultaneously has the call intensity ξY. According to the above, ξy is the mean number of calls of the selected traffic during the total time when the base traffic has the call intensity ξY. We then get

$$\int_0^\infty x f_y(x, \xi)\,dx = \xi y \tag{98 a}$$

Further we obviously have

$$\int_0^\infty f_y(x, \xi)\,dx = 1 \tag{98 b}$$

We further write $g_y(x)\,dx$ for the probability that the selected traffic has the call intensity x at a randomly chosen point of time. We must then have

$$g_y(x) = \int_{\xi=0}^\infty f_y(x, \xi)\,dH(\xi) \tag{99}$$

Finally, as usual,

$$\int_0^\infty g_y(x)\,dx = 1 \tag{100 a}$$

and

$$\int_0^\infty x g_y(x)\,dx = y \tag{100 b}$$

which can be derived from the relations above.

We now consider n traffic flows chosen from the same base traffic with the mean number of calls $y_1, y_2, \ldots y_n$. The probabilities that at a point of time when the base traffic has the intensity ξY these have the intensities $x_1, x_2, \ldots x_n$ are then, according to the foregoing, independent of each other. The probability that at a randomly chosen point of time the traffic flows have the intensities in question and at the same time the base traffic has the intensity ξY will then be

$$f_{y_1}(x_1, \xi)\, f_{y_2}(x_2, \xi) \ldots f_{y_n}(x_n, \xi)\, dx_1\, dx_2 \ldots dx_n\, dH(\xi)$$

By integrating this expression for all ξ-values we get the probability that at a randomly chosen point of time the n traffic flows will have the intensities in question independent of the value of ξ. The superposition function F_n according to (94) will then be

$$F_n = \int_{\xi=0}^\infty f_{y_1}(x_1, \xi)\, f_{y_2}(x_2, \xi) \ldots f_{y_n}(x_n, \xi)\, dH(\xi)$$

That the condition (95) is satisfied for the superposition function follows directly from (98 b). When further

$$f_{y_1+y_2}(z, \xi) = \int_{x=0}^{z} f_{y_1}(x, \xi)\, f_{y_2}(z-x, \xi)\,dx \tag{101}$$

then evidently the necessary and sufficient condition (96) for a traffic class is satisfied.

For a traffic class that is formed by means of the described procedure with random selection, the frequency functions $g_y(x)$ for the traffic call intensities are uniquely determined by two functions $H(\xi)$ and $f_y(x, \xi)$, for which the conditions (97) – (101) apply. A traffic class, expressed in this way by functions subject to the conditions set up here, is said to be of *Type A*.

Of great interest is the question how far the mathematical scheme used to derive classes of Type A expresses the properties of real telephone traffic. From the thinking at the beginning of this chapter this should be expected to occur to some extent. A definite answer to the question can be given only by measurements. However, it should be remarked that, even if the random selection is in reality realized only to a limited extent, and thus the justification put forward here for the occurrence of traffic classes of Type A should not be considered to be fully convincing, a satisfactory description of the superposing conditions for real telephone traffic on the basis of such traffic classes is not ruled out. As a matter of fact they give a very general description of intensity variations, which are partly of a common nature and partly independent of each other. Thus, regardless of the justification based upon the random selection, the traffic classes of Type A give a good picture of the properties that are generally expected to characterize real telephone traffic. The common intensity variations are derived from the similar habits of the subscribers regarding sleep, work times, meals, etc. The independent intensity variations appear because of more occasional occurrences in the subscriber group that originates or receives the traffic. For this reason we may expect that the influence of the independent intensity variations will be smaller the greater a traffic flow is. This is also the case for every traffic class of Type A, as will be shown in the following analysis of the conditions for the appearance of independent traffic variations.

In conclusion we can state that the traffic classes of Type A constitute a first and rather general specification of the general traffic class concept. We have also good reason to hope that this specification will not become too narrow to allow its application to normal telephone traffic. In this general view of the classes of Type A it is not necessary any longer to base our argument on the appearance of a base traffic, though really, as will be shown later, such a traffic can always be defined by means of the intensity distributions for the class. The parameter ξ, if one does not want to refer it to a base traffic, can be considered as a measure of the common intensity variations.

In order to investigate the properties of the traffic classes of Type A closer, a characteristic function for the distribution $f_y(x, \xi)$, is introduced, defined by

$$\vartheta_y(t, \xi) = \int_0^\infty e^{-xt} f_y(x, \xi)\, dx \tag{102}$$

By analogy with the expression (53) we realize that (102) is the distribution function for the next call during the time when the parameter of the common variations has the value ξ. (102) is an always decreasing, completely monotonic function of t. Further, based on (98 b), we get $\vartheta_y(0, \xi) = 1$. From (101) we thus get

$$\vartheta_{y_1+y_2}(t, \xi) = \int_{z=0}^{\infty} \int_{x=0}^{z} e^{-zt} f_{y_1}(x, \xi) f_{y_2}(z-x, \xi)\, dx\, dz$$

If a new variable $\varrho = z - x$ is introduced in the integral in the right-hand member and the integration order is reversed, we get

$$\vartheta_{y_1+y_2}(t, \xi) = \int_{x=0}^{\infty} \int_{\varrho=0}^{\infty} e^{-xt} f_{y_1}(x, \xi)\, e^{-\varrho t} f_{y_2}(\varrho, \xi)\, d\varrho\, dx$$

The right-hand side will now be equal to $\vartheta_{y_1}(t, \xi)\, \vartheta_{y_2}(t, \xi)$ because of the definition (102), and the relation

$$\vartheta_{y_1+y_2}(t, \xi) = \vartheta_{y_1}(t, \xi) \cdot \vartheta_{y_2}(t, \xi) \qquad (103)$$

is obtained. This also applies in reverse, so that if a function $\vartheta_y(t, \xi)$ satisfying the expression (103) is given, a function $f_y(x, \xi)$ which satisfies (101) is defined by (102).

The relation (103) should now be identically valid for all values of y_1 and y_2. Then as is known, if $\vartheta_y(t, \xi)$ is a limited function of y, we can write it in the form

$$\vartheta_y(t, \xi) = \{V(t, \xi)\}^y \qquad (104)$$

Later we shall touch upon the question of which conditions $V(t, \xi)$ must satisfy for the right-hand side of (104) to be a completely monotonic function of y for all positive values of y. We immediately see, however, that $V(t, \xi)$ as a function of t must be completely monotonic. Further $V(0, \xi) = 1$ since $\vartheta_y(0, \xi) = 1$. When we differentiate (104) with respect to t we get for $t = 0$

$$\frac{d}{dt}\vartheta_y(0, \xi) = y\frac{d}{dt}V(0, \xi)$$

Now from (102) and the relation (98 a) we get

$$\frac{d}{dt}\vartheta_y(0, \xi) = -\xi y \qquad (105\text{ a})$$

from which it follows that

$$\frac{d}{dt}V(0, \xi) = -\xi \qquad (105\text{ b})$$

The distribution function for the next call for a traffic flow whose frequency function for the call intensity is $g_y(x)$ is now, from (53)

$$\vartheta_y(t) = \int_0^\infty e^{-xt} g_y(x)\, dx$$

By means of (99) the right-hand side of this can be written

$$\int_{x=0}^{\infty}\int_{\xi=0}^{\infty} e^{-xt} f_y(x, \xi)\, dH(\xi)\, dx$$

which, with the use of (102), takes the form

$$\int_0^\infty \vartheta_y(t, \xi)\, dH(\xi)$$

By means of (104) we finally get

$$\vartheta_y(t) = \int_0^\infty \{V(t, \xi)\}^y\, dH(\xi) \qquad (106\text{ a})$$

From this we get the distribution function of the inter-arrival times by differentiating according to formula (8 b) in Chapter 2. This gives

$$\varphi_y(t) = \int_0^\infty \{-V'(t, \xi)\}\{V(t, \xi)\}^{y-1}\, dH(\xi) \qquad (106\text{ b})$$

where $V'(t, \xi)$ represents the derivative in respect of t. From (105 b) and (97 b) it follows that $\varphi_y(0) = 1$.

Now we have the means of showing a number of important properties of the inter-arrival time distribution function. By the differentiation of (106 b) we get

$$\varphi_y'(t) = -\int_0^\infty \{V''(t, \xi)\{V(t, \xi)\}^{y-1} + (y-1)\{V'(t, \xi)\}^2\{V(t, \xi)\}^{y-2}\}\, dH(\xi)$$

For $t = 0$, $V(0, \xi) = 1$ and $V'(0, \xi) = -\xi$. We then get

$$-\varphi_y'(0) = (y-1)\int_0^\infty \xi^2\, dH(\xi) + \int_0^\infty V''(0, \xi)\, dH(\xi) \qquad (107\text{ a})$$

The integral in (107 a)

$$m = \int_0^\infty \xi^2\, dH(\xi)$$

is always >1 except for zero variation of $H(\xi)$. To prove this we shall use the inequality rule derived from the expression (59) in Chapter 6. We then consider ξ^2 as a kernel function whereby the variable ξ from (97 b) has the mean value 1. The second derivative of the kernel function is now always positive. The inequality rule then states that the above m is always greater than the square of the mean value, thus >1. When $H(\xi)$ shows zero variation i.e. varies only in the point $\xi = 1$, m will be 1.

We now consider the integral in (107 a)

$$\int_0^\infty V''(0, \xi)\, dH(\xi)$$

First it will be shown that $V''(0, \xi) > \xi^2$. Since $V(t, \xi)$ as said above is a completely monotonic function of t, it may always be written in the form

$$V(t, \xi) = \int_0^\infty e^{-xt}\, dB(x)$$

in which $B(x)$ is a non-negative function of x and also dependent on ξ. According to (105 b) we then get

$$-V'(0, \xi) = \xi = \int_0^\infty x\, dB(x)$$

so that ξ is the mean value of the variables x. Further

$$V''(0, \xi) = \int_0^\infty x^2\, dB(x)$$

and, in the same way as above, it is proved that this integral is greater than the square of the mean value, i.e. greater than ξ^2. The only exception occurs for zero variation of $B(x)$ where $V(t, \xi) = e^{-\xi t}$. Hence, because of this inequality

$$\int_0^\infty V''(0, \xi)\, dH(\xi) > \int_0^\infty \xi^2\, dH(\xi)$$

If we put

$$\int_0^\infty V''(0, \xi)\, dH(\xi) = m(1 + \gamma)$$

in which γ is always positive, the expression (107 a) will be

$$-\varphi_y'(0) = m\gamma + my \qquad (107\text{ b})$$

For a traffic class of Type A, the derivative of the inter-arrival time distribution function for $t = 0$ is thus a linear function of y. The fact that the member independent of y in (107 b) is always >0 – except in a case to be studied later in greater detail – is of great interest. For this shows that also for $y = 0$ the probability of inter-arrival times of limited length is >0. This means that *the traffic flows in a traffic class of Type A for $y \to 0$ converge towards clustering.*

The only case in which the term independent of y in (107 b) is zero will then occur when $B(x)$ varies in only one point, i.e. $x = \xi$. Then, $V(t, \xi)$, as already remarked, has the form $e^{-\xi t}$ and from (104) it is clear that we then have $\vartheta_y(t, \xi) = e^{-\xi y t}$. By (102) $f_y(x, \xi)$ is now unambiguously determined by $\vartheta_y(t, \xi)$, and when the latter has the exponential form, $f_y(x, \xi)$ can only differ from zero in the point $x = \xi y$. This means that no random but only common intensity variations appear and the expressions (106 a) and (106 b) will be

$$\vartheta_y(t) = \int_0^\infty e^{-\xi y t}\, dH(\xi) \qquad (108\text{ a})$$

and

$$\varphi_y(t) = \int_0^\infty \xi e^{-\xi y t}\, dH(\xi) \qquad (108\text{ b})$$

The condition for the case that random intensity variations can occur is thus the convergence of the traffic flows towards clustering when $y \to 0$.

With clustering we must assume that the relatively dense call clusters are separated from each other by long call-free intervals. Correspondingly, the call intensity is generally zero but within certain shorter intervals it is greater than zero. When $y \to 0$, a general continuous decrease of the call intensity towards zero does not take place, but instead the intensity maintains a value greater than zero within certain areas, while the time distances between these areas increase without limit. These circumstances can also be seen in the superposition conditions (101). This is only a form of the usual formula for adding independent random variables. If a random variable is formed by summation of a number of primary variables and their number grows indefinitely while the mean value of the summed variables remains constant, then, as is well known, under certain conditions the standard deviation of the variable $\to 0$, which means that the variable can only assume a value equal to the mean value. These conditions imply that none of the primary variables may dominate. For a traffic flow to show random intensity variations, it is thus necessary that certain of the included partial traffic flows dominate. In the scheme used above to form a traffic of Type A by random selection from a base traffic formed by a great number of partial traffic flows, we must imagine that at each point of time the call intensities of a great majority of the partial traffic flows are very close to zero while a relatively small number of partial traffic flows have call intensities clearly different from zero. These will then exercise a dominating influence on the intensity of a selected traffic and, by the constantly renewed random selection, random intensity variations might arise. Such variations could also arise because of the fact that the intensities of the partial traffic flows vary at random between values close to zero and values of dominating magnitude. This condition has a great interest as it turns out that one does not necessarily need to assume a constantly renewed random selection to obtain random intensity variations. This disposes of one of the objections that could be raised against the lack of consistency between the assumptions for traffic of Type A and the properties of real telephone traffic.

Decisive for the question whether random intensity variations could be expected to appear in telephone traffic is, of course, to what extent the intensity distribution tied to clustering is present in reality. Experience seems to indicate that a clustering tendency appears in most small traffic flows. The traffic from a single subscriber, which in most cases is very small compared to the traffic flows present in the groups of a telephone plant, often shows a tendency towards clustering during short periods. In between there are longer periods during which the intensity must be considered to be zero, as the subscriber terminal is not available for calls, for instance because nobody is on the premises. Thus there are great possibilities that normal telephone traffic exhibits properties that to a considerable degree coincide with the assumptions for convergence towards clustering.

Apart from the boundary conditions for $y \to 0$, the conditions are interesting when $y \to \infty$. In order to investigate this, we consider the variance of the distribution $f_y(x, \xi)$. This is defined by

$$\int_0^\infty (x - \xi y)^2 f_y(x, \xi)\, dx$$

since ξy according to (98 a) is the mean value of the distribution. The expression above is clearly equal to

$$\int_0^\infty x^2 f_y(x, \xi)\, dx - (\xi y)^2$$

Now we get from (102)

$$\int_0^\infty x^2 f_y(x, \xi)\, dx = \vartheta_y''(0, \xi)$$

By differentiation of (104) we get

$$\vartheta_y''(t, \xi) = y(y-1)\{V'(t, \xi)\}^2 \{V(t, \xi)\}^{y-2} + yV''(t, \xi)\{V(t, \xi)\}^{y-1}$$

which for $t = 0$, and based upon (105 b), will be

$$\vartheta_y''(0, \xi) = y(y-1)\,\xi^2 + yV''(0, \xi)$$

We then get finally

$$\int_0^\infty (x - \xi y)^2 f_y(x, \xi)\, dx = y\{V''(0; \xi) - \xi^2\} \tag{109}$$

If we now consider the variable $\eta = x/y$, we find its variance to be

$$\frac{1}{y}\{V''(0, \xi) - \xi^2\}$$

As $y \to \infty$, this expression $\to 0$, which means that the probability of a deviation of the variable η from the mean value ξ will be zero. This implies that the relative magnitude of the random variations decreases when y increases and that the distribution function of the intensity approaches

$$H\left(\frac{x}{y}\right)$$

as $y \to \infty$. Even if, by the definition of a traffic class of Type A, we do not start from a base traffic, such a traffic will always be defined by the conditions at $y \to \infty$.

Finally, we shall also in some detail treat the question concerning the conditions under which the function $V(t, \xi)$ satisfies the condition that (104) is a completely monotonic function of t for all positive values of y. As this must also be valid for $y = 1$, naturally $V(t, \xi)$ itself must be completely monotonic. From the derivation of (103) it follows that the product of two completely monotonic functions is always completely monotonic. If $V(t, \xi)$ is completely monotonic, (104) will therefore also be completely monotonic for all integer values of y. On the other hand it seems uncertain whether the same is valid for non-integer values of y. However, the following theorem can be proved:

When V is a completely monotonic function of t and in addition

$$f = \frac{-V'}{V}$$

is completely monotonic as a function of t, then

$$\vartheta = V^y$$

is a completely monotonic function of t for all positive values of y.

A function is completely monotonic when successive derivatives in every point all have alternate signs. If we differentiate ϑ we now get

$$\vartheta' = -y\,\vartheta f$$

which is always negative. Further we get

$$\vartheta'' = y^2\,\vartheta f^2 - y\,\vartheta f'$$

Since by assumption f' is negative, ϑ'' becomes positive. It is clear that an arbitrary derivative can always be written as a sum of terms of the form

$$K\,\vartheta f^{n_0}(f')^{n_1}(f'')^{n_2}(f''')^{n_3}\ldots \tag{110}$$

where K is a constant and $n_0, n_1,\ldots$ are positive integers or 0. On differentiation of this expression we get

$$\begin{aligned}
&-K\,y\,\vartheta f^{n_0+1}(f')^{n_1}(f'')^{n_2}(f''')^{n_3}\ldots\\
&+K\,n_0\,\vartheta f_0^{n_0-1}(f')^{n_1+1}(f'')^{n_2}(f''')^{n_3}\ldots\\
&+K\,n_1\,\vartheta f_0^{n_0}(f')^{n_1-1}(f'')^{n_2+1}(f''')^{n_3}\\
&+\ldots\ldots
\end{aligned}$$

As according to the assumption of the completely monotonic character of the function the derivatives of f have alternate signs, every term of the sum gets an opposite sign compared with (110). When now all terms in the derivative of ϑ, in (110) which represent a term, have the same sign, then all terms in the next derivative of ϑ get the opposite sign. As already in ϑ'' above both terms are positive, it follows by induction that all derivatives of ϑ have alternate signs. From that it follows that ϑ is completely monotonic.

The theorem does not express a necessary condition, as f is not completely monotonic if V is a simple exponential function.

The traffic classes of Type A certainly constitute only a special case of the general traffic class concept. In spite of that they have a very general character. This will appear from the fact that they are determined by two (with only minor restrictions) arbitrary functions $H(\xi)$ and $V(t, \xi)$, of which the latter even contains two variables. In order to arrive at advantageous applications, it seems necessary to specialize the general form of these functions by means of further assumptions. Such assumptions may not be chosen too arbitrarily as otherwise difficulties might arise in applying the results; rather they must be based upon the ideas we may have regarding the properties of real telephone traffic.

Before any attempts are made in this direction, we shall look at the very special simplifications that will occur when either the random or the common intensity variations are completely lacking. For the case where there is no random variation and thus $f_y(x, \xi)$ differs from zero only for the value $x = \xi$, the formulas (108 a) and (108 b) have already been derived. In this case both distribution functions are only functions of the product yt and thus invariant for values of y and t that maintain this product constant. It is easy to observe that this is an absolute criterion, thus a *necessary and sufficient condition for the class to show only common intensity variations*. That the condition is necessary follows from (108 a) and (108 b). That it is sufficient is clear from the condition that, if the distribution function for the next call is to be a completely monotonic function of $t_1 = yt$ alone, it can always be brought into the form

$$\int_0^\infty e^{-\eta t_1}\, dH_1(\eta)$$

in which H_1 is independent of y and t. However, this is the general form for a traffic flow with the relative intensity distribution $H_1(\eta)$, and as this is independent of y, the relative variation of the intensity is equal for all traffic flows; in other words, there are only common intensity variations. The same applies for the distribution function of the inter-arrival times.

There is then a very simple method of finding out by measurements whether a series of traffic flows shows only common intensity variations. We realize besides that the proposed criterion is not tied to the assumption of a traffic class of Type A or even of a traffic class or a complete traffic group. If two traffic flows have the same relative intensity variations, the distribution functions of call arrivals must be invariant for constant values of yt.

If the common intensity distribution $H(\xi)$ varies in only one point, i.e. for $\xi = 1$, we get, evidently, the special case of pure random traffic flows and the distribution function for the inter-arrival times will be e^{-yt}.

Further, the special case will be considered that there are only random intensity variations. As no common variations occur, $H(\xi)$ will vary only in the point $\xi = 1$ and we get from (99)

$$g_y(x) = f_y(x, 1)$$

and from (106 a)

$$\vartheta_y(t) = \{V(t, 1)\}^y \qquad (111)$$

Also this form is generally valid and is not tied to any assumptions regarding traffic class or complete traffic group. Its necessity follows as a matter of fact from the general formula (10) in Chapter 2 for superposing mutually independent traffic flows and the generally valid theeorem that the mean values of calls are added together on superposing. On the other hand the condition that the distribution function for the next call has the form (111) is sufficient to guarantee that no common intensity variations exist. If, namely, two traffic flows which satisfy the formula (111) and which have the mean values of calls y_1 and y_2 are superposed and the resulting traffic also satisfies the formula (111), then

$$\vartheta_{y_1+y_2}(t) = \vartheta_{y_1}(t)\,\vartheta_{y_2}(t)$$

which means that the traffic flows are superposed at random.

Regarding $V(t, 1)$ which is now a function only of t, the same condition as before applies, viz. that for every positive value of y (111) shall be a completely monotonic function of t. Regardless of this condition, valid for all traffic flows with slow intensity variations, it is obvious *that the necessary and sufficient condition for the case of a series of traffic flows to be independent of each other is that the distribution functions of the next call arrival can be written in the form* (111).

Also in this case the pure random traffic appears as a special case, i.e. $V(t, 1) = e^{-t}$, which happens if $g_y(x)$ only differs from 0 when $x = y$.

With the simplification considered up till now most of the characteristic properties of the Type A traffic classes, which generally show random as well as common intensity variations, vanish. There is, however, a possibility of another specialization of the expression for the Type A classes, which implies a considerable simplification in a mathematical respect, thereby facilitating the applications, without deleting any of the characteristic properties found up till now. The starting point for the justification of this specialization is given by the formula (98 a). As the integral here is dependent only on the product ξy, not on the individual values of ξ and y, the simplest specialization of $f_y(x, \xi)$ is to consider this function as a function only of x and the product ξy but not of the individual values of ξ and y. The same result can, however, be obtained for a formally much more general assumption. We need only assume that $f_y(x, \xi)$ can be written as a function of x and ϱ where ϱ is a function of ξ and y. In the computation of the integral in the left member of (98 a) the result will be a function only of ϱ, but as this should be equal to ξy, ϱ must itself be a function only of ξy. From this it follows that $f_y(x, \xi)$ is a function of only x and ξy, thus the same as for the aforementioned more special assumption.

From this assumption it now follows from the relation (104) that $\vartheta_y(t, \xi)$ can be written in the following form:

$$\vartheta_y(t, \xi) = \{v(t)\}^{\xi y} \tag{112}$$

where $v(t)$ is a function of t independent of ξ and y. We then have

$$V(t, \xi) = \{v(t)\}^{\xi}$$

If we imagine a traffic class of Type A formed by random selection from a great base traffic constituted by a great many very small partial traffic flows, then the assumption just introduced will be satisfied on condition that the intensity variations of partial traffic flows are random in relation to each other and that the number of partial traffic flows in the base traffic varies. The variations in common in the intensity of the selected traffic flows then depend entirely on the variations in the number of partial traffic flows comprised in the base traffic and not on any variations in common of the intensity of the partial traffic flows. The value of the parameter ξ at a particular point of time is a measure of the number of the partial traffic flows that at the same time constitute the base traffic. The intensity variations of a selected traffic then depend on the random variations of the partial traffic flows which are independent of the value of ξ and on the selected number of partial traffic flows which, based upon (98 a), are measured by ξy. From that it follows that $f_y(x, \xi)$ becomes a function of ξy but not of the individual values of ξ and y.

The reasoning pursued may be considered to be somewhat vague but should be sufficient in this context, as it has an importance only for the judgement of the extent to which the pure mathematical assumption leading to (112) can be considered to correspond to any facts in real telephone traffic. At first this might be considered to be less probable, as the number of subscriber instruments from which a certain telephone traffic originates is generally considered to be very constant. If the traffic from every subscriber terminal is considered as a partial traffic which takes part in the constitution of a greater base traffic, then the number of partial traffic flows in the base traffic must also be considered constant. Nothing prevents us, however, from considering a traffic flow whose call intensity for the time being is zero as non existent on the occasion in question. Now as has been said before, there occur long time intervals during which the call intensity from the individual subscribers must be considered to have the value zero, as the instruments are not available for calls during these intervals. We can in this way, for real telephone traffic, imagine a picture of a base traffic in which the number of partial traffic flows varies with time.

In an attempt to compare the mathematical scheme set up to verify the assumption leading to (112) with the properties of real telephone traffic, we are led it appears to a rather subtle reasoning. This may be felt to be a weakness, but should in reality be a strength, as it turns out that the assumption in question cannot imply any considerable limitation regarding the applicability of the results to real telephone traffic. Therefore, in what follows it is generally assumed that $f_y(x, \xi)$ is a function only of x and ξy. A traffic class with this property is said to be of Type *A a*.

If we now enter (112) in (106 a) we get for the distribution function for the next call arrival

$$\vartheta_y(t) = \int_0^\infty \{v(t)\}^{\xi y}\, dH(\xi) \qquad (113\text{ a})$$

For the distribution function for the inter-arrival times (106 b) we further get

$$\varphi_y(t) = -v'(t) \int_0^\infty \{v(t)\}^{\xi y - 1}\, \xi dH(\xi) \qquad (113\text{ b})$$

A traffic class of Type *A a* is fully determined by two functions $H(\xi)$ and $v(t)$, each with only one variable. For the non-negative function $H(\xi)$ the conditions (97 a) and (97 b) apply. For the likewise non-negative function $v(t)$ the condition applies that $\{v(t)\}^y$ should be completely monotonic for all positive values of y. Further $v(0)$ should be 1 and, according to (105 b), also $v'(0) = -1$. In other respects the functions can have arbitrary properties.

As always, the two distribution functions $\vartheta_y(t)$ and $\varphi_y(t)$ are completely monotonic functions of t. It should be remarked that they are completely monotonic functions also of y. This is true also for the general case i.e. for every traffic class of Type *A*. From the nth derivative of (106 a) with respect to y we get

$$\int_0^\infty \{V(t, \xi)\}^y \{\ln V(t, \xi)\}^n\, dH(\xi)$$

For $t > 0$, $V(t, \xi) < 1$ and thus the function $\ln V(t, \xi)$ is negative. The nth derivative in respect of y has the same sign as $(-1)^n$ which is the criterion of a completely monotonic function. The same applies to (106 b) as is easily seen.

If we write (113 a) in the form

$$\vartheta_y(t) = \int_0^\infty e^{\xi y \ln v(t)}\, dH(\xi) \qquad (113\text{ c})$$

we see that $\vartheta_y(t)$ is an unambiguous, completely monotonic function of $-y \ln v(t)$. It can be shown that this is a criterion of the case that a traffic class of Type *A* belongs to the special class Type *A a*. This leads to the following statement:

A necessary and sufficient condition that a traffic class of Type A should be of Type A a is that there exists a function $v(t)$ such that the distribution function for the next call $\vartheta_y(t)$ will be an unambiguous completely monotonic function of $-y \ln v(t)$. Furthermore the func-

tion $\{v(t)\}^y$ *should be completely monotonic for all positive values of* y *and satisfy* $v(0) = 1$ *and* $v'(0) = -1$.

That the condition put forward is necessary has just been shown. That the condition is sufficient is clear from the following. If $\vartheta_y(t)$ is a completely monotonic function of $-y \ln v(t)$ it must be possible to express it as a *Laplace* integral of the form (113 c). This can therefore be rewritten in the form (113 a). The traffic class considered was now of Type A according to the assumption. From that it follows that $\vartheta_y(t)$ can be expressed in the form (106 a). The function $V(t, \xi)$ thus has in this case the form $\{v(t)\}^\xi$ and from that it follows according to (104) that $\vartheta_y(t, \xi) = \{v(t)\}^{\xi y}$. The relation (102) will then have the form

$$\{v(t)\}^{\xi y} = \int_0^\infty e^{-xt} f_y(x, \xi)\, dx$$

According to the inversion law for the *Laplace* integral (further treated in Chapter 10), $f_y(x, \xi)$ will then be univocally determined by $\{v(t)\}^{\xi y}$, but as the latter is a function only of x and ξy, the same thing must be valid for $f_y(x, \xi)$, which is exactly the definition previously given for a traffic class of Type $A\ a$.

From the aforementioned criterion it is evident how the function $v(t)$ can be determined if we know the distribution function for the next call $\vartheta_y(t)$ as a function of both y and t. We investigate the value pairs y, t that satisfy the equation

$$\vartheta_y(t) = c_0$$

where c_0 is a positive constant < 1. These value pairs should according to the criterion above also satisfy the equation

$$y \ln v(t) = -c_1$$

where c_1 is another positive constant, whose value also depends on c_0. In order to satisfy the criterion above, the two curves for y as a function of t obtained for any two different c_0-values must then have the same ratio between the y-coordinates for all t-values. Besides it is quite evident that these curves are permanently decreasing and have both the y- and the t-axis as asymptotes. We can then determine $v(t)$ from one of the resulting curves, as from the equation above it follows that

$$v(t) = e^{-\frac{c_1}{y}}$$

In order to determine the constant c_1, we may use the expression $v'(0) = -1$ (the condition $v(0) = 1$ is already satisfied). We choose then an arbitrary constant k and compute a function $v_0(t)$ from

$$v_0(t) = e^{-\frac{k}{y}}$$

Thus we have

$$v(t) = \{v_0(t)\}^{\frac{c_1}{k}}$$

and find from that

$$v'(t) = \frac{c_1}{k}\{v_0(t)\}^{\frac{c_1}{k}-1} v_0'(t)$$

and thus

$$\frac{c_1}{k} v_0'(0) = -1$$

from which c_1 is determined and also $v(t)$. Among the simplifications that appear in the former formulas when the traffic class is of Type $A\ a$ it should be mentioned that (107 b) gets the form

$$-\varphi_y'(0) = v''(0) - 1 + my \qquad (107\text{ c})$$

The results obtained up to now in this chapter could be summarized as follows: In order to achieve a systematic treatment of the superposition problem, it has proved useful to introduce the concept of traffic classes. Traffic flows belonging to the same traffic class have a common form for the frequency function $g_y(x)$ of the intensity distribution and, on superposition, traffic flows are formed which always belong to the same class as the traffic flows from which they are made up. By specifying the properties of the function $g_y(x)$ we obtain firstly traffic classes of Type A for which the intensity variations are separated into variations in common for the class and random variations within the class and, secondly, the classes of Type $A\ a$ for which the random variations are determined by a function of a single variable $v(t)$ subject to certain conditions. An investigation of the significance of these mathematical specifications has proved that we need not fear that they imply any limitations regarding the possibility of applying the theoretical results to real telephone traffic. This investigation has been made so that probability schemes, satisfying the proposed properties of $g_y(x)$ e.g. the process of random selection from a base traffic, have been compared with the properties which we may believe to be present in real telephone traffic.

By such comparisons we can, as a matter of fact, only get a certain guidance for the theoretical analysis. Even if this guidance sometimes seems to be uncertain, it is of the greatest importance for economizing in the theoretical work. As a first result of the above considerations it has turned out that it must be possible to derive the frequency function $g_y(x)$ (dependent upon two variables) for the call intensity of a traffic in a definite way from two functions of one variable, $v(t)$ and $H(\xi)$. Of these functions, which are subject to certain conditions, $v(t)$ represents the random intensity variations and $H(\xi)$ those in common. However, this representation is still too general to be successfully applied to a practical classification of real traffic flows. Hitherto, we have been used to describe each traffic flow by a single parameter, the traffic intensity. If we now want to supplement this description, which is unsatisfactory for the treatment of most traffic problems, it is of course desirable for practical reasons that its character is still expressed in the simplest possible way, viz. by introducing a limited number of new parameter values. This means that we must limit ourselves to certain presumed forms for the functions $v(t)$ and $H(\xi)$. The question then arises whether it is still possible to reach results applicable with good approximation to all, or at least to the great majority, of the telephone traffic flows occuring in reality. This question can only be answered by experience, but it is clear that already the consideration of the intensity variations of the traffic flows which are obtained by introducing one or two characteristic constants must offer possibilities for computations having a considerably better agreement with real conditions than is possible when every traffic flow is described only by the traffic intensity.

Concerning the task of setting up suitable forms for the functions $v(t)$ and $H(\xi)$, it seems to be more difficult than has been the case earlier to find any guidance from discussions of the probable properties of the traffic. The forms must clearly satisfy the conditions that have been set up above and, in accordance with the recent remarks, also contain a limited number of parameters. There are many known probability distributions which could be adapted to these conditions. In the choice between these we have as sole guidance the desire that the results obtained from the treatment of traffic problems should to the greatest possible extent have a simple mathematical form. As these results are mostly derived in a simple way from the distribution functions $\vartheta_y(t)$ and $\varphi_y(t)$ for the calls, we should aim at a simple mathematical form for these functions. It has now been found that this desire probably can best be realized if we start throughout from distributions of the *Poisson* type.

For the earlier introduced frequency function $f_y(x, \xi)$ the following *normal form* is now chosen:

$$f_y(x,\xi) = \tau_0 \frac{(\tau_0 x)^{\tau_0 \xi y - 1}}{\Gamma(\tau_0 \xi y)} e^{-\tau_0 x} \tag{114}$$

where τ_0 is a constant with the dimension time. The expression in the denominator is the gamma function which, as is well known is defined by

$$\Gamma(\tau_0 \xi y) = \int_0^\infty x^{\tau_0 \xi y - 1} e^{-x} dx$$

The integral is convergent for all positive values of $\tau_0\xi y$. For integral values of this expression $\Gamma(\tau_0 \xi y) = (\tau_0 \xi y - 1)!$ as is well known. Further, the relation

$$\Gamma(\tau_0 \xi y + 1) = \tau_0 \xi y \, \Gamma(\tau_0 \xi y)$$

applies. From this relation and the definition of the gamma function it is immediately evident that the form (114) satisfies both the conditions (98 a) and (98 b). For the variance we obtain the expression

$$\int_0^\infty (x - \xi y)^2 f_y(x,\xi)\, dx = \frac{\xi y}{\tau_0}$$

As $\tau_0 \to \infty$ this expression tends to 0, which means that the probability is 0 for every value of x other than the mean value ξy.

From the derivative of (114) with respect to x

$$\frac{d}{dx} f_y(x,\xi) = \tau_0^2 (\tau_0 \xi y - 1 - \tau_0 x) \frac{(\tau_0 x)^{\tau_0 \xi y - 2}}{\Gamma(\tau_0 \xi y)} e^{-\tau_0 x}$$

it is clear that for $\tau_0\xi y > 1$, $f_y(x,\xi)$ has a maximum for $x = \xi y - \frac{1}{\tau_0}$.

In this case $f_y(x,\xi)$ starts with the value 0 for $x = 0$ and also tends towards 0 as $x \to \infty$. If on the other hand $\tau_0\xi y = 1$, $f_y(x,\xi)$ starts with the value τ_0 for $x = 0$ and then decreases exponentially towards 0 for increasing x-values. If finally $\tau_0\xi y < 1$, then $f_y(x,\xi) \to \infty$ as $x \to 0$. From the derivative it is evident that in this case the function is always decreasing for increasing x-values.

Thus as the value of ξ varies, the function (114) changes type when ξy passes the value $1/\tau_0$. It will soon be shown that this is connected with the appearance of clustering of the calls.

Instead of the constant τ_0 it is sometimes advantageous to introduce the inverted value $\eta_0 = 1/\tau_0$. The constant η_0 has then the dimension 1/time and can suitably be understood as a call intensity.

By the form (114) of $f_y(x,\xi)$ we now get for the distribution function introduced through (102)

$$\vartheta_y(t,\xi) = \frac{\tau_0^{\tau_0 \xi y}}{\Gamma(\tau_0 \xi y)} \int_0^\infty x^{\tau_0 \xi y - 1} e^{-x(\tau_0 + t)} dx$$

The integration can be performed by means of the definition of the gamma function and after some simplifications we then get

$$\vartheta_y(t,\xi) = \left(1 + \frac{t}{\tau_0}\right)^{-\tau_0 \xi y} \tag{115 a}$$

or by means of the recently introduced constant η_0

$$\vartheta_y(t,\xi) = (1 + \eta_0 t)^{-\frac{\xi y}{\eta_0}} \tag{115 b}$$

From this it is evident that the distribution introduced satisfies the conditions for a traffic class of Type A a and that the function $v(t)$ defined by (112) becomes

$$v(t) = \left(1 + \frac{t}{\tau_0}\right)^{-\tau_0} \tag{116 a}$$

or

$$v(t) = (1 + \eta_0 t)^{-\frac{1}{\eta_0}} \tag{116 b}$$

It is clear that both $v(t)$ and $\vartheta_y(t,\xi)$ are always completely monotonic functions of t. Besides $v(t)$ satisfies the sufficient condition set up on page 86 that $\{v(t)\}^k$ should be a completely monotonic function, as

$$-\frac{v'(t)}{v(t)} = \frac{1}{1 + \eta_0 t}$$

is completely monotonic. Further $v(t)$ satisfies the conditions $v(0) = 1$ and $v'(0) = -1$. Furthermore $v''(0) = 1 + \eta_0$, so that (107 c) gets the form

$$-\varphi_y'(0) = \eta_0 + my \qquad (107\text{ d})$$

If $\tau_0 \to \infty$ or $\eta_0 \to 0$ we get the limit value $v(t) = e^{-t}$ and $\vartheta_y(t, \xi) = e^{-\xi y t}$. This corresponds to the case when no random intensity variations are present and is fully concordant with statements on page 88 for this case. The The function $\vartheta_y(t, \xi)$ is, as has been mentioned earlier, the distribution function for the next call during the time when the parameter of the common variations has the value ξ. If we form the mean value

$$\int_0^\infty \vartheta_y(t, \xi)\, dt$$

this expresses the mean value of the distance between an arbitrarily chosen point of time and the next call (still during the time when the parameter of the common variations has the value ξ). This integral will now according to (115 b) be

$$\int_0^\infty \vartheta_y(t, \xi)\, dt = \frac{1}{\eta_0 - \xi y}\left[(1 + \eta_0 t)^{1 - \frac{\xi y}{\eta_0}}\right]_0^\infty$$

If $\eta_0 < \xi y$ this expression has the value $1 / (\xi y - \eta_0)$. If on the other hand $\eta_0 \geqq \xi y$ we get the value $+ \infty$.

The concept of clustering has in the foregoing been used in such a manner that a traffic flow is said to tend towards clustering when $y \to 0$ if there are finite t-values for which

$$\lim_{y \to 0} \varphi_y(t) < 1$$

Because of the conditions now shown it seems to be appropriate to use the concept in a somewhat different form. We may thus state that traffic with a distribution function for the next call $\vartheta(t)$ is clustered if

$$\int_0^\infty \vartheta(t)\, dt$$

does not converge. A traffic flow with the call intensity ξy and the distribution function for the next call (115 b) is thus clustered for all $\xi y \leqq \eta_0$. The constant η_0 should therefore be designated the *clustering limit* and may be considered as a call intensity, as has already been remarked.

Using the form (114) for the random intensity variations, we now obtain from (106 a) and (106 b) for the general distribution function for the next call

$$\vartheta_y(t) = \int_0^\infty \frac{dH(\xi)}{(1 + \eta_0 t)^{\frac{\xi y}{\eta_0}}} \qquad (117\text{ a})$$

and for the distribution function for the interarrival times

$$\varphi_y(t) = \int_0^\infty \frac{\xi\, dH(\xi)}{(1 + \eta_0 t)^{\frac{\xi y}{\eta_0} + 1}} \qquad (117\text{ b})$$

From the second formula we get the limit as $y \to 0$

$$\lim_{y \to 0} \varphi_y(t) = \frac{1}{(1 + \eta_0 t)}$$

which shows the convergence towards clustering.

Of interest is the special case when no intensity variations in common are present and thus $H(\xi)$ varies only in the point $\xi = 1$. We then get

$$\vartheta_y(t) = (1 + \eta_0 t)^{-\frac{y}{\eta_0}} \qquad (118\text{ a})$$

and

$$\varphi_y(t) = (1 + \eta_0 t)^{-\left(1 + \frac{y}{\eta_0}\right)} \qquad (118\text{ b})$$

The frequency function of the call intensity will in this case according to (99) and (114) be

$$g_y(x) = \frac{1}{\eta_0} \frac{\left(\frac{x}{\eta_0}\right)^{\frac{y}{\eta_0}-1}}{\Gamma\left(\frac{y}{\eta_0}\right)} e^{-\frac{x}{\eta_0}} \qquad (118\text{ c})$$

In some of the practical applications the description of the traffic properties so obtained is satisfactory though it disregards the intensity variations in common. It seems therefore suitable to introduce a special name for this description. A traffic flow whose properties can be described by the relations (118) is said to be expressed in the *normal form 1.*

If we now take into account the occurrence of common intensity variations, it seems appropriate to assume a frequency function of similar form to (114). For this purpose the following normal form for $dH(\xi)$ is introduced

$$\frac{dH(\xi)}{d\xi} = \frac{1}{\varkappa} \frac{\left(\frac{\xi}{\varkappa}\right)^{\frac{1}{\varkappa}-1}}{\Gamma\left(\frac{1}{\varkappa}\right)} e^{-\frac{\xi}{\varkappa}} \qquad (119)$$

where κ is a dimensionless constant as ξ itself is a relative number. Evidently, the conditions (97 a) and (97 b) are satisfied. Further the variance is

$$\int_0^\infty (\xi - 1)^2\, dH(\xi) = \varkappa$$

and thus κ is equal to the variance. How (119) varies as a function of ξ is clear directly from the previous discussion of (114) as a function of x. For $\kappa < 1$ (119) thus has a maximum for $\xi = 1 - \kappa$.

If we use the form (119) we obtain from (113 a)

$$\vartheta_y(t) = \frac{1}{\varkappa^{\frac{1}{\varkappa}} \Gamma\left(\frac{1}{\varkappa}\right)} \int_0^\infty \xi^{\frac{1}{\varkappa}-1} e^{-\xi\left\{\frac{1}{\varkappa} - y \ln v(t)\right\}} d\xi$$

The integration can be carried out directly and we get

$$\vartheta_y(t) = \{1 - \varkappa y \ln v(t)\}^{-\frac{1}{\varkappa}} \qquad (120\text{ a})$$

In a similar way we get from (113 b)

$$\varphi_y(t) = -\frac{v'(t)}{v(t)} \{1 - \varkappa y \ln v(t)\}^{-\left(1 + \frac{1}{\varkappa}\right)} \qquad (120\text{ b})$$

In these forms the function $v(t)$ can still be arbitrarily chosen. If we combine the form (119) for $dH(\xi)$ with the form (114) for $f_y(x, \xi)$ and with $v(t)$ from (116) we get

$$\vartheta_y(t) = \frac{1}{\left\{1 + \varkappa \frac{y}{\eta_0} \ln(1 + \eta_0 t)\right\}^{\frac{1}{\varkappa}}} \qquad (121\text{ a})$$

and

$$\varphi_y(t) = \frac{1}{(1 + \eta_0 t)\left\{1 + \varkappa \frac{y}{\eta_0} \ln(1 + \eta_0 t)\right\}^{1 + \frac{1}{\varkappa}}} \qquad (121\text{ b})$$

In this case we get from (107 d)

$$-\varphi_y'(0) = \eta_0 + (1 + \varkappa)\, y \qquad (107\text{ e})$$

By introducing the two normal forms (114) and (119) for the intensity variations, relatively simple mathematical expressions for the distribution functions of the calls have been obtained. They comprise two parameters of which η_0 represents the random intensity variations and κ the intensity variations in common for the class; because of that a good adaptation to curves obtained by measurements can be expected. A traffic flow whose properties are described by the relations (121) is said to be expressed in *normal form 2.* Evidently normal form 1 is a special case of normal form 2 for $\kappa = 0$.

If the values of the parameters η_0 and κ have been determined by measurements on a traffic flow, then the frequency functions of the intensity variations – random and in common – can be formed according to (114) and (119). A disadvantage is, however, that the frequency function $g_y(x)$ of the total intensity variations cannot be given any simple mathematical expression as the integral in (99) cannot be expressed in elementary terms. If we wish to use $g_y(x)$ we have to satisfy ourselves with a numerical integration.

Chapter 9

Properties of traffic flows in normal forms 1 and 2

For traffic flows that could be expressed in the normal forms 1 and 2, introduced in the preceding chapter, all traffic quantities were dependent only on the call intensity y, the mean occupation time s and the parameters η_0 and κ. These quantities could then be determined by measurements in a number of ways. Later on a series of special methods for carrying out the necessary computations will be treated. These methods are based upon certain basic properties of the normal forms. Some of them were already shown in Chapter 8. The remainder will be treated in more detail in this chapter.

By introducing forms of the intensity distribution function $G(x)$ connected with the normal forms, one gets in many cases relatively simple mathematical expressions for the more important traffic quantities. For a series of traffic quantities important in practice the expressions are, however, already so complicated for random traffic that any moderate assumption for the form of the distribution function $G(x)$ leads to extremely cumbersome formulas for exact numerical computations. One must in such cases rely on approximate formulas. For the forming of such formulas the expansion (58) in Chapter 6 for an integrated traffic quantity F_s becomes of basic importance. This expansion should therefore be investigated first for traffic flows in one of the normal forms. It shows then that also in this case the normal forms mostly lead to simple and readily calculated approximate expressions for the traffic quantities.

The expansion (58) applies for arbitrary variation limits a and b. In this case the boundaries $a = 0$ and $b = \infty$ are of foremost interest. They comprise all intensity values. Since $G(x)$ is a distribution function, then from (57 a) $m_0 = 1$. Since further $G(x)$ is the distribution function of the call intensity, its mean value $m = y$. The sum of the first ν terms of the expansion (58)

$$F(y) + \sum_{\sigma=2}^{\nu} \frac{F^{(\sigma)}(y)}{\sigma!} \int_0^{\infty} (x - y)^{\sigma}\, dG(x) \quad (122\text{ a})$$

could then be designated as the νth approximation for the traffic quantity F_s defined by

$$F_s = \int_0^{\infty} F(x)\, dG(x)$$

The first approximation is then $F(y)$, namely the value we get for random traffic with a call intensity y which is equal to the mean call intensity for slow intensity variations. This approximate value is the one which has so far been used generally.

If in (122 a) we introduce the quantities $m_{y,\nu}$ defined by (63) for the distribution $G(x)$ we get

$$F(y) + \sum_{\sigma=2}^{\nu} \frac{(-y)^{\sigma}}{\sigma!} m_{y,\sigma} F^{(\sigma)}(y) \quad (122\text{ b})$$

as an expression of the νth approximation of F_s. Now first the question arises as to which values the constants $m_{y,\sigma}$ have for traffic flows expressed in the normal forms. For normal form 1, $dG(x)$ is given by (118 c) and we get

$$(-y)^\sigma m_{y,\sigma} = \int_0^\infty (x-y)^\sigma \frac{\left(\frac{x}{\eta_0}\right)^{\frac{y}{\eta_0}-1}}{\Gamma\left(\frac{y}{\eta_0}\right)} e^{-\frac{x}{\eta_0}} \frac{dx}{\eta_0} \tag{123 a}$$

For normal form 2 we get from (99), (114) and (119)

$$(-y)^\sigma m_{y,\sigma} = \int_{x=0}^\infty \int_{\xi=0}^\infty (x-y)^\sigma \frac{\left(\frac{x}{\eta_0}\right)^{\frac{\xi y}{\eta_0}-1}}{\Gamma\left(\frac{\xi y}{\eta_0}\right)} \cdot e^{-\frac{x}{\eta_0}} \frac{\left(\frac{\xi}{\varkappa}\right)^{\frac{1}{\varkappa}-1}}{\Gamma\left(\frac{1}{\varkappa}\right)} e^{-\frac{\xi}{\varkappa}} \frac{d\xi\, dx}{\varkappa\, \eta_0} \tag{123 b}$$

The integrals in (123 a) and (123 b) can easily be carried out by considering the definition of the gamma function. For higher values of σ, however, we get more and more complicated expressions, especially for normal form 2. For the smallest σ-values we get:

For normal form 1

$$\left.\begin{aligned} m_{y,2} &= \frac{\eta_0}{y} \\ -m_{y,3} &= 2\left(\frac{\eta_0}{y}\right)^2 \\ m_{y,4} &= 3\left(\frac{\eta_0}{y}\right)^2 + 6\left(\frac{\eta_0}{y}\right)^3 \\ -m_{y,5} &= 20\left(\frac{\eta_0}{y}\right)^3 + 24\left(\frac{\eta_0}{y}\right)^4 \end{aligned}\right\} \tag{124 a}$$

For normal form 2

$$\left.\begin{aligned} m_{y,2} &= \varkappa + \frac{\eta_0}{y} \\ -m_{y,3} &= 2\left(\varkappa + \frac{\eta_0}{y}\right)^2 - \varkappa\frac{\eta_0}{y} \\ m_{y,4} &= 3\left(\varkappa + \frac{\eta_0}{y}\right)^2 + 6\left(\varkappa + \frac{\eta_0}{y}\right)^3 - \\ &\quad - 6\varkappa\frac{\eta_0}{y}\left(\varkappa + \frac{\eta_0}{y}\right) - \varkappa\left(\frac{\eta_0}{y}\right)^2 \\ -m_{y,5} &= 20\left(\varkappa + \frac{\eta_0}{y}\right)^3 + 24\left(\varkappa + \frac{\eta_0}{y}\right)^4 - \\ &\quad - 10\varkappa\frac{\eta_0}{y}\left(\varkappa + \frac{\eta_0}{y}\right) - 36\varkappa^3\frac{\eta_0}{y} - \\ &\quad - 74\varkappa^2\left(\frac{\eta_0}{y}\right)^2 - 46\varkappa\left(\frac{\eta_0}{y}\right)^3 \end{aligned}\right\} \tag{124 b}$$

From a comparison between the two expressions for $m_{y,2}$ it is clear that if we do not consider it necessary to have a more accurate value for the computation of an integrated traffic quantity F_s than the one obtained from the second approximation, then the same result is obtained for a traffic in normal form 2 with the parameters η_0 and κ as for a traffic in normal form 1 with the parameter $\eta_0 + y\kappa$. This condition is in itself rather trivial as the second approximation is always the same for all traffic flows for which the distribution function of the intensity has the same variance. However, we find that also the higher quantities $m_{y,\sigma}$ will be rather close to each other in normal form 2 and normal form 1 if in the latter case the parameter is put equal to $\eta_0 + y\kappa$. Thus for this case according to (124 a)

$$-m_{y,3} = 2\left(\varkappa + \frac{\eta_0}{y}\right)^2$$

and according to (124 b)

$$-m_{y,3} = 2\left(\varkappa + \frac{\eta_0}{y}\right)^2 - \varkappa\frac{\eta_0}{y}$$

The relative difference between these expressions is

$$\frac{\varkappa \frac{\eta_0}{y}}{2\left(\varkappa + \frac{\eta_0}{y}\right)^2}$$

This has a maximum for $\kappa = \eta_0/y$ and then takes the value 0.125. If we investigate $m_{y,4}$ in a similar way, we find that the relative difference has a maximum for $\kappa = 0.927\ \eta_0/y$ with the value

$$\frac{1{,}627\left(\varkappa + \frac{\eta_0}{y}\right)}{3 + 6\left(\varkappa + \frac{\eta_0}{y}\right)}$$

For practical applications we now find that $\kappa + \eta_0/y$ has the order of magnitude of 0.1 or less. In such a case the maximum relative difference for $m_{y,4}$ will be around 0.045 or less.

Also for $m_{y,5}$ the relative difference generally has a maximum near $\kappa = \eta_0/y$. For $\kappa + \eta_0/y = 0.1$ the order of magnitude of this maximum will be 0.16.

If the ratio of κ to η_0/y differs considerably from 1, then the relative differences will be considerably smaller than in the maximum points and they will be 0 if any of the parameters is 0. This follows, moreover, from the fact that normal form 2 passes into normal form 1 if any of the parameters is 0.

The above investigation now shows that for such integrated traffic quantites F_s, which can be expressed with sufficient accuracy by means of any of the lower approximations (122 b), we can usually without greater error apply the formulas which pertain to normal form 1, even if the traffic is expressed in normal form 2. In this case the parameter $\eta_0 + y\kappa$ should be used where η_0 and κ are the parameters in normal form 2. This condition is of great importance for numerical computations. In many cases normal form 1 gives so simple explicit expressions for the integrated traffic quantities that they can with advantage be used instead of the approximations (122). On the other hand the corresponding explicit expressions for normal form 2 will often be more complicated. Because of that, this normal form will then mainly be of theoretical interest which depends upon the fact that it can be interpreted as an expression for common as well as random intensity variations.

In Chapter 6 we discussed the properties of the most important integrated traffic quantities. The starting point was the second approximation according to (122). When the traffic is expressed in one of the normal forms it is now possible to carry the discussion considerably further, as in this case we can also directly start from the explicit expressions for the traffic quantities. First we consider the distribution function of the inter-arrival times $\varphi(t)$ according to (118 b) that is for normal form 1. This is a function of the time t and the parameters y and η_0. A discussion of this function does not involve any restriction if we put $y = 1$; this only implies that the previously undetermined time unit is put equal to the mean inter-arrival time. The function that should be investigated is then

$$\varphi_1(t) = (1 + \eta_0 t)^{-\left(1 + \frac{1}{\eta_0}\right)} \qquad (125\text{ a})$$

For comparison with the exponential distribution function with the same mean value, that is e^{-yt}, the *difference curve of the inter-arrival times* (introduced in Chapter 6) will also be considered. It is here designated

$$D_{\eta_0}(t) = \varphi_1(t) - e^{-t} = (1 + \eta_0 t)^{-\left(1 + \frac{1}{\eta_0}\right)} - e^{-t} \qquad (125\text{ b})$$

The general course of this function is discussed in Chapter 6 where it has been shown that besides in the point $t = 0$ it also has a zero point around $t = 2$, a minimum around $t = 0.6$, and a maximum around $t = 3{,}4$. As $m_{y,2}$ now, according to (124 a), will be equal to

η_0, the minimum according to (66 c) is approximately $-0.23\,\eta_0$, and the maximum close to $+0.08\,\eta_0$. Further the form factor that was defined in (67) will in this case be $\varepsilon = 2/(1-\eta_0)$ and is finite only for $\eta_0 < 1$.

A more thorough discussion can be obtained by forming the expressions for higher approximations than the second by means of (124 a). However, the difference curve according to (125 b) has such a simple form that it can be investigated directly without introducing approximations.

We consider first the derivative of (125 b) with respect to t

$$D'_{\eta_0}(t) = -\frac{1+\eta_0}{1+\eta_0 t}\varphi_1(t) + e^{-t} \qquad (126\text{ a})$$

From this we get $D'_{\eta_0}(0) = -\eta_0$. Furthermore for $t = 1$

$$D'_{\eta_0}(1) = -D_{\eta_0}(1)$$

If $D_{\eta_0}(1)$ is negative then $D'_{\eta_0}(1)$ will be positive. For $t = 1$ one is already on the increasing branch and from that follows that the minimum will occur for $t < 1$. The other alternative, that $D_{\eta_0}(1)$ is positive, means that the total negative part of $D_{\eta_0}(t)$ must occur for $t < 1$.

In the two points where the derivative (126 a) is zero, obviously

$$D_{\eta_0}(t) = \frac{\eta_0}{1+\eta_0}(t-1)\,e^{-t} \qquad (126\text{ b})$$

which is a formula that can be useful for some computations.

We now consider $D_{\eta_0}(t)$ as a function of η_0. When $\eta_0 \to 0$ then $D_{\eta_0}(t) = 0$ and when $\eta_0 \to \infty$ then $D_{\eta_0}(t) \to -e^{-t}$. The derivative with respect to η_0 is

$$\frac{d}{d\eta_0}D_{\eta_0}(t) = \frac{\varphi_1(t)}{\eta_0^2}\left\{\ln(1+\eta_0 t) - \eta_0 t\,\frac{1+\eta_0}{1+\eta_0 t}\right\} \qquad (127)$$

The expression within brackets in (127)

$$K(\eta_0) = \ln(1+\eta_0 t) - \eta_0 t\,\frac{1+\eta_0}{1+\eta_0 t}$$

is zero for $\eta_0 = 0$ and goes evidently towards negative values for large η_0-values. As

$$\frac{d}{d\eta_0}K(\eta_0) = \frac{\eta_0 t}{(1+\eta_0 t)^2}(t-2-\eta_0 t)$$

is always negative for $t \leqq 2$ and $\eta_0 > 0$, it is apparent that $K(\eta_0)$ in this case is always negative and consequently the derivative (127) is negative. From this follows again that for $t \leqq 2$, $D_{\eta_0}(t)$ is never positive. The positive root of $D_{\eta_0}(t) = 0$ thus always appears for $t > 2$. This is not a general property, valid for all distribution functions for the inter-arrival times. For instance it is not valid for the distribution considered in Chapter 6, Fig. 6, (where $y = 0.5$).

Since (127), as has just been shown, has no zero point for positive $t \leqq 2$, there is at most one positive η_0-root of the equation $D_{\eta_0}(t)$ = a constant for every positive value of $t \leqq 2$. The curves for $D_{\eta_0}(t)$ as a function of t thus cannot cross within this area.

For $t > 2$

$$\frac{d}{d\eta_0}K(\eta_0) \text{ is positive for } 0 < \eta_0 < 1 - \frac{2}{t}$$

$$\frac{d}{d\eta_0}K(\eta_0) \text{ is negative for } \eta_0 > 1 - \frac{2}{t}$$

As $K(0) = 0$ and $K(\eta_0)$ assumes negative values for great η_0-values, $K(\eta_0) = 0$ must have one and only one positive η_0-root which evidently is greater than $1 - 2/t$. The same is true for the derivative (127). From this follows that $D_{\eta_0}(t)$ for $t > 2$ is positive at least within the area $0 < \eta_0 < (1 - 2/t)$. The curve can evidently have only one maximum which occurs for $\eta_0 > (1 - 2/t)$. For increasing η_0 it thereafter decreases continuously towards $-e^{-t}$.

Fig. 10 shows the course of $D_{\eta_0}(t)$ as a function of η_0 at one value of $t < 2$ and at another value of $t > 2$. For every value of $t > 2$ there is a finite maximum value of $D_{\eta_0}(t)$. These maximum values constitute together a *positive limiting curve* for $D_{\eta_0}(t)$ as a function of t defined within the area $t \geqq 2$. This limiting curve evidently constitutes an envelope for $D_{\eta_0}(t)$ as a function of t for different η_0-values. There are obviously, for every value of $t > 2$, two positive η_0-roots of the equation $D_{\eta_0}(t) = k$, where k is a positive constant, which is smaller than the maximum value of $D_{\eta_0}(t)$. Obviously, in each point inside the positive limiting curve, two $D_{\eta_0}(t)$-curves as functions of t cross each other.

For numerical computation of the envelope the zero points of (127) are determined and then the corresponding $D_{\eta_0}(t)$-values are computed. One starts suitably from different values of the product $\eta_0 t$ and computes the corresponding η_0-value from $K(\eta_0) = 0$, that is from

$$1 + \eta_0 = \left(1 + \frac{1}{\eta_0 t}\right) \ln(1 + \eta_0 t)$$

and thereby t. In this way one avoids a troublesome iterative procedure. Then the wanted maximum value is obtained from

$$D_{\eta_0}(t) = e^{-\frac{(1+\eta_0)^2}{1+\eta_0 t} t} - e^{-t}$$

A similar method can be used for the numerical computation of the point where the difference curve crosses the t-axis. If we put $D_{\eta_0}(t)$ according to (125 b) equal to zero, we get a relation that can be written in the form

$$1 + \eta_0 = \frac{\eta_0 t}{\ln(1 + \eta_0 t)}$$

For given values of $\eta_0 t$, η_0 is computed and from that t will be determined. With the methods mentioned the positive limiting curve and the crossing point t_0 have been computed. Fig. 11 shows t_0 as a function of η_0. Further the difference curve $D_{\eta_0}(t)$ has been computed for a number of η_0-values. In Fig. 12 some typical courses for the difference curve are shown together with the envelope. Diagrams 1 A and 1 B (at the end of this work) show a tighter family of curves of the function $D_{\eta_0}(t)$ which is limited to such magnitudes of

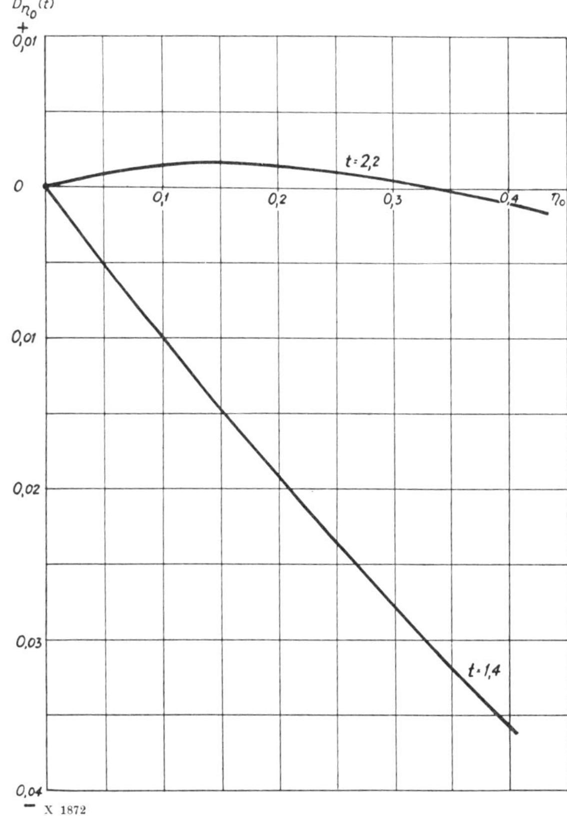

Fig. 10. $D_{\eta_0}(t)$ as a function of η_0 for one value of $t < 2$ and for another value of $t > 2$.

η_0 that are mostly present for practical applications. For the same purpose, in Diagrams 2 A – D, curves are presented for $D_{\eta_0}(t)$ as a function of η_0 for some values of t. For smaller η_0-values these curves are close to

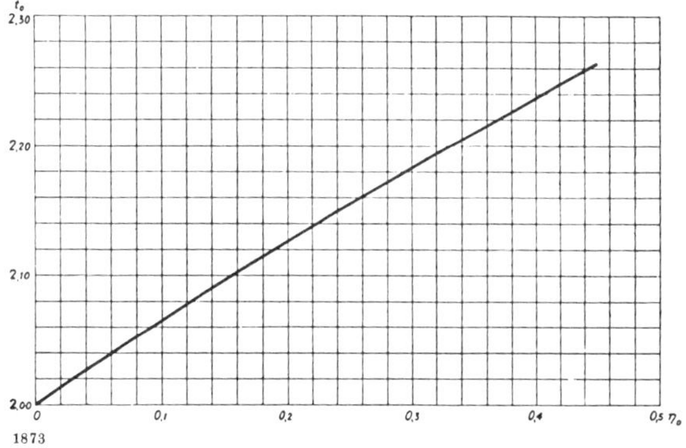

Fig. 11. The root t_0 of $D_{\eta_0}(t) = 0$ as a function of η_0.

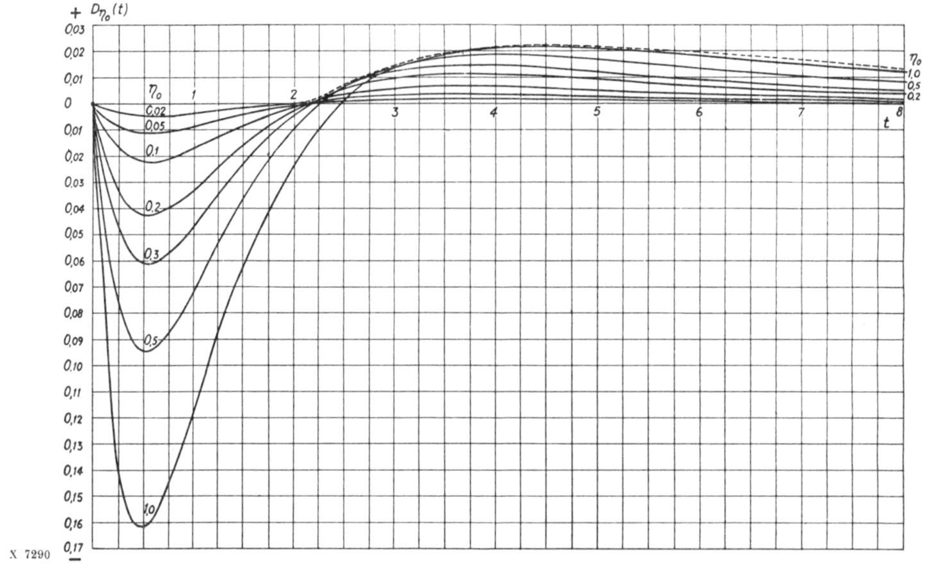

Fig. 12. D_{η_0} as a function of t for some different values of η_0. The dotted line is the envelope.

straight lines with the equation

$$D_{\eta_0}(t) \approx \frac{1}{2} t(t-2) e^{-t} \eta_0 \qquad (128)$$

which follows from the second approximation according to (122) for $\varphi_1(t)$. From that follows for instance

$$\eta_0 \approx -2e\,D_{\eta_0}(1) \qquad (128\text{ b})$$

a formula that can be useful for the estimation of the order of magnitude of η_0.

It has become evident from the computation of $D_{\eta_0}(t)$ that the approximation (128) is sufficiently accurate for most of the actual applications. The computation can then be performed by means of Table 1 at the end of this work. The table gives the coefficient

$$k(t) = \frac{1}{2}\, t\,(t-2)\; e^{-t}$$

for a number of t-values and the upper permitted limit for η_0, for a computation error not greater then 0.0001.

Further the more general distribution function for the inter-arrival times (121 b), valid for normal form 2, will now be considered. Also in this case we can put $y = 1$ and thus choose the time unit equal to the mean inter-arrival time. The difference curve will then get the form

$$D_{\eta_0,\varkappa}(t) =$$

$$= \frac{1}{1+\eta_0 t}\left\{1 + \frac{\varkappa}{\eta_0}\ln(1+\eta_0 t)\right\}^{-\left(1+\frac{1}{\varkappa}\right)} - e^{-t} \qquad (129)$$

As this besides t comprises two parameters η_0 and κ, a direct discussion will be considerably more difficult than for the curve (125 b). We can then make use of the condition shown earlier in this chapter during the investigation of the approximations (122), that a traffic quantity for normal form 2 can be approximated by means of normal form 1. From this result follows that as an approximation of (129) we have

$$D_{\eta_0,\varkappa}(t) \approx D_{\varkappa+\eta_0}(t) \qquad (130\text{ a})$$

A series of approximations of another kind can be obtained from the expansion (122 b) and the coefficients (124 b). The derivatives can in this case be obtained from (65) in

Chapter 6. For instance the third approximation will be

$$\left.\begin{aligned} D_{\eta_0,\varkappa}(t) \approx \frac{1}{2}\, t\,(t-2)\, e^{-t}(\varkappa+\eta_0) - \\ -\frac{1}{3}\, t^2 (t-3)\, e^{-t} (\varkappa+\eta_0)^2 + \\ +\frac{1}{6}\, t^2 (t-3)\, e^{-t} \varkappa\eta_0 \end{aligned}\right\} \quad (130\text{ b})$$

A better approximation can generally be obtained by combining (130 a) and (130 b). If one writes

$$D_{\eta_0,\varkappa}(t) \approx D_{\varkappa+\eta_0}(t) + \frac{1}{6}\, t^2 (t-3)\, e^{-t} \varkappa\eta_0 \quad (130\text{ c})$$

one has corrected (130 a) for the deviation that occurs by the term $\kappa\eta_0$ in $m_{y,3}$ according to (124 b). As an example of the accuracy of the approximations, the following is given: At $\kappa = \eta_0 = 0.05$ according to (129) $D_{\eta_0,\kappa}(1)$ will get the value -0.01765. According to (130 a) it will be -0.01739 and according to (130 b) -0.01625. From (130 c) finally we get -0.01770 which is closest to the real value. The approximations put up for (129) generally seem to give a good accuracy compared with the order of magnitude of the extreme values of $D_{\eta_0,\kappa}(t)$. In contrast, the relative accuracy can sometimes be unsatisfactory. This is especially the case for greater t-values. For those both $D_{\eta_0,\kappa}(t)$ and its approximation are very small by comparison with the extreme values mentioned but the quotient between the approximate and the real values can differ considerably from 1. That this must be the case is evident from the fact that the form factor for the distribution function (121 b) is always infinitely great. If one tries to compute this according to the definition (67) in Chapter 6 for normal form 2 one obtains by means of (99), (114) and (119) in Chapter 8 an integral that is not convergent for $\eta_0 > 0$. The reason for this is easy to realize. For normal form 1 the form factor is finite, as has been shown earlier, only if the mean call intensity is greater than η_0. Now normal form 2 implies a summation of traffic flows in normal form 1 where the summation also comprises indefinitely small mean call intensities. If $\eta_0 > 0$ there are therefore always time intervals for a traffic in normal form 2 for which the mean call intensity is smaller than η_0 and from this follows that for normal form 2 the form factor must always be indefinitely great.

As long as $\kappa + \eta_0 < 1$ all approximations (130) evidently correspond to inter-arrival time distribution functions with a finite form factor. Then obviously the approximations concerned for great t-values must represent curves with a considerably different nature from (129).

An idea of the conditions of (129) for great t-values can be obtained by introducing a new variable

$$\tau = \frac{1}{\eta_0} \ln (1 + \eta_0 t) \quad (131)$$

and inversely

$$t = \frac{e^{\eta_0\tau} - 1}{\eta_0} = \tau + \frac{1}{2}\eta_0\tau^2 + \frac{1}{6}\eta_0{}^2\tau^3 + \ldots$$

From this it is evident that $t > \tau$ although t and τ nearly coincide for small values of τ and η_0. From (129) we get

$$D_{\eta_0,\varkappa}(t) = \frac{1}{1+\eta_0 t}(1+\varkappa\tau)^{-\left(1+\frac{1}{\varkappa}\right)} - e^{-t}$$

If we now introduce (125 b), which by means of τ gets the form

$$D_{\eta_0}(t) = \frac{1}{1+\eta_0 t}\, e^{-\tau} - e^{-t}$$

we obtain

$$D_{\eta_0,\varkappa}(t) = \frac{1}{1+\eta_0 t}\left\{(1+\varkappa\tau)^{-\left(1+\frac{1}{\varkappa}\right)} - e^{-\tau}\right\} + D_{\eta_0}(t)$$

or

$$D_{\eta_0,\varkappa}(t) = e^{-\eta_0\tau} D_\varkappa(\tau) + D_{\eta_0}(t) \quad (132\text{ a})$$

For small values of η_0 and t we get the approximation

$$D_{\eta_0,\varkappa}(t) \approx D_\varkappa(t) + D_{\eta_0}(t) \qquad (132\text{ b})$$

For greater t-values the deformation of the time scale τ in relation to the time scale t, caused by the transformation (131) will be very noticeable since τ becomes considerably smaller than t. For very large t-values the second term in the right member of (132 a) can be totally neglected and we get

$$D_{\eta_0,\varkappa}(t) \sim e^{-\eta_0\tau} D_\varkappa(\tau) \qquad (132\text{ c})$$

For large t-values (129) has then the same form as a difference curve for normal form 1 with the parameter κ where, however, the time scale is distorted according to (131) and an attenuation factor $e^{-\eta_0\tau}$ is added.

The next type of integrated traffic quantities, which will be considered here more closely, are the moments of the intensity distribution, for which some theorems have been formulated already in Chapter 6. Later on it will be shown that we can define, directly as well as indirectly by measurements, quantities which for pure random traffic have the form $(sy)^n$ or A^n where A is the traffic intensity. If the traffic has slow intensity variations with the distribution function $G(x)$ for the call intensity we get the expression

$$M_n = \int_0^\infty (sx)^n \, dG(x) \qquad (133\text{ a})$$

As the distribution function of the inter-arrival times in this case has the form (52) from Chapter 5, we generally obtain the expression

$$M_n = (-1)^{n-1} y s^n \varphi^{(n-1)}(0) \qquad (133\text{ b})$$

If now the traffic is expressed in normal form 1, so that the distribution function for the inter-arrival times has the special form (118 b) then

$$\varphi_y^{(n-1)}(t) = (-1)^{n-1} \frac{(y+\eta_0)(y+2\eta_0)\ldots(y+(n-1)\eta_0)}{(1+\eta_0 t)^{n+\frac{y}{\eta_0}}}$$

From (133 b) we then obtain

$$M_n = s^n y(y+\eta_0)(y+2\eta_0)\ldots(y+(n-1)\eta_0) \qquad (134\text{ a})$$

If we here introduce $A = sy$ and put $y = 1$ and thus make the time unit equal to the mean inter-arrival time we get

$$M_n = A^n(1+\eta_0)(1+2\eta_0)\ldots(1+(n-1)\eta_0) \qquad (134\text{ b})$$

For the considerably more general case where the distribution function for the common intensity variations has the general form $H(\xi)$ we get from the expression (117 b) for the distribution function of the inter-arrival times

$$\varphi_y^{(n-1)}(t) = (-1)^{n-1} \int_0^\infty \frac{(\xi y+\eta_0)(\xi y+2\eta_0)\ldots(\xi y+(n-1)\eta_0)}{(1+\eta_0 t)^{n+\frac{\xi y}{\eta_0}}} \, \xi \, dH(\xi)$$

and from that according to (133 b)

$$M_n = s^n \int_0^\infty \xi y(\xi y+\eta_0)(\xi y+2\eta_0)\ldots(\xi y+(n-1)\eta_0)\, dH(\xi) \qquad (135\text{ a})$$

or if as earlier we put $y = 1$

$$M_n = A^n \int_0^\infty \xi(\xi+\eta_0)(\xi+2\eta_0)\ldots(\xi+(n-1)\eta_0)\, dH(\xi) \qquad (135\text{ b})$$

In order to compute this expression it is necessary to know the integrals

$$b_\nu = \int_0^\infty \xi^\nu \, dH(\xi) \qquad (136\text{ a})$$

In order to expand the product in the integrand in (135 b), a number of numerical coefficients $L_{n,\mu}$ are introduced, defined by

$$\xi(\xi+\eta_0)(\xi+2\eta_0)\ldots(\xi+(n-1)\eta_0) = \\ = \xi^n \sum_{\mu=0}^{n-1} \left(\frac{\eta_0}{\xi}\right)^\mu L_{n,\mu} \qquad (137\text{ a})$$

Evidently we here have the following starting values and the recursion formulas

$$\left.\begin{aligned} &L_{n,0} = 1 \\ &L_{n,n-1} = (n-1)! \\ &L_{n,\mu} = L_{n-1,\mu} + (n-1)\,L_{n-1,\mu-1} \end{aligned}\right\} \qquad (137\text{ b})$$

From (135 b) we obtain

$$M_n = A^n \sum_{\mu=0}^{n-1} \eta_0{}^\mu L_{n,\mu}\, b_{n-\mu} \qquad (135\text{ c})$$

For a traffic in normal form 2 where $H(\xi)$ has the form (119) we get b_ν according to (136 a)

$$b_\nu = (1+\varkappa)(1+2\varkappa)\ldots(1+(\nu-1)\varkappa) \qquad (136\text{ b})$$

In Table 2 at the end of this work some coefficients $L_{n,\mu}$ are tabulated. In addition, for normal form 2, viz. the form (135), the expressions for M_n are explicitly presented for low n-values.

For a numerical computation of M_n the form (134 b) for normal form 1 is very suitable for direct use. For normal form 2 the computations will be somewhat more difficult and it could be desirable to have access to simpler approximate formulas. To this can be remarked that the expansion (135 c) in increasing powers of η_0 might very often be discontinued at lower values of μ and the upper limit $n-1$. Another possibility might present itself in the approximations (122). We easily find, however, that also these give an expansion in increasing powers of η_0, which, besides, will be more complicated than (135 c). For greater n-values and not too small η_0 one must include rather many terms in (122) to get a reasonably good approximation for M_n. The reason for this is that under such conditions M_n differs considerably from the value A^n valid at zero variation. Generally M_n seems to be one of the traffic quantities which is influenced most strongly by the intensity variations. For instance for $\eta_0 = 0.1$ the value of M_{20} is about 240,000 times greater than at zero variation.

Of importance for the applications are finally also the expressions (71) in Chapter 6, which indicate the mean value of the part of the total time when n devices and no more are busy at the same time in a full availability group. In (71) the mean occupation time s is put equal to 1. If this is not the case we have

$$P_n = \int_0^\infty \frac{(sx)^n}{n!} e^{-sx}\, dG(x) \qquad (138\text{ a})$$

Next some relations between P_n and M_n not previously mentioned must be remarked upon. From the expansion of e^{-sx} in powers of sx we get directly the relation

$$P_n = \frac{1}{n!} \sum_{\mu=0}^{\infty} (-1)^\mu \frac{M_{n+\mu}}{\mu!} \qquad (139\text{ a})$$

Inversely evidently

$$M_n = \sum_{\mu=0}^{\infty} (\mu+1)(\mu+2)\ldots(\mu+n)\, P_{n+\mu} \qquad (139\text{ b})$$

For P_n an expression analogous to (133 b) can also be found. If we differentiate the distribution function of the inter-arrival times in the general form (52) in Chapter 6 altogether $n-1$ times and after that put $t = s$, then we get

$$P_n = (-1)^{n-1}\, y\, \frac{s^n}{n!}\, \varphi^{(n-1)}(s) \qquad (138\text{ b})$$

If now the traffic is expressed in normal form 1 so that the distribution function of the inter-arrival times has the form (118 b), we obtain for $n > 0$

$$P_n = \frac{s^n}{n!} \frac{y(y+\eta_0)(y+2\eta_0)\ldots(y+(n-1)\eta_0)}{(1+\eta_0 s)^{n+\frac{y}{\eta_0}}} \qquad (140\text{ a})$$

By comparison with (134 a) the relation

$$P_n = \frac{M_n}{n!}(1+\eta_0 s)^{-\left(n+\frac{y}{\eta_0}\right)} \tag{140 b}$$

is obtained. For the more general case, when the distribution function for the common intensity variations has the general form $H(\xi)$ we obtain, as is easily seen, instead of (140 a)

$$P_n = \frac{s^n}{n!}\int_0^\infty \frac{\xi y(\xi y+\eta_0)(\xi y+2\eta_0)\ldots(\xi y+(n-1)\eta_0)}{(1+\eta_0 s)^{n+\frac{\xi y}{\eta_0}}}\,dH(\xi) \tag{141 a}$$

In order to determine this expression we can use a similar method to the one used to get the formula (135 c) for M_n. If we put $y = 1$ and introduce the notation

$$c_\nu = \int_0^\infty \frac{\xi^\nu}{(1+\eta_0 s)^{\frac{\xi}{\eta_0}}}\,dH(\xi) \tag{142 a}$$

we obtain by analogy with (135 c)

$$P_n = \frac{A^n}{n!}(1+\eta_0 s)^{-n}\sum_{\mu=0}^{n-1}\eta_0^{\mu} L_{n,\mu} c_{n-\mu} \tag{141 b}$$

and here the L-quantities are those defined by (137 a). If the traffic is expressed in normal form 2 and $H(\xi)$ has the form (119), we get for c_ν by means of the expression (136 b)

$$c_\nu = b_\nu\left\{1+\frac{\varkappa}{\eta_0}\ln(1+\eta_0 s)\right\}^{-\left(\nu+\frac{1}{\varkappa}\right)} \tag{142 b}$$

Numerical computations according to the formulas obtained are sometimes rather cumbersome. In this case. however, use of the approximations (122) offers greater possibilities than computations of the moments. The ratio between P_n for intensity variations and zero variation is commonly relatively close to 1 (except for large values of n) and less than the corresponding ratio for the moments. The second approximation according to (122) is already expressed in (71) in Chapter 6. From the third approximation according to (122) we get

$$\frac{P_n}{\frac{A^n}{n!}e^{-A}} \approx 1+\frac{1}{2}\{(n-y)^2-n\}\,m_{y,2}-\frac{1}{6}\{(n-y)^3-3n(n-y)+2n\}\,m_{y,3} \tag{143}$$

where $m_{y,2}$ and $m_{y,3}$ are obtained from (124 a) and (124 b). The treatment of the measurement results carried out in Chapter 14 gives an idea of this approximation for different cases.

A number of other methods could be given with the purpose of permitting a simpler treatment of the numerical computations for the applications of the normal forms. These will be elaborated on in Chapter 12.

Chapter 10

Fundamentals of calculating intensity variations from results of measurements

By use of the theory of slow intensity variations and the related superposition theory, there would seem to be a possibility of carrying out more accurate computations of the traffic conditions in telephone plant than hitherto. This hope is based upon considerations of the physical background of the course of traffic events, starting from general ideas on the relation between the causes of the calls or on the lack of causes in common. Thereby valuable guidance is obtained for establishing suitable probability schemes. One has, however, to be cautious of believing that these discussions constitute sufficient proof to guarantee the applicability of the mathematical schemes in practice. Such a guarantee can only be obtained by comparisons between the theoretical results and the real traffic conditions based upon measurements.

The theories that have been established in Chapter 5 and the following are based upon the assumption that for every traffic flow and for every point of time there exists a fully determined traffic intensity and that the probability of a call is a function only of this instantaneous intensity value. The justification of such an assumption can only be verified indirectly by investigating whether the mean values of different traffic quantities pertaining to real telephone traffic coincide with the theoretical results. There does not seem to be any possibility of a more direct examination as one can never establish the instantaneous value of the intensity of real traffic.

For the further assumption introduced in Chapter 5 about the slowness of the intensity variations, the conditions will be somewhat different. The investigations that were made in Chapter 7 show, as a matter of fact, that the traffic expressions which are valid for slow variations are unchanged in their form even when the assumptions of slow variations are abandoned. In this case, however, the expressions for the intensity distribution, which are a part of the results, do not apply to the real traffic but to traffic modified in a certain way, the equilibrium traffic. From this follows that the assumption that was introduced for the theoretical investigations, about the slowness of the intensity variations, does not imply any significant limitation of the applicability of the formulas to real telephone traffic. It is found, however, that one can expect somewhat different intensity distributions for different types of traffic quantities.

The theories based upon slow intensity variations now result in traffic expressions in the form of mean values of the corresponding traffic expressions for random traffic, that is, in the form of integrals of type (50 a). The intensity distribution function $G(x)$ appearing here can, as was just remarked, vary somewhat for different types of traffic expressions, since the intensity variations are not ideally slow. Regarding the inter-arrival times one thus gets an intensity distribution which probably agrees very closely with the real one. For states and traffic quantities derivable from these, an intensity distribution applies which corresponds to in a certain manner attenuated intensity variations. Besides it should be observed that, as has already been remarked, variations of the mean occupation time could also cause a certain difference between the intensity distributions mentioned.

One can now put the question as to which properties the traffic quantities obtained by measurement should have, in order that the probability scheme based upon slow intensity variations can be considered satisfactory for description of the traffic. It is evidently satisfactory if all traffic quantities could be expressed in the form (50), that is in the form (50 a) or (50 b) depending on whether it concerns a time mean value or a call mean value. That a certain traffic quantity can always be expressed in this form is trivial, as the possibility exists of choosing the function $G(x)$ to a high degree arbitrarily. On the other hand it is not trivial that it should be possible to express all traffic quantities by one and the same distribution function $G(x)$. According to what has just been pointed out, this is not rigorously necessary, as $G(x)$ can for many reasons be expected to have different forms for different types of traffic quantities. For one and the same type, for instance for the states and traffic quantities derived from these $G(x)$ must, however, have the same form. It is out of the question to investigate all imaginable traffic quantities, as an unlimited number of them exist. For measurements we have to restrict ourselves to those which are important for dimensioning, or are especially available for measurement, or in other ways interesting and suitable. In order to investigate whether a number of these quantities could be derived from a function $G(x)$ in common, it is often in principle not necessary to carry out any explicit determination of $G(x)$. For different types of traffic quantities it is in many cases, as will be shown later, possible to establish criteria which permit a direct decision whether the description in question is possible or not.

In order to justify the theories based upon slow intensity variations it is, however, not sufficient that it can be proved that the results describe the reality with wanted accuracy; it mut also be possible to determine explicitly the intensity distribution function $G(x)$ from the results of measurements. It is then necessary to analyze the important principal question, under which conditions the intensity distribution is uniquely determined and which traffic quantities are suitable for this determination. From an investigation later in this chapter it appears that $G(x)$ can be uniquely determined from several types of traffic quantities. However, the numerical computation of $G(x)$ is very difficult in the general case as one cannot in advance start from a certain form of the function. One would therefore in practice always be referred to the introduction of one such form. This must, besides, be considered suitable also taking into account the need for access to formulas which are only moderately complicated. For applications in practice there exists no possibility of letting each traffic flow be characterized by a function that must be determined individually in its total course. If the theories put forward are to be of any practical use it is evidently necessary that each traffic flow can be characterized by a small number of parameters. Obviously, one should then investigate the possibilities of using one of the normal forms introduced in Chapter 8. These are, however, derived from discussions regarding the superposition conditions, but this should not prevent them being of use independently of the extent to which the superposition conditions for real traffic can be expressed by the theories in Chapter 8.

Of the two normal forms, form 1 gives a considerably simpler formula system than form 2. Now it has been shown, among other things by the investigations in Chapter 9, that the normal forms 1 and 2 do not give considerably different values for the traffic quantities of most interest. It is, therefore, hardly probable that a traffic flow that cannot be described satisfactorily in normal form 1, could generally be considerably better expressed in normal form 2. It should therefore generally be sufficient to investigate the possibilities of representing the traffic flows in normal form 1. This seems then to be the primary task for measurements aiming at an investigation of the usefulness of the theories for slow intensity variations. As a matter of fact it is evident that, if the properties of a traffic flow can be represented with sufficient accuracy by the formula system derived from normal form 1, it follows that all basic assumptions for the theory of slow intensity variations are justified for the traffic in question.

To this measurement program the following remark should be made. The theory of slow intensity variations does not claim to explain satisfactorily all deviations that could occur between the properties of real telephone traffic and theoretically random traffic. As a matter of fact for real telephone traffic there is, as already has been remarked, a certain after-effect of the blocking that causes a special correlation between the calls in the vicinity of the blocking conditions. This phenomenon, that really cannot be described as a slow intensity variation, has been totally left aside in this work, as it has not been possible to tackle this problem theoretically with any success. If it is a question of verifying the basic hypotheses of the theory for slow variations by comparison of the results obtained it is therefore necessary to suppress the influence of such phenomena as are irrelevant to the case. In the mentioned example of after-effect of the blocking, this can be done by always carrying out the measurements in so abundantly dimensioned groups that blocking practically does not occur.

If now the measurements show that the results of the theories based upon the principle of slow intensity variations can be used to compute the traffic conditions in device groups of a telephone plant, this brings up the question to what extent the superposition theories in Chapter 8 could also be applicable. In principle, measurements for this purpose could be carried out by investigating whether a family of traffic flows satisfies the conditions of Chapter 8 for different types of traffic classes. From these conditions it is easy to find out if the family belongs to one and the same traffic class of Type A or the special Type $A\ a$ or if the family only shows random or common intensity variations. For such measurements it is best to select the studied traffic flows in such a way that they represent traffic for which some traffic flows are formed by superposing. If one investigates three traffic flows a, b and c then one should also investigate the traffic flows that are formed by superposing $a + b$, $a + c$, $b + c$, $a + b + c$. As a matter of fact the superposition need not be realized in any group of devices. It can be carried out fictitiously in the measuring equipment.

Measurements of the type outlined seem to be rather complicated and could therefore presumably be carried out only to a limited extent. If most traffic flows could be expressed in normal form 1 with sufficient accuracy – and there seems to be reasons to rely on this being the case – the entire superposition question will, however, assume another and more advantageous aspect. The properties of a traffic flow are then dependent upon two parameters only, the traffic intensity and the clustering limit η_0. By superposition the traffic flows are added directly and one need only investigate by measurements how the clustering limit varies by superposition. It was shown in Chapter 8 that if the traffic flows are expressed in normal form 2, then the two parameters η_0 and κ belonging to this normal form become invariant when superposing. If again a traffic is expressed in normal form 2 then it may, as already has been repeatedly remarked, be expressed in normal form 1 with good approximation with the clustering limit $\eta_0 + y\kappa$. One can thus expect, for traffic flows that can be expressed in normal form 1 and which belong to the same traffic class, that the clustering limit should vary linearly with the mean call intensity. In such a case, in order to be able to apply the theories put forward in this work to full extent, one need only carry out relatively limited measurements with the purpose of determinining the parameters for different types of telephone traffic.

After this short survey of the measurements desirable for the application of the theories, the principles for carrying out measurements intended to determine the presence in telephone traffic of those properties which characterize random traffic modified by slow intensity variations will be treated in more detail. The other kind of desirable measurements, namely those which concern the superposition conditions, can on the contrary not be treated in any great detail in this work, as there has not been any occasion to carry through any systematic measurements of this kind.

Of basic importance for the possibility of carrying out measurements of the kind mentioned is, according to the foregoing, the question to what extent there are traffic quantities available for measurements that uniquely determine the intensity distribution $G(x)$, and which conditions the traffic quantities in such a case should satisfy, in order that they can be represented by means of such a function. First it should again be pointed out that it does not seem possible to determine $G(x)$ directly, as one can never establish by measurements the magnitude of the intensity at a given moment. The only possibility for the determination of $G(x)$ is an indirect computation from integrals of type (50), which express some of the quantities available for measurements. Now evidently, as $G(x)$ is a function of a variable x, it is not possible to determine the entire function only from some isolated measurement values, that is, from some isolated integrals of type (50). In order to compute the entire function $G(x)$ one must have access at least to so great a number of measurement values for traffic quantities that they can be interpolated to a continuous curve. One can thus expect that theoretically one needs a series of unlimitedly many integrals of type (50) in order to have $G(x)$ uniquely determined. Examples of series of such integrals have appeared many times in the foregoing. Thus, an integral of the wanted form for each value of the parameter t is obtained from the distribution function of the inter-arrival times in the general form (52). The distribution function (52), thus representing an unlimited number of values of type (50), should be called a *measuring function* and the parameter its *measuring variable*. It might occur for certain conditions that such a measuring function uniquely determines $G(x)$. Now it appears that the question whether this occurs is closely connected with the properties of the *Laplace* integral transform and thereby also with the *Stieltjes* moment problem. By the *Laplace* transform, which has been touched upon several times in the foregoing, a function $f(z)$ is determined by another function $G(x)$

through the relation

$$f(z) = \int_0^\infty e^{-zx}\, dG(x) \qquad (144\text{ a})$$

One considers only such cases when $G(x)$ never decreases. The function $f(z)$ is then evidently completely monotonic. The question is now under which circumstances a function $f(z)$ can be represented in the form (144 a) and to what extent $G(x)$ then is determined by $f(z)$, that is to what extent the relation (144 a) can be inverted. This problem and other related questions have been studied in recent years by several mathematicians as *Bernstein, Hausdorff, Widder* and *Feller*. For the literature the reader is referred to *Doetsch.*[13] A summary of most results that are pertinent in this connection has been given by *Feller.*[5] The main result is *Bernstein-Widder*'s theorem:

The necessary and sufficient condition that a function $f(z)$ can be represented in form (144 a) for $z > 0$, where $G(x)$ never decreases and is limited in every finite interval, is that $f(z)$ is completely monotonic so that for $z > 0$ it has finite derivatives of every order satisfying the condition

$$(-1)^n f^{(n)}(z) \geq 0 \qquad (144\text{ b})$$

It should be remarked that the restriction to the domain $z > 0$ is formal and does not imply any essential restriction of the general validity.

Further it has appeared that if the criterion put forward is valid, then $G(x)$ will be uniquely determined by $f(z)$. Here then the uncertainty implied by the integration constant is disregarded. In the following the constant is always determined so that $G(0) = 0$. *Dubordieu*[2] and *Feller*[5] have set up the following inversion formula

$$G(x) = \lim_{\varrho \to \infty} \sum_{n=0}^{[x\varrho]} \frac{(-\varrho)^n}{n!} f^{(n)}(\varrho) \qquad (144\text{ c})$$

where $[x\varrho]$ designates the greatest integer that is $\leqq x\varrho$. Neither this inversion formula nor the criterion (144 b) have very suitable forms for use in such cases when $f(z)$ is only a numerically determined curve, which of course is always the case when it has been obtained by measurements. It is here difficult to determine numerically the values of the derivatives with any greater accuracy. However, the formulas mentioned could be transformed with unchanged validity domain so that they contain only the values of $f(z)$ in a number of distinct points. These results are mainly due to *Feller* who in the work just quoted has investigated the so-called interpolation problem. The most important results in this connection are the following:

Let z_n, where $n = 0, 1, 2, \ldots$, represent an infinite number of positive numbers which increase towards infinity and for which it is assumed that the series

$$\sum_{n=0}^{\infty} \frac{1}{z_n} \qquad (145\text{ a})$$

is divergent. Then every completely monotonic function $f(z)$ is uniquely determined by the values in the points z_0, z_1, z_2, etc., that is, by $f(z_0)$, $f(z_1)$, etc. Thus, in order to determine a completely monotonic function one need not know its value in every single point but only in an unlimited number of distinct points. The condition that the series (145 a) should be divergent means that the distance between the points may not increase too fast.

The derivatives in the criterion (144 b) are now substituted by the so-called difference quotients for the function values in the points z_0, z_1, etc. These difference quotients are defined as follows:

$$f(z_r, z_{r+1}) = \frac{f(z_{r+1}) - f(z_r)}{z_{r+1} - z_r}$$

$$f(z_r, z_{r+1}, z_{r+2}) = \frac{f(z_{r+1}, z_{r+2}) - f(z_r, z_{r+1})}{z_{r+2} - z_r}$$

etc. so that one generally has the recursion formula

$$f(z_r, z_{r+1}, \ldots z_{r+n}) = \\ = \frac{f(z_{r+1}, \ldots z_{r+n}) - f(z_r, \ldots z_{r+n-1})}{z_{r+n} - z_r} \qquad (145\text{ b})$$

These formulas are well known from the mathematical theories for differences, for instance Reference 19, p. 8. The number series, formed by $f(z_0)$, $f(z_1)$, etc. is now said to be completely monotonic if for all non-negative integer values of r and n

$$(-1)^n f(z_r, z_{r+1}, \ldots z_{r+n}) \geq 0 \qquad (145\text{ c})$$

is valid. It can now be shown that all number series that are formed in the manner mentioned from a completely monotonic function must be completely monotonic. Further, *Feller* has shown that every completely monotonic number series that satisfies the condition that the series (145 a) is divergent, uniquely determines a completely monotonic function $f(z)$, the values of which coincide with number series in the points z_1, z_2, etc. In the work just quoted, *Feller* has also shown how $f(z)$ and its corresponding distribution function $G(x)$ can be determined from the difference quotients (145 b).

The foregoing gives an absolute criterion, convenient in applications, for deciding whether a function can be represented in the form (144 a) where $G(x)$ is a never decreasing limited function. It is sufficient to investigate a single arbitrarily chosen series of numbers provided, however, that the condition (145 a) is satisfied. It is then often advantageous to chose equidistant points so that $z_n = nk$, where k is a constant. As the series

$$1 + \frac{1}{2} + \frac{1}{3} + \frac{1}{4} + \ldots$$

is divergent, the condition (145 a) is satisfied. In this case, the difference quotients (145 b) could simply be expressed as normal differences. These are generally designated

$$\left.\begin{aligned} \underset{k}{\triangle} f(z) &= \frac{f(z+k) - f(z)}{k} \\ \underset{k}{\overset{2}{\triangle}} f(z) &= \underset{k}{\triangle}(\underset{k}{\triangle} f(z)) \\ &\vdots \\ \underset{k}{\overset{n}{\triangle}} f(z) &= \underset{k}{\triangle}\left(\underset{k}{\overset{n-1}{\triangle}} f(z)\right) \end{aligned}\right\} \qquad (146\text{ a})$$

and one easily obtains the formula

$$\underset{k}{\overset{n}{\triangle}} f(z) = k^{-n} \sum_{\sigma=0}^{n} (-1)^{n-\sigma} \binom{n}{\sigma} f(z + \sigma k) \qquad (146\text{ b})$$

The following relation evidently exists for equidistant points between these differences and the difference quotients defined by (145 b)

$$f(z, z+k, \ldots, z+nk) = \frac{1}{n!} \underset{k}{\overset{n}{\triangle}} f(z) \qquad (146\text{ c})$$

The condition for a number series to be completely monotonic (145 c) can in this case also be expressed by

$$(-1)^n \underset{k}{\overset{n}{\triangle}} f(z) \geq 0 \qquad (146\text{ d})$$

Feller has derived the following inversion formula for (144 a)

$$G(x) = \lim_{N \to \infty} \sum_{n=0}^{[N(e^{kx}-1)]} (-1)^n \binom{N+n-1}{n} \cdot \\ \cdot k^{1-n} \underset{k}{\overset{n}{\triangle}} f(Nk) \qquad (147)$$

Hausdorff and *Widder* have derived similar inversion formulas, but which contain other types of differences. It seems, however, as though (147) is the one most suitable for numerical computations. In spite of this it proves to be of little use for computations where one starts from the values of $f(z)$

obtained from measurements. It has appeared in the actual test computations that the values for $f(z)$ must be given with extremely high accuracy in order to have $G(x)$ determined to two or three reliable digits. There do not seem to exist any other methods more suitable for numerical computations of $G(x)$.

If one disregards the difficulty just mentioned of computing $G(x)$, the other above-mentioned theoretical results are of great importance in the actual case. If one has, as a matter of fact, been able by measurements to determine a function $f(z)$, which according to the theory of slow intensity variations should be possible to express in form (144 a), then the arithmetically convenient criterion (146 d) gives good information about the accuracy with which such a description is possible and this without the necessity of determining $G(x)$ explicitly.

It now appears that there exist several measuring functions which could be represented in the form (144 a) and because of that could be investigated by means of the above-mentioned criterion. In many important cases, however, one gets measuring functions of another type which is closely connected to the *Laplace* transform. One has then instead the moments

$$M_n = \int_0^\infty x^n \, dG(x) \qquad (148\text{ a})$$

which except for the signs are equal to the derivatives in the zero point of $f(z)$ according to (144 a). On account of this one can expect that all these moments should uniquely determine $f(z)$ and then also $G(x)$. This is also the case under certain conditions.

The question under which conditions an infinite positive series of numbers M_0, M_1, M_2, etc. uniquely determines a never negative and never decreasing finite function $G(x)$, so that (148 a) is satisfied for every value of n has first been treated by *Stieltjes* and is called *Stieltjes' moment problem.* During later years several mathematicans have concerned themselves with this problem; the results given below are taken from *Bernstein.*[1]

It appears that the necessary and sufficient conditions for the solution of the given problem are closely connected with the properties of the *Hankel* determinants,[17] based upon the number series $\{M_n\}$. Generally the simplified representation

$$(M_r, M_{r+1}, \ldots M_{r+2n}) = \begin{vmatrix} M_r & M_{r+1} \ldots M_{r+n} \\ M_{r+1} & M_{r+2} \ldots M_{r+n+1} \\ M_{r+2} & \cdots\cdots\cdots \\ \vdots & \\ M_{r+n} & \cdots\cdots \; M_{r+2n} \end{vmatrix} \qquad 149\text{ a)}$$

is used. A necessary condition for the existence of a never decreasing finite function $G(x)$ for which (148 a) is satisfied for every non-negative integer value of n, is that the two conditions

$$\left.\begin{array}{l} (M_0, M_1, \ldots M_{2n}) \geq 0 \\ (M_1, M_2, \ldots M_{2n+1}) \geq 0 \end{array}\right\} \qquad (148\text{ b})$$

are satisfied for all n. This condition, however, is not sufficient for the existence of one and only one such function $G(x)$. First the following condition by *Carleman* applies: if there is a function $G(x)$ that satisfies the conditions just set up, then the necessary and sufficient condition that this shall be the only function that satisfies these condition is that the series

$$\sum_{n=0}^{\infty} M_n^{-\frac{1}{n}} \qquad (148\text{ c})$$

is divergent. Further one must distinguish between the cases when the expressions (148 b) are greater than or equal to zero. If the conditions

$$\left.\begin{array}{l} (M_0, M_1, \ldots M_{2n}) > 0 \\ (M_1, M_2, \ldots M_{2n+1}) > 0 \end{array}\right\} \qquad (148\text{ d})$$

are satisfied for all n, then there is always a function $G(x)$ that has the mentioned properties. The other case, when some of the expressions (148 b) are equal to zero, occurs for such functions $G(x)$ that vary only in a finite number of points and where thus the related completely monotonic function $f(z)$ is a sum of a finite number of exponential expressions. It is then necessary and sufficient that both expressions (148 b) are greater than zero up to a certain value of n and equal to zero for all greater values of n. In this case, naturally, the condition (148 c) is satisfied. For zero variation, that is when $G(x)$ varies in only one point, all expressions (148 b) are equal to zero for $n \geqq 1$.

The pertinent conditions (148 b) for the case of moments are somewhat more cumbersome for a numerical test than the corresponding conditions (146 d) for the case when the function $f(z)$ is given. To facilitate the computations some well known properties of the *Hankel* determinants[17] should be mentioned. First it should be remarked that evidently

$$(M_r, kM_{r+1}, \ldots k^{2n} M_{r+2n}) = \\ = k^{n(n+1)} (M_r, M_{r+1}, \ldots M_{r+2n}) \quad (149\text{ b})$$

Further it is clear, that if in the *Hankel* determinants (149 a) we substitute every element M_{r+p} by the difference

$$\overset{p}{\triangle} M_r = \sum_{\sigma=0}^{p} (-1)^{p-\sigma} \binom{p}{\sigma} M_{r+\sigma} \quad (149\text{ c})$$

then the determinants remain unchanged, so that we have

$$(M_r, M_{r+1}, \ldots M_{r+2n}) = (M_r, \triangle M_r, \ldots \overset{2n}{\triangle} M_r) \quad (149\text{ d})$$

Finally it should be remarked that the conditions (148 b) and (148 d) respectively mean as a matter of course that every determinant (149 a) for an arbitrary r must satisfy the same conditions. This follows immediately from the fact that every moment M_{r+n} of the function $G(x)$ also can be considered as the nth moment of a function $x^r G(x)$. When investigating measuring results the inequalities of the formula (60 b) in Chapter 6 can also be of interest. They can be written in the form

$$1 < \left(\frac{M_2}{M_1^2}\right)^{\frac{1}{2}} < \left(\frac{M_3}{M_1^3}\right)^{\frac{1}{3}} < \ldots < \left(\frac{M_n}{M_1^n}\right)^{\frac{1}{n}} \quad (149\text{ e})$$

from which can be seen that the quantities M_n / M_1^n always form a series of increasing numbers greater than one. The inequalities mentioned are necessary but not sufficient conditions for a distribution function $G(x)$ to exist and not to have zero variation.

As in the case considered earlier when $f(z)$ is given in the form (144 a), it seems that, also at given moments according to (148 a), no methods at present exist that can be used with advantage for numerical computations of $G(x)$ from measured values. This is, however, as was already mentioned in the introduction of this chapter, of minor importance for the possibility of using the results of the theories for dimensioning computations in practice, as one is then also for other reasons restricted to presumed forms of $G(x)$, which comprise only few parameters. It is then only necessary to find methods for the computation of these parameters.

From the phenomena of statistical nature that occur in a group of devices handling telephone traffic one can produce a practically infinite number of traffic quantities which could be of importance for the determination of $G(x)$. Some of these quantities could be merged into common types which form measuring functions and from the foregoing it is evident that they are of interest in the first place for carrying out measurements. It now appears to be possible in principle to produce a great number of measuring functions. Of these, however, only those are of greater interest that directly or after transformations can be related to any of the forms (144 a) or (148 a). Although these seem to form the majority of the possible measuring functions, the necessary transformations imply such difficulties concerning the computations that the treatment of the measuring results will be very cumbersome. For many other measuring functions considerable technical measurement difficulties arise. It seems, therefore, that there are only relatively few measuring functions that can with advantage be used for the actual investigations. In the following a short survey is given of the advantages and disadvantages of the different measuring functions.

According to Chapter 5 p. 47 the traffic quantities present in a group of devices can be divided into call mean values and time mean values. These two types of means can both be used to provide measuring functions. In a group there is usually a multitude of call quantities. In the first place, and in this connection the most important, are the incoming calls to the group. By the handling of the traffic within the group the incoming calls are then divided in different ways and each such way represents a call quantity. These quantities may, in contrast to the incoming calls, be named the internal call quantities of the group as they are generated by the internal conditions in the group. Examples of such internal call quantities are the blocked calls from the group and the blocked calls from each separate device or the calls that will occupy the device. One can also define further internal call quantities related to the different traffic states in the group. An infinite number of measuring functions can now be obtained from every call quantitiy. One has first the distribution function of the inter-arrival times and the distribution for the next call. Both quantities are functions of a time parameter t, which is the measuring variable of the measuring function. Further one has the distribution functions $\varphi_n(t)$ introduced in Chapter 6 p. 59 which give the probability for courses of successive calls. These form two-dimensional measuring functions as one can use both t and n as measuring variable. One can also obtain measuring functions if one considers the mean values of the inter-arrival times for series of different call quantities. For instance for an infinitely great group with sequential hunting, the blocked traffic from every simple device forms a measuring function with the order number of the device as measuring variable.

Also for the time mean values there are ample possibilities for forming measuring functions. One has then in the first place the different states of the group, that is, the mean of those times when n and only n of the devices are busy at the same time in the group. Here n is the measuring variable. Further one can by special measuring arrangements divide the devices in the group in different subgroups and look at the states within these subgroups. As a special case one can look at the quantities which indicate the time mean value when n specific devices in the group are busy at the same time and again as a special case of this the busy times of the separate devices which, however, only give a useful measuring function if the hunting is sequential. Finally, one can get additional measuring functions by considering the distribution functions for the durations of the different states. Some of these coincide very closely with the internal call quantities.

Suppose that in the mathematical treatment of the measuring values for a certain measuring function, one starts from the assumption that the intensity distribution $G(x)$ has a given form, so that only a small number of parameters obtained in this form need be computed. It is nevertheless necessary, in order to avoid too complicated a treatment, that the kernel function $F(x)$, which according to expression (50 a) or (50 b) appears in the measuring function, has a relatively simple form as a function of x and also as a function of the measuring variable. This is unfortunately not the case for most of the possible measuring functions. Practically all internal call quantities $F(x)$ seem to have too complicated a form. The kernel function $F(x)$ is obtained, as a matter of fact, from formulas valid for random traffic and these formulas are rather complicated for internal call quantities, which is clear from the results obtained in Chapters 3 and 4. Also the expressions for the call intensities for most internal call quantities seem to be too complicated to produce measuring functions that are easy to treat.

Regarding the measuring functions formed by time mean values, one must, in order to carry out the treatment, refrain from using functions that are obtained from distributions of the duration of times, as to the extent they are known these have turned out to be too complicated. The same thing is true of certain of the state mean values in subgroups (mentioned above), for instance for the busy times of the separate devices.

The demand for easily treated measuring results has thus to a high degree limited the number of suitable measuring functions. Those remaining are of two kinds: functions that are based upon the distribution of the incoming calls and functions that start from the mean values of the different states of the group. Both possibilities will be scrutinized here.

Of the measuring functions which are based upon the distribution of the incoming calls the most important is the inter-arrival time distribution function, which has the form (52) for slow variations. It is thus given in the form of a *Laplace* integral and according to the theorems for these integrals just mentioned becomes $dG(x)$, or in this case rather $x\,dG(x)$, uniquely determined by the measuring function $\varphi(t)$. The investigations in Chapter 9, regarding the special forms the distribution function takes when the traffic is expressed in one of the normal forms, also give indications of several methods for computing the parameters in the normal forms from $\varphi(t)$. The distribution function of the inter-arrival times thus forms a measuring function that has favourable properties regarding the treatment of the measuring material. However, the disadvantage that has already been mentioned earlier is that the difference between the distribution function in question and the purely exponential distribution is usually very small and hence the measurement must be performed with high accuracy. This necessitates among other things a great amount of measurement material. This need not, however, imply any serious difficulties as the inter-arrival times are mostly rather small, and because of that the total measuring time need not be especially long to give sufficient measuring material. A more serious inconvenience is the necessary accuracy for the determination of the measuring variable t. The measuring equipment for this purpose can be rather complicated.

As already shown in Chapter 6, the inconveniences mentioned can be reduced if as measuring function one uses instead the distribution function expressed in (69) for a course of $n+1$ inter-arrival times where $n > 0$. A question of principal importance is then whether the distribution function $G(x)$ is uniquely defined by means of $\varphi_n(t)$ according to (69). This is easy to show. By differentiating (69) one gets

$$\varphi_n'(t) = -\frac{1}{y}\frac{t^n}{n!}\int_0^\infty x^{n+2}e^{-xt}\,dG(x) \quad (150\text{ a})$$

The integral here is a *Laplace* integral. The function

$$-y\frac{n!}{t^n}\varphi_n'(t)$$

then uniquely determines the function $x^{n+2}\,dG(x)$. Further the following relations, easily obtained from (69), should be observed

$$\varphi_n(t) - \varphi_{n-1}(t) = -\frac{t}{n}\varphi'_{n-1}(t) \quad (150\text{ b})$$

and

$$\varphi_n(t) - \varphi_{n-1}(t) = \frac{1}{y}\frac{(-t)^n}{n!}\varphi^{(n)}(t) \quad (150\text{ c})$$

In the latter $\varphi(t)$ is the distribution function according to (52). From the formulas mentioned one can derive several very convenient methods for determining the parameters of the normal forms. From the point of view of treatment it is very advantageous to use distribution functions for courses of successive inter-arrival times as measuring functions. A disadvantage is, however, that the measuring equipment will inevitably be considerably more complicated than that for measuring the distribution function of the inter-arrival times only.

Besides the measuring functions mentioned one more can be considered and that is the distribution function for the next call. Here, however, technical difficulties arise, as the measuring equipment must generate points of time chosen at random. It is thereby difficult to avoid systematic faults of such a magnitude that there is a risk of totally ruining the results.

For all measuring functions which are based on the distribution of the incoming calls, one gets a function $G(x)$ which is solely related to the slow variations of the call intensity and not influenced by a possible variation of the mean occupation time. Further it should be observed that according to the investigations of rapid intensity variations in Chapter 7 the influence of the transition courses is extremely insignificant. The function $G(x)$ determined by measurements on incoming calls can thus be considered as practically independent of the rapidity of the intensity variations.

Another factor of greatest importance for the carrying out of the measurements on incoming calls is the following: One of the basic assumptions for the theories put forward in this work is the notion that the individual durations of the busy times do not show any direct correlation with the inter-arrival times occurring at the same time. If now the calls arrive distributed at random, then the same must be the case for the end points of the busy times and this is independent of the distribution function that applies for the busy times. Now, the theory for the slow variations is based upon the assumption that the calls arrive distributed at random, although the call intensity can experience slow changes. From this follows, that if the theoretical assumptions are satisfied then exactly the same distribution functions should apply to the end points of occupation times as for the incoming calls. The practical importance of these conditions will be shown in Chapter 11. It should be remarked that, in measuring end points of occupation times, one must check that all occupations will continue undisturbed and not be subject to any systematic displacements in relation to the calls, which means that the groups where the measurements are made must be so abundantly dimensioned that blocking is practically out of the question.

Among the time mean values that according to the foregoing discussion could be expected to be suitable for the production of measuring functions, the state mean values of the group should first be noticed. These indicate the mean for the part of the total time when a certain number of devices are busy at the same time. If one has a full availability group that is so abundantly dimensioned that the blocking can be neglected, then these time mean values for slow intensity variations are expressed by (138 a) where n represents the number of devices busy at the same time. If one imagines the group infinitely great, then $G(x)$ is uniquely determined by all P_n-quantities. This is clear from the fact that by means of the relation (139) and the knowledge of the different P_n one can determine all moments M_n. These moments uniquely determine $G(x)$ according to what has been mentioned earlier regarding the inversion of the *Laplace* integral. For the practical computation the greatest interest is connected with the question to what extent one can get a simple determination of the parameters of the normal forms. In this respect P_n-quantities are relatively advantageous. The determination of state mean values by measurements does not seem to offer any great difficulties.

It has been mentioned earlier that by dividing device groups into subgroups and by investigation of the states in these subgroups one might obtain further measuring functions based upon time mean values. In an investigation of these conditions it has been found possible to design a measuring function that for certain conditions is very advantageous to work with. One is here concerned with such groups, where the hunting is random, so that every device has the same probability of receiving an incoming call. $a_{p,n}$ may denote the mean of the part of the total time when in a group of n devices in a certain subgroup of p devices all p devices are busy. Because of the assumed random hunting and from the resulting symmetry it follows that $a_{p,n}$ must have the same value for all subgroups of p devices. If the main group is so abundantly dimensioned that the blocking can be neglected and further the ratio p/n is sufficiently small, then for random traffic the following applies, where A is the incoming traffic intensity

$$a_{p,n} = \frac{A^n}{n(n-1)(n-2)\ldots(n-p+1)} \quad (151\text{ a})$$

For the derivation of this expression see Reference 12. For traffic with slow intensity variations one now according to the general formula (50 a) obtains instead the expression

$$a_{p,n} = \frac{s^n}{n(n-1)(n-2)\ldots(n-p+1)} \int_0^\infty x^n \, dG(x) \quad (151\text{ b})$$

From the quantities $a_{p,n}$ a measuring function is thus obtained that directly makes it possible to determine the moments of the intensity distribution. For computing the parameters of the normal forms this is of great importance since as has been shown in Chapter 9 the expressions of the moments as functions of these parameters have a relatively simple form. Measurements of such so called *combinations of occupations of devices* $a_{p,n}$ can usually be carried out rather simply. Sometimes considerable difficulties can arise because the assumed random hunting is not realized with sufficient accuracy in practice and must therefore be improved by special arrangements.

All the quantities have now been run through which seem to offer the best possibilities for measurements to investigate the applicability in practice of the theories put forward in this work. It should be remembered, however, that for measurements which are based upon time mean values of the above treated types, one gets an intensity distribution $G(x)$ that for two reasons may differ from the intensity distribution which applies for incoming calls. Firstly, the time mean values are dependent on the variations of the call intensity as well as the variations of the mean occupation time. The latter, which probably are rather insignificant, do not influence the distribution of the incoming calls. Secondly, as has been shown in Chapter 7, the speed of the intensity variations influences the call distributions far less than the time mean values. From this it may be expected that a clear influence appears regarding measured intensity distributions. It should be observed, however, that according to the derivations in Reference 12 the intensity distribution that applies for the state mean values must be identically the same as that for the combinations of occupations of devices.

REPORT ON MEASUREMENTS CARRIED OUT

Chapter 11

Measurement methods and equipment

In connection with the development of the theories presented in the foregoing account, a number of measurements were carried out during the years 1937−1942 at one of the large telephone exchanges in Stockholm. These measurements, which have been performed under the auspices of the Royal Swedish Telegraph Board, have been relatively comprehensive. For instance, the number of measured inter-arrival times amounted to more than 400,000. The measuring methods and the equipment used have undergone several modifications, since the experience gained in the course of the measurements has, of course, provided indications of possibilities for considerable improvement. There are no reports in the literature of measurements of this kind and degree of accuracy. It is therefore natural that the methods used at the outset could not meet all the requirements. Therefore only some of the measurements made will be reported, viz. a series which was carried out in 1941−1942 for a final check of the applicability of the results derived form the theory of slow intensity variations. These measurements also seem to meet high demands as regards accuracy.

For measurements of telephone traffic and its treatment in groups of devices, it is possible to use two methods, in principle different. By one of these methods the traffic process is determined and recorded in such detail by the measuring equipment that the traffic process in question can be completely reconstructed afterwards. The record can be made, for instance, by writers having a pen for each device in the group. The pen records the occupations of the device. The advantage of this method is that it is possible to study the traffic process in every respect. On the other hand, analysis of measurements involving determination of the traffic process would create considerable work. This would in practice prevent application of the method on a large scale. Furthermore, for instance for determining inter-arrival times, it is difficult to obtain sufficiently accurate values by this method. Finally it has become apparent that the unavoidable manual analysis of the records sometimes causes systematic errors which might jeopardize the results. In the first of the mentioned measurements for investigation of the slow intensity variations, this method of determining the traffic process was applied. The method was abandoned, however, on account of the disadvantages mentioned. That, of course, does not prevent the method from being useful in other kinds of measurements.

The method that has turned out to be more suitable than the detailed determination of the traffic process may be named *the direct summation method.* Here the measuring equipment immediately determines the desired quantities, but generally in such a way that no subsequent reconstruction of the actual traffic process is possible. It is therefore necessary to determine in advance which quantities it is desirable to know. It is not possible to supplement the result afterwards in respect of any other quantities. As a rule, however, this would probably involve only a small inconvenience and is counterbalanced by the considerable advantages of the direct summation method. As a rule one can obtain considerably more accurate measurements by this method than by the method of detailed determination of the traffic process. It also requires considerably less analysis work as this is to a great extent performed by the measuring equipment itself. On the other hand the method of direct summation as a rule requires rather extensive and complicated equipment.

The measurements reported below – carried out by the direct summation method – are intended to give an idea of the extent to which the basic prerequisites for the slow intensity variation theory are satisfied in real telephone traffic. Unfortunately it has not been possible to give the measurements such a scale and form that the applicability of the theories established previously, regarding superposing of traffic flows, can be checked. It has become evident, however, that the normal forms established in connection with these theories offer a fair chance of describing the properties of the investigated traffic.

In accordance with the results of the investigations carried out in Chapter 10, the measurements made have comprised determination of distribution functions for inter-arrival times, traffic states and certain combinations of device occupations. The equipment designed for these measurements has throughout been made up of components common in telephony such as telephone relays, switches, and subscriber meters.

The problems arising when designing the recording equipment concern mainly two questions; (1) the determination of the length of various time intervals with sufficient accuracy, and (2) the most suitable circuits for obtaining the desired summation with the simplest possible equipment. Also as regards the time recording, the use of common telephony components has turned out to be preferable. The most difficult phase in the measurements arose when connecting the recording equipment to the telephone exchange. The functioning of the automatic telephone exchanges is rather simple in principle, but on a closer study often exhibits extremely complicated processes. For measurements of the desired accuracy this may cause serious complications owing to the difficulties of defining the traffic elements, i.e. the calls and the holding times, sufficiently precisely. Further, many phenomena arise which are of secondary significance for the traffic conditions in general but could appreciably influence the quantities to be measured. Through the influence of the operating times of the switches, the points of time that represent the arrivals in the measurement might be exposed to mutual systematic displacements. This deformation has no noticeable influence on the traffic capacity of the devices but can cause considerable deformation of the distribution function of the inter-arrival times obtained in the measurement. In order to avoid the influence of such secondary phenomena it must be ensured that suitable circuits are used as far as possible. This can, however, often bring about considerable complications in the measuring equipment.

In the following, a description is given of the principles for the design and functioning of the measuring equipment.

The telephone exchange, Östermalm in Stockholm, where the measurements were performed, is automatic and equipped with 500-point switches of Ericsson design. In

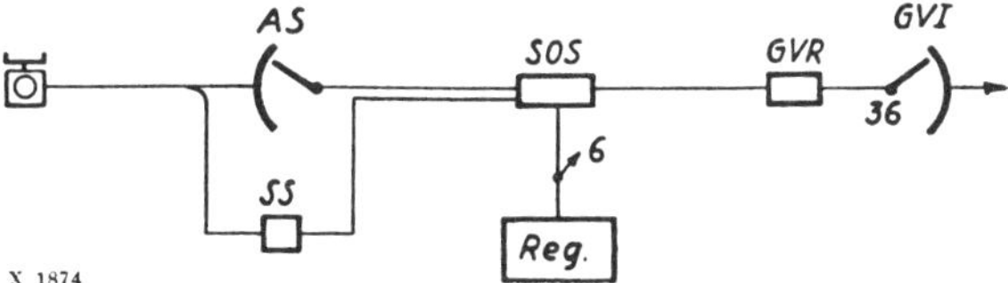

Fig. 13. Excerpt of the traffic routing plan of the Östermalm exchange

Fig. 13 an excerpt of the traffic routing plan of the exchange is shown. The exchange is equipped with line finders AS, allotter SS and sequence switches SOS (which also include control devices for each line finder) first group selectors GV1 and their control relays GVR. Further the exchange is provided with registers REG, which can be connected to the cord circuits constituted by the line finders, first group selectors and their control devices in a so called common group connection. During the measurements this grouping was designed in such a way that each register served six cord circuits, one in each of six 500-groups of subscriber circuits. The incoming subscriber lines are combined in groups of 500 each. The traffic from each such group is served by a number of line finders which are part of the cord circuits of the said kind. The line finders in a 500-group then form a group of devices that under certain conditions – as will be discussed below – can be considered to work under full availability. As it was deemed desirable, considering the scope of the measurements, that the investigated traffic should not be exposed to deformation by possible blocking in previous device groups, it was considered appropriate to carry out the measurements on the traffic in a line finder group. As a matter of fact, this traffic can always be considered "fresh".

The line finder group for the investigated traffic consisted of 36 line finders, which turned out to be quite sufficient to avoid blocking during the measurement period. As has been repeatedly mentioned before, it is necessary in the carrying out of these measurements to avoid blocking as it might deform the properties of the original traffic. Owing to the abundant dimensioning, the line finder group considered could be regarded as fully available. This is not generally the case with common group connection, as described above. Since the registers serving the line finders in a group are common to several groups of line finders, blocking in a group might occur although all line finders are not engaged on occupations caused within their own 500-group. The remaining free line finders might be incapable of receiving calls as their registers are engaged on connecting calls in other groups. If all line finder groups served by the same register group are so abundantly dimensioned that no blocking occurs, the common group connection does not cause any disadvantage in this respect.

The measurements reported here were carried out in the 56th 500-group of the exchange. The registers concerned were also connected to the 500-groups 53, 54, 55, 57, and 58. In each of the six 500-groups, 36 cord circuits with line finders were in service during the measurement period. The number of connected registers was thus also 36. The composition of the subscriber population of the investigated 500-group is reported in Chapter 14.

From various contact points in the cord circuits, shown in principle in Fig. 13, electrical potentials can be obtained which mark the states of the line finders as busy or free and, without any effect on the functioning of the exchange, can be used directly to operate the relays in the measuring equipment. Such contact points can advantageously be obtained, for instance, in the sequence switches SOS, which are provided with a cylindrical contact field with 12 contact rows, each having 12 contact points, i.e. totally 144 possible contact points. Most of them could be used for connection to the measuring equipment. It might be mentioned in this context that positions 1 and 2 are home positions (when the line finder is free), positions $3-5$ are passing positions during the hunting for the calling subscriber line by the line finder, and positions $11-12$ are passing positions during the release of the connections. It is possible by means of contacts in the sequence switches to obtain potentials during the busy time of the line finders and also short impulses marking the points of time for the start and termination of occupations. These impulses have a length corresponding to the passing of a contact point, which in this case is about 80 ms.

In spite of the many alternatives that the contact field of the sequence switches present, it is rather difficult to obtain a satisfactory connection to the measuring equipment. Among the reasons for this is the difficulty of accurately defining the points of time when the devices of the group become busy or free and thus unavailable or available for new calls. Further, anomalous switching processes which are rather infrequent could appear, for instance simultaneous tests, the correct recording of which causes certain inconveniences.

Here should only be mentioned a phenomenon of more common interest which has an important effect on the measuring process. The distribution function of the inter-arrival times in the earlier measurements had a rather unexpected form for small t-values. As a result of an investigation carried out by *T. Frisk* M.Sc. (as diploma paper, Royal Institute of Technology, Stockholm 1941), made on account of the observation mentioned, the background to this phenomenon turned out to be a certain correlation between the operating times of the line finders. The hunting for a calling subscriber is normally done as follows: When the operation of a line relay of a subscriber has indicated that the subscriber seeks a connection, a number of free line finders, usually around 5, are started. The line finder that during its rotary movement first reaches the frame in which the calling subscriber line is situated moves into the frame and hunts for the line in question whereby the line finder becomes occupied. Thus there is as a rule a time interval – which might vary within rather wide limits – between the point of time for a call and the point of time when a line finder becomes busied by the call. If the calls are now distributed at random and the lengths of the time intervals in question are not dependent upon the positions of other calls, then also the points of time when the line finders become engaged are distributed at random. These points of time can be used as basic material for the determination of the distribution function of the inter-arrival times. However, it appears that in instances when several calls arrive close to one another, an increase of the length of the time intervals can occur up to the points of time when the calls seize line finders. The result is that the frequency of the short intervals between the times of seizure of line finders is less than the frequency of the same intervals between call arrivals. This has no noticeable influence on the mean values of the states in the group, but it causes considerable deformation of the distribution function of the inter-arrival times. In order to avoid this complication it might be possible to carry out the measurement directly on the call arrival times marked by the operations of the line relays instead of using the starting times of the occupations of the line finders. On account of the great number of subscribers in relation to the number of line finders this would, however, result in a considerable complication. For this reason the most convenient method seems to be to base the measurement on the termination points of the occupations. In this case there are no similar, mutually dependent time displacements as in the case of the starting times of the occupations. Further, as has already been pointed out, under the assumptions valid for random traffic, modified by

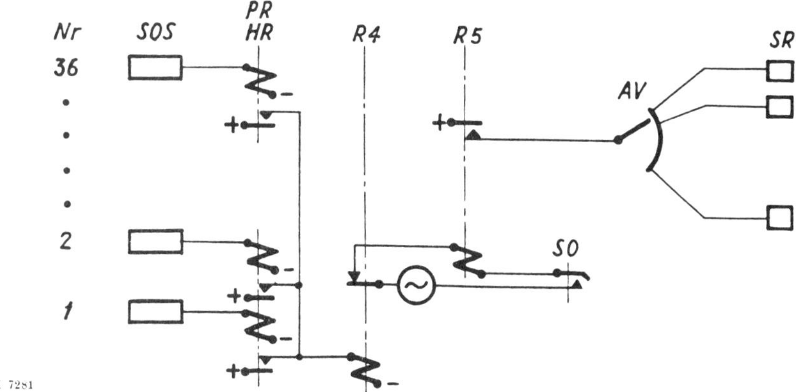

Fig. 14. Schematic diagram of the interval meter

intensity variations in non-blocking groups, the distribution function of the intervals between the times of terminations will be the same as that of the inter-arrival times. For investigating the validity of the assumptions in question, either of the two methods can be chosen. The measurements reported here have been based on the times of the terminations.

The measuring equipment contains four meters, i.e. one interval meter, one state meter, one occupation meter and one traffic volume meter. Below is given a description in principle of the functions of these meters.

Time interval meter

The time interval meter records the time intervals between the times of terminations of occupation in the investigated group. These are obtained as impulses at the homing of the sequence switches. The impulses have a length of around 80 ms. The interval meter should deliver material for determination of the distribution function of the intervals between these impulses. They should thus render it possible to determine a sufficient number of points on this curve. Hence, the interval meter does not need to measure the real interval between all impulses. It is sufficient that it records the number of all intervals between impulses that are longer than certain predetermined times indicating the desired measuring points on the distribution curve. The arrangement in principle of an interval meter, functioning as indicated, is shown schematically in Fig. 14. In this meter 36 relay sets, HR, PR, are adapted to deliver – in cooperation with contacts in the associated sequence switch SOS – a current impulse of about 80 ms. The impulse is generated at the release of each connection established through the line finder and cord circuit associated with the sequence switch. These impulses are fed into a relay R4, common to the entire 500-group, which in this way is made to perform a short armature operation at every termination of an occupation. After the operation of a starting switch SO the relay R4 operates an AC driven impulse relay R5, which is the driving means for a switch AV to the various positions of which a total of 45 counters SR are connected.

The switch AV is brought back to its home position by the short operation of R4, which marks the starting time of each interval. When this operation is finished, switch AV is started by the AC-driven impulse relay R5. This relay steps the switch forward from position to position. In each of these positions the connected counter SR is moved one step forward. The end of the interval is marked by a new short operation of R4, which starts the recording of the next interval in the same way. Through this mode of operation every single counter SR will indicate the number of intervals longer than a certain time, equal to the time of setting switch AV to the position where the counter is connected, plus the operating time of relay R4. The ratio between the recording on the counters and the total number of intervals directly indicates corresponding points of the distribution function of the interval lengths.

The technical design of a meter working on this principle is obviously highly influenced by the accuracy required in the times involved. This is determined by the mean value of the time intervals to be recorded, and, therefore, by the mean calling rate of the traffic investigated. In the present measurements the mean lengths of the intervals are 6 – 7 s. Consequently for the determination of the times it is desirable that the inaccuracy should not exceed 1 – 2 ms. Fortunately, as will be seen from Chapter 13, the occurrence of considerably greater occasional errors does not necessarily deform the distribution curve in a detrimental way, provided that the mean of the recorded times can be determined with the aforementioned accuracy. As a result it has been possible to design the interval meter with standard telephony components such as relays, mechanical switches, and subscriber meters, although in this case the requirements of correct functioning will be higher than normal. Therefore, without altering the principles used for the technical design, it might be difficult to reach a considerably higher accuracy. If the task is to measure traffic at considerably greater volumes, and consequently with increased requirements of accuracy in the time determinations, it would be necessary to introduce other components, for instance relays and switches based on electron tubes. This would cause unavoidable complication and higher costs of the measuring equipment.

For an understanding of the factors underlying the design of the interval meter, reference is made to Fig. 15, showing a time diagram of the functioning of the equipment during a time interval. The time span 1 – 3 represents the operation of relay R4 caused by an impulse generated by the sequence switch SOS at the end of an occupation and repeated by relay sets PR, HR. Switch AV starts to home at time 1, after the release of relay R4, and at time 2 the switch has reached the home position. At time 4 switch AV starts again causing the first counter to move one step at time 5. At time 1′ relay R4 is operated again and the above described sequence is repeated.

For the correct functioning of the interval meter as described, it is necessary that switch AV has sufficient time to home during the time that relay R4 is operated. As this time, due to the desired accuracy of measurement, cannot be considerably longer than the length of the impulse from the sequence switch, i.e. 80 ms, it has been necessary to design switch AV with relays, thereby allowing for the short homing time necessary. It should be remarked that all relays in the measuring equipment are provided with twin contacts in order to avoid contact failures.

With the release of relay R4 at time 3, switch AV should start. It has turned out, however, that it is not possible to establish exact synchronization without too many complications, the reason being as follows: In order to bring about the necessary accurate stepping speed of switch AV, the driving relay R5 has to be governed by a sufficiently exact frequency source. For this purpose an alternating current controlled by an accurate frequency standard is used. Thus the switch can be started only at points of time corresponding to a certain phase of the AC mentioned. As the frequency of this AC cannot be too high owing to the greatest possible stepping speed of the switch, a noticeable time difference may arise between the release time 3 of relay R4 and the starting time 4 of switch AV. This time difference may vary by

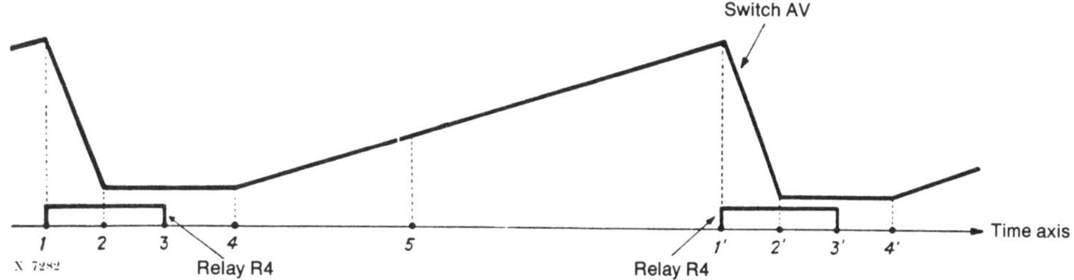

Fig. 15. Time diagram for the function of the interval meter

Point of time 1: Relay R4 operates and starts the homing of AV
Point of time 2: Switch AV arrives at its home position
Point of time 3: Relay R4 releases
Point of time 4: Switch AV moves on
Point of time 5: The first counter SR records
Point of time 1': Relay R4 operates again and the switch AV starts its homing
Point of time 2': Switch AV arrives at its home position
Point of time 3': Relay R4 releases
Point of time 4': Switch AV moves on

maximum 40 ms, depending on when time 3 occurs in relation to the phase of the AC. A more detailed account of the process is presented below in connection with the description of the time source.

The interval between the operating time 1 of relay R4 and time 5 for the recording by the first counter has been found by measurement and computation to be on average 182 ms. This time includes also the working time of the counter, i.e. the impulse time required for the counter to record. As mentioned, the time between 1 and 5 may vary due to variable distance between 3 and 4. It also varies somewhat since the impulse times of the sequence switches are variable. Likewise the times between 1 and 3 are variable. This variation is at most 15 ms. Therefore there is a possible maximum variation of 55 ms in the entire time span between 1 and 5. In Chapter 13 it will be shown that such a variation does not have any detrimental effect on the form of the distribution function as produced by the measurements. The same variation in time as between times 1 and 5 is present also between time 1 and the recording times of all other counters, since, when switch AV has started, it works completely synchronously with the controlling AC. Hence no variation in time can appear between the recording time 5 and the recording times of the other counters. The first five counters represent the following mean interval lengths:

0.182 s
0.482 s
0.986 s
1.481 s
2.452 s

Furthermore, there are counters for each additional complete second, i.e. 3.452 s, etc. up to 21.452 s. From there on there are counters with differences of 2 s up to 41.452 s. Finally, there are counters with differences of 5 s up to 96.452 s. Longer intervals appear very rarely in the measurement in question. Altogether 45 counters are connected to switch AV. Their placing is chosen with regard to the volume of the investigated traffic and the general form of the expected distribution curve. This placing has been found to be suitable.

If the interval between time 3 for the release of relay R4 and time 1′ for its next operation is not sufficiently long, no recording will occur on the first counter of switch AV. In order to determine the distribution curve of the intervals, however, it is necessary to know the total number of seizure terminations. To determine this, a counter could be introduced to record the number of operations of relay R4. It is obvious, however, that this is not sufficient, as impulses might appear from the sequence switches which are simultaneous or so close to one another that relay R4 will not have time to release between them. In this case the termination impulses are said to overlap one another. For the determination of the total number of occupations there is therefore special equipment belonging to the interval meter. This is connected to relay set PR, HR (Fig. 14) which includes a relay for each sequence switch. The relay is held in the operated position at the start of each occupation. These relays are scanned by a special relay switch at fixed time intervals. In this scanning the number of operated relays is recorded on counters and at the same time the relay is released. In this way a reliable count is obtained of the total number of occupations during the measurement period. For circuitry reasons these are recorded on four counters, whose aggregate recordings give the total number of occupations. It became necessary to have the aforementioned relays continuously connected to enable the measuring equipment to register also those occupations that are in force at the time when the measurement is started.

It is clear that overlaps will not influence the measurement of intervals between impulses positioned at such a distance that the counters of switch AV will have time to function. In order to have an idea of the number of overlaps, a counter for the number of operations of relay R4 has been connected in the measurement. There is also a special device whereby the number of intervals between times 3 and 1′ (see Fig. 15) not exceeding 200 ms will be recorded by means of a slow releasing relay not connected to switch AV.

As regards the technical details of the interval meter it should finally be mentioned that the starting and stopping devices should be of a special design. In order to record only the real interval lengths during a continuous measurement period, the start device has been designed in such a way that, after the start key has been thrown, the interval meter does not start until the first following termination of an ongoing occupation. For the same reason the interval meter is constructed in such a way that, by this arrangement, the recording of the total number of occupations will show one unit too many. A correction of the result must therefore be made.

The interval meter designed for the measurements contains a total of 115 relays, 51 subscriber meters and a 25-point rotary step switch.

State meter

The task of the state meter is to record the total length of the parts of the measurement period when any n of the 36 devices in the investigated group are busy at the same time. This should be done for thirtyseven possible values of n, i.e. 0, 1, 2, ... 36. As this is a matter of pure time measurement, a fairly obvious solution might be to consider the use of electrically controlled clocks, one for each state, which run only when the corresponding state prevails. The totalling of the different time intervals would then be done directly by the clocks. The application of such a measuring principle, however, implies great circuitry difficulties. Besides, it is also difficult to provide sufficiently accurate and suitable clocks at a reasonable cost. For control of the clocks the meter must be equipped with a circuit for each state, the electrical condition of which indicates whether the traffic state in question prevails or not. Now the n occupations which form the state n can exist in any of the 36 devices in the group. They can thus form

$$\binom{36}{n}$$

combinations of occupation, each of which will cause the same effect on the circuit of the state clock. This, of course is not impossible to bring about but necessitates an exceedingly

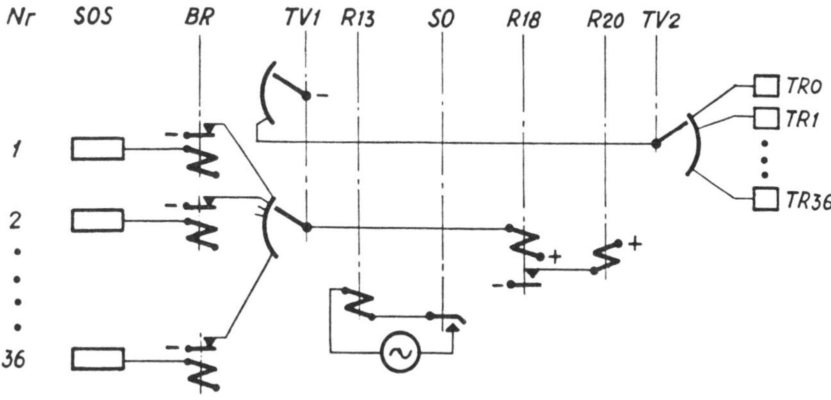

Fig. 16. Schematic diagram of the state meter

large contact field. To avoid this, it is possible to equip every device with a resistor. The resistors of the busy devices are connected in parallel to a circuit, the total (parallel) resistance of which will be a unique function of the prevailing state. (The individual resistors are all alike. The design is not uncommon in traffic volume meters.) By means of an automatic device, the clock of the state corresponding to the momentary resistance of the circuit is connected. It is difficult, however, to make such a device sufficiently accurate and fast. A state meter based on this principle cannot therefore be recommended for measurements of the type in question.

The state meter which has been designed for the present measurements is based on a totally different mode of operation which is schematically shown in Fig. 16. For each cord circuit, i.e. device in the group (in the figure represented by the sequence switch SOS), there is a relay BR, which indicates the state of the cord circuit by being non-operated when the cord circuit is busy, and vice versa. A switch TV1 is stepped during the measurements at an exactly set speed, controlled by the AC-driven impulse relay R13. In each of its first 36 positions switch TV1 connects relay R18 to a state-indicating contact of the BR relay of a cord circuit. The relay R18 is therefore briefly operated every time a busy cord circuit is passed. Another switch TV2 is driven by a relay R20 one step every time relay R18 is operated. When switch TV1 has passed its first 36 positions, switch TV2 has thereby been moved as many steps as the number of busy cord circuits that switch TV1 has come across during its search. Switch TV2 has 37 positions including the home position. A counter TR is connected in each position of switch TV2. When switch TV1 has moved past its 36th position, it closes a brief operating circuit through switch TV2 for the counter TR that is connected to the position assumed by switch TV2. Thereafter switches TV1 and TV2 move instantaneously to their home positions and a new work cycle can start. By means of the described apparatus a recording is obtained on one of the counters for every complete search of the 36 cord circuits in the group. This counter indicates by its position the prevailing state in the group during the search. An additional counter gives the total number of complete searches during the measurement. The time of each complete work cycle is exactly 2 s, providing 1800 searches per hour.

The switches TV1 and TV2 are in actual fact made up of relays only. In this way the greatest possible operating reliability is obtained. The components of the state meter are 36 BR relays, 29 other relays and 38 counters.

One may ask to what extent the recordings on the counters in the state meter make it possible to compute the total durations of the states appearing during the measurement. As a matter of fact there is no exact time measurement but the function of the state meter can be characterized as a sampling method. It is evident, however, that the time of each search cycle – 2 s – multiplied by the number of recordings on a counter, should give at least a rough measure of the total time of the corresponding state during the measurement period. Actually it can be proved that the time obtained with the hunting speed used gives a sufficiently accurate value. For a closer investigation, reference is made to the treatment of the measurement results in Chapter 13. There also the question is touched upon whether the fact that the different devices of the group are not observed at exactly the same time has an influence on the result, for which reason the state meter in reality gives no recording of occupations occuring at exactly the same point of time. In an earlier work by the author, reported in some detail in Chapter 13, it is proved that this condition has no influence on the accuracy of the results.

The occupation meter

The task of the occupation meter is to record the total duration during the measurement period of different combinations of device occupations (subgroup occupancy), i.e. the periods when certain selected devices are occupied at the same time. For this measurement it seems to be suitable to use the measurement method of summation clocks, discussed in the first part of the description of the state meter. In the occupation measurement it is not particularly difficult to realize the necessary circuits for the control of the clocks recording the periods for the subgroup occupancy. The clocks have been made very simple as impulse-driven counters and provide fully adequate accuracy.

The principal functioning of the occupation meter is shown schematically in Fig. 17 by an example of the measuring circuit for the determination of subgroup occupancy. The subgroup consists of three devices, i.e. the line-finders 1, 2 and 3. In this measuring circuit the following components are included: contacts of the BR relays (used also for the state meter), a counter UR allotted to the group to be measured and an AC-driven impulse relay IR, common to several groups.

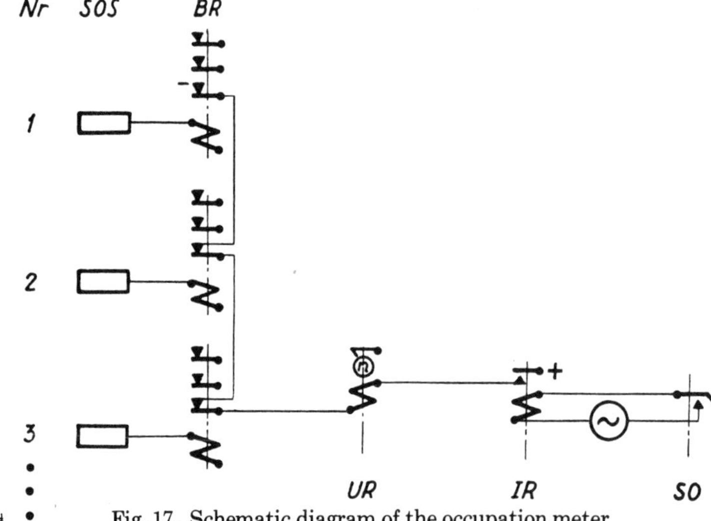

Fig. 17. Schematic diagram of the occupation meter

The three contacts of relay BR, closed when the corresponding line-finders are busy, are connected in series in a circuit comprising also the counter UR and a contact of the impulse relay IR. As long as the three line-finders are busy, the counter UR steps with a speed determined by the impulse frequency. The recording by the counter will be a direct, although somewhat approximate, measure of the total duration of the state when all three devices in the group have been busy.

To obtain the greatest possible material, the occupancy of the largest feasible number of subgroups has been recorded. However, it has not been deemed suitable to have any particular line-finders in more than one subgroup of each kind. It has been considered sufficient to investigate groups of five sizes, i.e. groups of 3, 4, 6, 9 and 12 line-finders each. As the total number of line-finders in the group was 36, the following number of mutually independent subgroups could be formed: 12 groups of 3 line-finders, 9 goups of 4 each, 6 of 6 each, 4 of 9 each and finally 3 groups of 12 line-finders each. One of the possible groups of 9 cord circuits could not be realized for circuitry reasons; the unavoidable complication could hardly be compensated by the gain of having this group included in the measurements. The occupation measurement has thus comprised 33 groups. Besides the contacts of the relays BR, the occupation meter comprises 33 subscriber meters, 8 relays, and 3 rotary step switches with 12 positions.

As with the state meter, the time recording of the occupation meter can be considered a sampling method giving a statistical measure of the time to be determined. The accuracy of such a method is studied in detail in Chapter 13. It turns out that the accuracy is dependent upon the ratio between the impulse frequency for the counters and the mean duration of the occupation states. Thus the ratio must be as small as possible. For the three groups of 12 line-finders each, which obviously have the shortest occupancy, a frequency of 10 impulses/s is used. This should give adequate accuracy. In order to obtain sufficiently reliable recording at this high frequency, 12-step switches are introduced as auxiliary equipment to the counters, giving a recording only for each complete revolution of the switches. For other subgroups it was considered sufficient to have a frequency of 5 impulses/s. At this speed the counters follow the impulses directly with adequate reliability.

The two impulse frequencies have been generated by a frequency-dividing relay which is driven by 50 Hz AC with accurate constancy of the frequency.

In order to obtain results from measurements of combinations of devices, which are of value for the study of the moments of the intensity distribution, it is essential, as has been repeatedly mentioned previously, that the hunting within the group is in principle at random. That means that the probability of occupation of a device by an incoming call has to be equal, or at least very nearly equal, for all free devices. To fulfil this requirement it was necessary to make a slight change of the line-finders' method of hunting for calling subscribers. Because of the common group arrangement, similar changes had to be made in all six 500-groups served by the same register group. These changes mainly concern the connection and function of the allotter. The allotter for a 500-group comprises, in addition to control relays, two step switches working as a unit, and causing an incoming call to start at the most five free line finders. Such changes were now made that the step switches of all allotters were continuously stepped during the measurements. The two step switches of each 500-group were stepped synchronously and maintained in a predetermined relative position. This results in a relation between the hunting of the different line-finders so that the equivalent of random

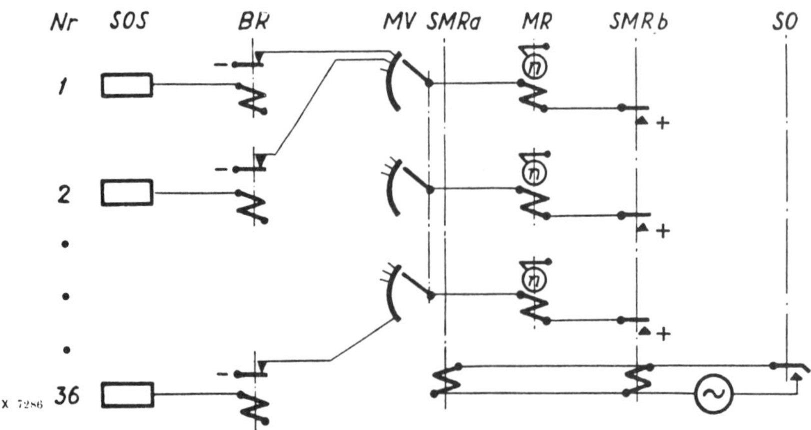

Fig. 18. Schematic diagram of the traffic volume meter

hunting could be considered fulfilled, or nearly fulfilled.

Traffic volume meter

A measurement, as far as possible correct, of the traffic volume of the 500-group under study was also necessary for the purpose of the measurements. A special meter was designed for the traffic volume. Its principle is shown in Fig. 18.

The traffic volume meter contains a switch MW with three wipers moving in parallel, each with 12 positions. The field contacts of the switch are connected to a contact on each of the 36 relays BR, mentioned in connection with the state meter. The contacts on relays BR indicate the free or busy state of the device in question. The three wipers are each connected to a counter MR. During the measurement the switch is stepped at an accurately defined speed by an AC-controlled relay SMRa. Each time the switch AV reaches a new position, another AC-driven relay SMRb closes the circuit for the three counters MR.

A recording is then obtained each time a busy line-finder is passed. The traffic volume meter is arranged to complete a work cycle within 1.6 s and the 36 line-finders are thus searched 3600/1.6 = 2250 times/h. By dividing the switch into three parts, each with its counter, a satisfactory reliability is achieved also with this high search speed. For the same reason the switch AV is made up of relays. The traffic volume meter comprises 12 relays in addition to the three counters and the contacts of the BR relays.

The functioning of the traffic volume meter can also be considered as a sampling procedure. The accuracy of measurement, which is studied in Chapter 13, seems to be very good and certainly sufficient for the purpose. It should be pointed out, however, that the principle of traffic volume meters working according to the scanning method is already well known. Traffic volume measurements can be carried out in several other ways, but none of the methods at present in use seems to offer such good possibilities for reliable and accurate measurement as those offered by the present scanning method.

It should be observed that the traffic volume meter used and the state meter described above show considerable analogies in their functioning. Actually the state meter also measures traffic volumes. Clearly, if the recording of each of the state meter counters TR is multiplied by the corresponding state number, and the results of all counters are totalled, a measure of the traffic volume of the group is obtained. The measurement has thus in principle been performed by a scanning method.

It seems superfluous under these circumstances to use a special meter for the traffic volume. However, computation of the traffic volume from the readings of the state meter turns out to be rather complicated. Actually the computations are facilitated by the introduction of a special traffic volume meter. Besides, the latter is designed in such a way as to give somewhat more accurate results than the state meter. By comparison of results of the two meters, a valuable check can be obtained, which also permits some verification of formulas applied in Chapter 13 for estimating the accuracy of measurement.

Time source

All the measuring apparatus described above comprises switches whose stepping must be accurately time-controlled for correct functioning. The interval meter is particularly critical in this respect. The simplest technique for time control of switches of the kind in question is to use relays which can operate and release following the oscillations of an applied AC. The measurement programme was therefore based on the use of the frequency of an alternating current with highly constant frequency as source for the time indications. It turned out, however, to be impossible to use AC from the mains, as the frequency is continually exposed to irregular variations, amounting to a few percent of the mean value (50 Hz). Instead an AC was used from a frequency standard belonging to the Radio Section of the Royal Board of Telecommunications. This frequency standard, which is crystal-controlled, generates an AC with a frequency of 50 kHz and is kept at this value with an accuracy of 2×10^{-7}. The Radio Section has designed frequency-dividing equipment for the traffic measurements with an output of 50 Hz. This AC has the same high degree of accuracy as the frequency standard. The AC is transmitted to the measuring site on an ordinary telephone line and amplified by an amplifier belonging to the measuring equipment, with an output of about 20 W, which is sufficient to drive all time controlled relays in the measuring apparatus.

In order to obtain time control as reliably as possible by means of the generated AC, a special circuit, containing a polarized relay, directly fed from the 50 Hz, has been introduced in the equipment. Its armature moves backwards and forwards with a frequency of half of the feeding frequency, i.e. 25 Hz. The 50 Hz sinousoidual AC is transformed into 25-cycle DC impulses which can be used to operate ordinary telephone relays. These relays work quite satisfactorily at this speed. A schematic drawing of the polarized relay in its circuit is shown to the left in Fig. 19. A diagram of the armature movement as a function of the applied AC is shown to the right.

The polarized relay has an armature 1, which pivots on a knife-edge bearing on a permanent magnet 2. The armature has a contact spring 3 which can be brought into alternate contact with the two fixed contacts 4 and 5. The armature 1 is situated between pole pieces 6 and 7 of a soft iron yoke. One of the poles of the permanent magnet rests on the middle part of the yoke. The two legs of the yoke have their respective windings 7 and 8. The relay built up in this way is controlled in the circuit shown by the 50 Hz source *e* in series with a resistor *m*.

The function of the relay will be clear from the right-hand side of Fig. 19, in the upper part of which the AC curve is shown. The movement of the armature spring between the fixed contacts 4 and 5 is shown in the lower part of the figure. The functioning can be described as follows.

The relay armature is assumed to be positioned as shown with contact between

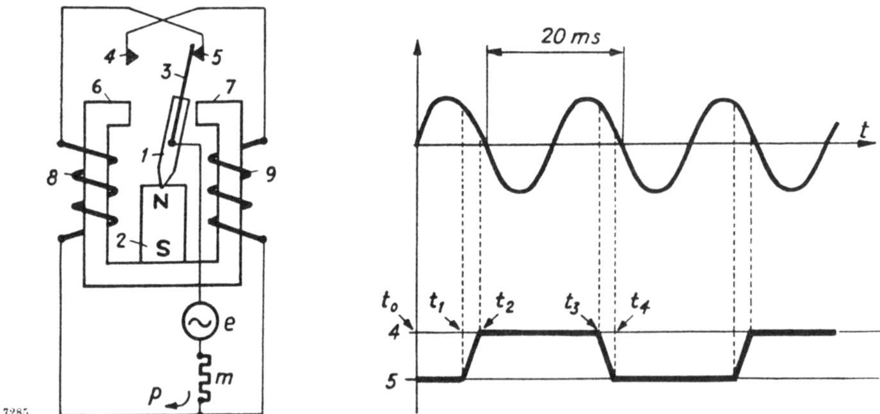

Fig. 19. Frequency division (by two) with a polarized relay

spring 3 and contact 5 at a point of time t_0, when a first half period of the AC starts in the direction indicated by the arrow *p*. The winding 8 is now passed by a current the field of which causes attraction between armature 1 and pole piece 6. At a point of time t_1, after the major part of this half period, spring 3 leaves contact 5. At a somewhat later point of time t_2 spring 3 comes into contact with contact 4, thus closing the circuit through winding 9. However, t_2 is so close to the termination of the first half period that the current of the first period through winding 9 will not have time to cause any armature attraction.

During the second half period, current is fed through winding 9 in such a direction that the field of this winding supports the field of the permanent magnet in its effort to keep armature 1 against pole piece 6. The armature therefore remains still during this half period.

During the third half period the current direction in winding 9 is reversed and its field therefore causes attraction between armature 1 and pole piece 7. As a result spring 3 at a point of time t_3, close to the termination of the third period, will leave contact 4. At a somewhat later point of time t_4 spring 4 comes into contact with contact 5, causing winding 8 to be connected. Winding 8 will not cause any field effect during the remaining part of the third half period. During the fourth half period its field contributes to keeping the armature against pole piece 7.

The process during the fifth half period will be the same as during the first, and so on. The relay armature will thus oscillate at half of the frequency of the input AC voltage.

The relay equipment described has been made completely reliable within rather wide limits of the feeder voltage by making a choice of suitable series resistors and by parallel capacitors (Fig. 19). The polarized relay is also provided with another contact of the same kind operating the time-controlled switches. Other relay sets are in some cases inserted between the polarized relay and the switches for further frequency division. In this case, ordinary telephone relays can be used as they function quite satisfactorily on DC impulses with a frequency of 25 Hz.

Two polarized relays of this kind are included in the interval meter and one such relay in each of the other three meters. The circuit which drives switch AV in the interval meter is particularly interesting, Fig. 15. It is a typical example of the design of the frequency dividing relay sets in the measuring equipment. In describing the apparatus it is sufficient to consider the eight relays

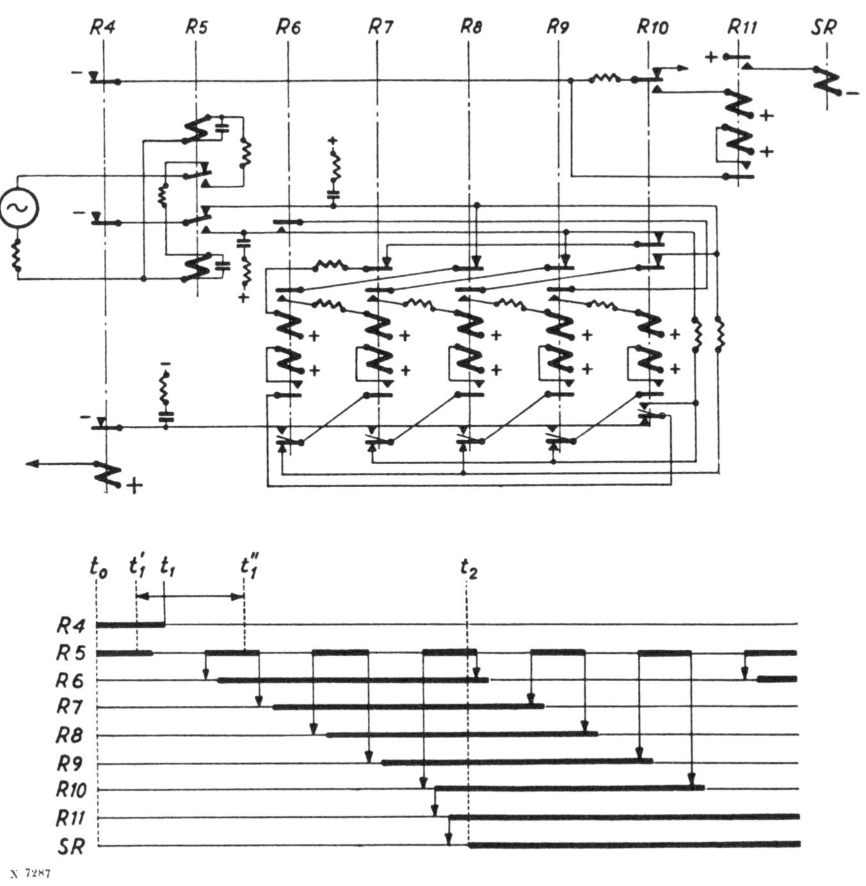

Fig. 20. Example of the use of a polarized relay for frequency division (by two) in the measuring equipment. – The upper part shows a detail of the circuit of the interval meter that contains a polarized relay R5, and the relay chain R6 – R10, the latter being a frequency divider (by two) and the first part of switch AV. The lower part of the figure is a time diagram for the operation of the relays in the upper part.

R4 – R11 in Fig. 20. Relay R4 is described in connection with Fig. 15. It is momentarily operated at the termination of each occupation. R5 is a polarized relay of the kind just described. Relays R6 – R11 are common telephony relays. The functioning of the equipment is shown by the time diagram in the lower part of Fig. 20. It contains in the ordinary way a time axis for each relay, on which the intervals are marked by thick lines when the relay is operated. Relay R5 is regarded as released in the state shown in the upper part of the drawing and operated when its change-over contacts are in the other position. The scale of the time diagram is clear from the operating time of relay R5 which, according to the foregoing, is operated for 20 ms, alternating with release for 20 ms (the short change-over times disregarded). Relays R6 – R10 have the same operating and release time, 5 ms.

For study of the equipment a process will be considered from the point of time t_0, when relay R4 is operated and relays R6 – R11

X 7276

Fig. 21. The entire equipment

released. Relay R4 is presumed to release at a point of time t_1, at which relay R5 is released. Time t_1 then corresponds to the termination of an impulse which is generated at the termination of an occupation. In the next operation of relay R5 an operate circuit (easily found in the figure) is closed through the upper winding of relay R6. Relay R6 operates and for the time being is locked through its lower winding. Relay R5 then releases, causing operation and holding of relay R7. The next operation of R5 likewise causes operation and holding of relay R8. Relay R9 is operated at the next release of R5 and, when R5 operates again, relay R10 is operated. All relays R6 – R10 are now operated.

When the next release of relay R5 occurs, the holding circuit of relay R6 is broken and this releases at the next operation of R5. At the next operation of relay R5 relay R7 releases. In the same manner relay R8 releases at the next release of R5, whereupon R9 releases at the next operation of R5. Finally, relay R10 releases at the next release of R5. As all relays R6 – R10 have now released, the same starting state prevails as before the operation of relay R6 (just described). As long as no change of the state of relay R4 occurs, relays R6 – R10 will work with an impulse frequency exactly one fifth of the impulse frequency of relay R5. This frequency-dividing property of the relay set has been made use of by letting relay R10, with its sketched contacts in the figure, control other frequency-dividing apparatus in its turn.

The frequency-dividing relay set described is the first part of switch AV in the interval meter. The first counter connected to the switch is then arranged to be operated by the operation of relay R11. The release of R11 is caused by the first operation of relay R10.

It is evident from the circuit diagram in Fig. 20 that all relays R6 – R11 and also counter SR are connected in such a way that they release as soon as relay R4 operates. If a new impulse, recording the termination of an occupation, arrives at such a point of time, after time t_1, that the operation of the counter SR (occurring normally at time t_2) cannot take place, no recording of the interval in question will appear on counter SR. In all other cases such a recording takes place at least on the counter SR.

The points of time t_1' and t_1'' indicated in the diagram limit the interval within which time t_1 (for the termination of the operation of relay R4) can occur anywhere without changing the course of events in the diagram. If time t_1 occurs before time t_1', relay R6 will have time to operate during the first period indicated in the figure when relay R5 is operated. In this case the periodic work cycle of the relay set starts one cycle earlier (i.e. 40 ms) than is shown in the diagram. If on the other hand t_1 occurs after t_1'', relay R6 will not have time to operate during the remaining time of the operation of relay R5. The work cycle of the relay set then starts only during the third of the intervals in the diagram. On account of this the interval between times t_1 and t_2 can vary up to 40 ms, as has already been mentioned in the description of the interval meter. The mean time between times t_1 and t_2 which is determined by the fact that the termination of the impulses from relay R4 occurs completely independently of the work cycle of relay R5, is now important for the computation of the recorded times of the interval meter. This has been determined by measurements, giving 104 ms as the mean value of the shortest interval between the termination of an impulse and the start of the next impulse giving a recording on counter SR.

Installation and operation

The various measuring apparatus described is installed together with the amplifier in the switchroom on the 4th floor of the Östermalm Telephone Office. An overall view of the equipment is shown in Fig. 21. All counters are brought together in a rack together with a date indicator and a synchronous electric clock for recording the days and hours of the measurements. The synchronous clock is driven by the AC used in the measurements.

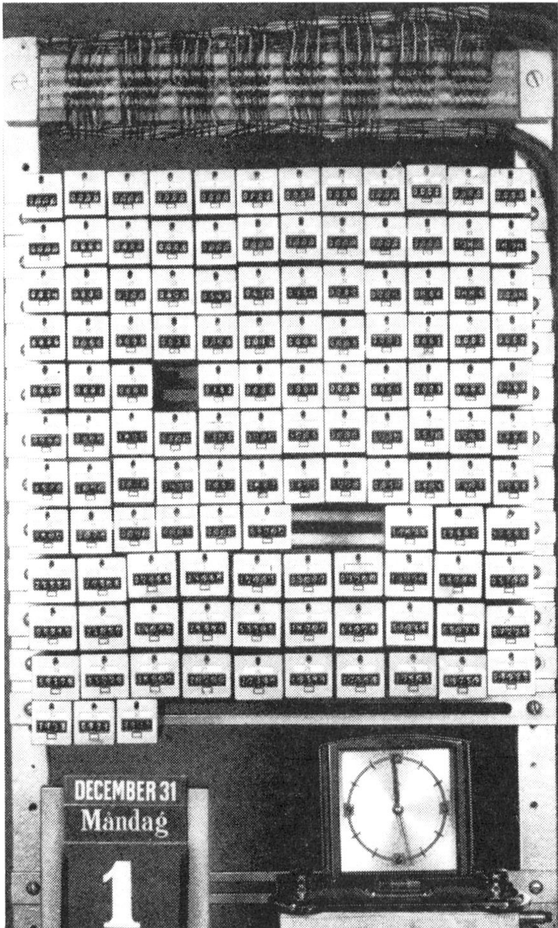

x 7868 Fig. 22. The counter rack

To obtain readings from the equipment, the counter rack was photographed at certain points of time. The counter recordings could then be read from the photos. A photographic procedure is necessary for several reasons in measurements of this kind. Regarding the counters, it may be mentioned that in most cases it was sufficient to use counters with only four figures. It has been necessary, however, to equip about one-third of the counters with five-figure capacity.

The counter rack is shown in Fig. 22. In the reading of the photos the use of a projector has been of advantage. Copying and enlarging of the negatives have thus been avoided. The measuring equipment has been operated every weekday during the measurement period – around 2 months – except days before holidays, during the hours 9 AM – 4.30 PM. The photographing has taken place before and after the daily measurements and at 10 and 11 AM.

The greatest care has been taken to check the reliability of the equipment during operation. Continuous watch has been kept. Every week the entire measuring equipment as well as the equipment in the 500-groups concerned has been thoroughly run through. The surveillance and control have required an average of two technicians.

Chapter 12

Methods of treating the measured results

In accordance with the principles put forward in Chapter 10, the purpose of the measurements carried out with the equipment described in the previous chapter is to investigate the following conditions:

1 To what extent the criteria in Chapter 10 of random traffic, as modified by slow intensity variations, are fulfilled by real telephone traffic.

2 To what extent the formulas for traffic according to normal form 1 are applicable to real telephone traffic. If the result turns out not to be satisfactory, the alternative should also be investigated as to whether a considerable improvement of the representation of the properties of real traffic could be obtained from the formulas valid for normal form 2.

In the foregoing, especially in Chapters 9 – 10, methods and formulas which are needed for studies of the kind mentioned have been developed. In the present chapter the methods and formulas have been compared and systematized. In some cases they have been transformed to facilitate the treatment as much as possible. The arithmetical treatment of the measurements is also touched upon. A more detailed description of this will be found in Chapter 14 together with a report on the analysis of the measurements.

The criteria put forward in Chapter 10 give necessary and sufficient conditions for the distribution function of the inter-arrival times to be represented by a *Laplace* integral and thus fulfil the requirements of the theory of intensity variations. Only one condition is important in practice in judging the distribution curve given in a table or graphically, this being that every series of values of the distribution function should form a completely monotonic series of numbers so that the difference quotients according to (145 b) will satisfy the condition (145 c). The absolute magnitude of the difference quotients decreases, however, very rapidly with increasing order. By an investigation of the validity of the condition mentioned for the distribution curve based on measurements, difference quotients are very soon obtained of such a magnitude that the uncertainty of the measured values becomes dominant. There is then no reason to expect that (145 c) will still always be satisfied. A pertinent question is how to separate the two phenomena:

1 Deviations that are due to the accuracy, always limited, of the measuring equipment.

2 Deviations that are of a fundamental nature owing to the fact that the traffic does not exhibit to a sufficient degree the assumed properties of the theory of slow intensity variations.

The only possibility is to be guided by the extent to which the deviations due to random errors show up as irregular steps in the values of the differences. Every considerable deviation of the measured distribution curve from the completely monotonic form should, on the other hand, show up in a sufficiently detailed analysis as similar deviations of the differences in certain domains. Evidently only such deformations of the distribution curve as are sufficiently great in relation to the random measuring errors can be traced. It is quite natural, however, that such minor but fundamental deformations as do not appear in the analysis of a distribution curve based on measurements, cannot correspond to properties of the traffic of such importance that their influence on the traffic condition is considerable compared with the random statistical variations. Such minor deformations of the distribution curve should then not be of any importance for judging the properties of the traffic investigated.

It should be remarked that possible fundamental deformations, that might occur, in a distribution curve based on measurements do not necessarily suggest deviations of traffic from the prerequisites of the theory of intensity variations. They might also be caused by systematic errors in the measurements. Therefore the first thing to investigate is the extent to which such errors might be the reason for any deviations.

From a theoretical point of view it is sufficient that a unique series of numbers derived from a given distribution function satisfies the condition (145 c) in order to check that the function is completely monotonic. In analysing a distribution curve based on measurements, it is necessary on the other hand, due to the influence of measuring errors, to look at several different series of numbers. In this way it will be easier to trace possible deformations of a regular character. The question arises as to which series should be chosen, i.e. which points of the curves to start from. The closer these points are placed to one another, the greater the number of points that can be included, and consequently differences of a still higher order can be calculated. On the other hand, the absolute magnitude of the differences decreases the more closely the points are placed and the sooner irregularities of the differences caused by measuring errors appear. If, for example, differences of first order already have varying signs, the formation of the differences only gives the information that the distribution curve is decreasing, and that is trivial. The best information on the character of the curve is clearly obtained from points which are so positioned as to give sign irregularities first at the highest possible order. It would seem difficult to create special rules in this context. In practice, it would be appropriate therefore to study several series of numbers with successively increasing distances between the points.

In selecting the points of the distribution curve intended for the difference study, it would be preferable to use the points that are directly determined by the interval meter. Another possibility is to draw a continuous curve through the points given by the interval meter and select points on this curve. These points can then be chosen equidistantly, thus simplifying the computations. It is, however, more important to design the curve so that a certain smoothing of the random measuring errors is obtained. This smoothing should therefore give a greater regularity of the differences.

Examples of the above methods for checking the completely monotonic character of a measured function are shown in Chapter 14. For this purpose direct computation from the values of the interval meter are used as well as the smoothing method, whereby the so-called difference curve is used for the graphical construction.

The next step in the study of the distribution curve for the inter-arrival times obtained by measurements is a comparison with the type of distribution function valid for traffic in normal form 1. The tables and graphs of the course of the difference curve for different values of the parameter η_0, drawn up in conjunction with the studies in Chapter 9 of normal form 1, are in this case helpful. It is advisable to start from a number of points on the difference curve computed from the measurement results to find the values of η_0

that correspond to these points in normal form 1. The mean of the quantities η_0 found are then calculated and the difference curve for this mean is computed. Finally this curve is compared with the difference curve derived from the measurements. It might be of advantage in forming the mean value to endeavour to allot to the different η_0 values different weights, taking into account the estimated measuring accuracy at different points of the measured difference curve. It is, however, difficult to set up any general rules in this respect.

The η_0 value obtained from the difference curve very close to $t = 0$ is of special interest. As is well known from Chapter 6, we have according to formula (62 a) or (63) and (64)

$$D'(0) = -\frac{1}{y}\int_0^\infty (x - y)^2 \, dG(x)$$

so that the variance of the intensity distribution is directly obtained from the derivative of the difference curve in the point $t = 0$. The following applies for normal form 1

$$D'_{\eta_0}(0) = -\eta_0$$

Unfortunately, this method for determining η_0 gives rather poor accuracy. The relative errors in the difference curve derived from measurements are often rather great for very small t-values due to the influence of the errors in determination of the t-quantities.

Some additional methods for deriving η_0 from the difference curve based on measurements are given below.

If a difference curve according to normal form 1 can have been constructed by means of η_0-values, determined in different ways, agreeing so closely with the difference curve based on measurements that the remaining deviations can be regarded as completely dependent on measuring errors, no further study of the completely monotonic character of the distribution curve is necessary and therefore the comparison with normal form 1 should be made first. It is possible, however, that the measurement-based and the constructed difference curves have deviations of a more regular character regardless of how the parameter η_0 is chosen. In such a case it should be investigated whether a better approximation of the measured difference curve could be obtained with the aid of difference curves constructed by use of normal form 2. It is necessary then to determine two parameters η_0 and κ.

The investigations in Chapter 9 regarding normal form 2 gave the result that the difference curve of normal form 2, with the parameters η_0 and κ for small t-values, asymptotically approaches the difference curve of normal form 1 with the parameter $\eta_0 + y\kappa$. Values of the sum $\eta_0 + y\kappa$ are thus obtained from the course of the measurement-based difference curve for small t-values. In order to separate the two parameters, the difference curve for higher t-values should be studied, as in this area the curves for normal form 1 and 2 differ more noticeably. Lacking tables of the difference curve for normal form 2, which would be very extensive because of the two parameters, it might be possible to use a trial-and-error method by testing different values of the parameters giving the previously determined value of the sum $\eta_0 + y\kappa$. A task of this kind can be facilitated by the fact that there are several methods by which, at least in principle, the parameters can be determined separately. The accuracy of practical calculations will, however, as a rule be so poor that only a rough estimation is possible.

The simplest of these possibilities for determining the two parameters is to apply the approximate formula (130 c) in Chapter 9. (It should be observed that in this formula $y = 1$.) This applies for small t-values but always up to higher t-values than the approximate formula (130 a). By applying (130 c) within a t-area near $t = 0$, where the measurement-based difference curve practically coincides with a difference curve according to normal form 1, it is possible to obtain an approximate value of the product $\eta_0\kappa$. By this method it should be possible to separate the two parameters as the sum $\eta_0 + y\kappa$ is determined beforehand.

Other ways of determining the two parameters can be obtained by computation of the distribution function $\vartheta(t)$ for the next call arrival. For reasons mentioned above it seems to be difficult to measure this function directly. It can, however, be derived instead from integration of the distribution function of the inter-arrival times. From formula (8 a), Chapter 2, we obtain

$$\vartheta(t) = 1 - y \int_{x=0}^{t} \varphi(x)\, dx$$

For a measurement-based distribution curve of the inter-arrival times the integration is carried out numerically or graphically. It is feasible to start from the difference curve

$$D(t) = \varphi(t) - e^{-yt}$$

whereby

$$\vartheta(t) = e^{-yt} - y \int_{x=0}^{t} D(x)\, dx \qquad (152\text{ a})$$

is obtained.

For normal form 1 the expressions (118 a) and (118 b) in Chapter 8 are valid for the distribution functions of the next call and of the inter-arrival times respectively. We then obtain

$$\frac{\vartheta(t)}{\varphi(t)} = 1 + \eta_0 t \qquad (152\text{ b})$$

Thus, for normal form 1 the ratio between the distribution functions of the next call arrival and of the inter-arrival times is a linear function of t. Hereby quite adequate means are obtained for checking – at least in principle – whether normal form 1 is valid and also for determining the value of parameter η_0.

For normal form 2 we instead obtain from the expressions (121 a) and (121 b)

$$\frac{\vartheta(t)}{\varphi(t)} = 1 + (\eta_0 + y\varkappa)\, t + $$
$$+ y\varkappa \left\{ \left(t + \frac{1}{\eta_0} \right) \ln(1 + \eta_0 t) - t \right\} \qquad (152\text{ c})$$

This is no longer a linear function of t. For $\eta_0 t < 1$ we obtain the series expansion

$$\frac{\vartheta(t)}{\varphi(t)} = 1 + (\eta_0 + y\varkappa)\, t + $$
$$+ \frac{1}{2} y\varkappa\eta_0 t^2 - \frac{1}{6} y\varkappa\eta_0^2 t^3 + \ldots \qquad (152\text{ d})$$

Differentiating both members of (152 c) with respect to t, we obtain

$$\frac{\vartheta'(t)\,\varphi(t) - \varphi'(t)\,\vartheta(t)}{\{\varphi(t)\}^2} = \eta_0 + y\varkappa + y\varkappa \ln(1 + \eta_0 t)$$

If we observe that $\vartheta'(t) = -y\varphi(t)$ and introduce the difference function $D(t)$, we obtain, after some transformation

$$\{ye^{-yt} - D'(t)\} \frac{\vartheta(t)}{\{\varphi(t)\}^2} =$$
$$= \eta_0 + y(1 + \varkappa) + y\varkappa \ln(1 + \eta_0 t) \qquad (152\text{ e})$$

If the left-hand member of (152 e) can be calculated with the desired accuracy from the distribution functions obtained by the measurements, this relation offers adequate means of determining the values of η_0 and κ.

Finally, an interesting quantity for the computations is the integral

$$\int_0^\infty t\,\varphi(t)\, dt \qquad (152\text{ f})$$

which is connected with the form factor ε, introduced in Chapter 6. For normal form 1 this integral equals $1/[y(y - \eta_0)]$ for $y > \eta_0$ otherwise being infinite. For normal form 2 the integral is always infinite. Obviously, if we endeavour to determine the integral from the distribution function obtained from measurements, it always acquires a finite value, as all inter-arrival times involved are finite. It is clear, however, that if a distribution function, determined by measurements, coincides more closely with the form valid for normal form 2 than to that for normal form 1, the integral (152 f), computed from the measurement results, will be considerably greater than $1/[y(y - \eta_0)]$. The value of the integral therefore becomes extremely interesting in deciding which type comes closest to a measurement-based distribution function of inter-arrival times.

The moments of the intensity distribution can be determined in two ways by measurements with the equipment described in the previous chapter. Certain moments are obtained directly according to formula (151 b) by measuring combinations of device occupations. Unfortunately, the accuracy of higher moments will be poor, among other reasons because the formula mentioned is not absolutely exact for groups with a limited number of devices. Further, the moments can be determined by means of formula (139 b) from the state quantities. The accuracy of lower moments is reasonably good, but the accuracy of higher moments is poor. This is clear from a closer study of the structure of formula (139 b). As a matter of fact the main part of the contribution to the sum in the right-hand member derives from state quantities P_r, whose r-values increase with increasing n. States with large r-values relative to the traffic intensity A occur very seldom and cannot therefore be so accurately determined. In each measurement performed during a limited time there is furthermore a maximum value of r, such that no states with a greater number of occupations have appeared during the measurement. These states then do not give any contributions to the sum in (139 b), which should be regarded as a measuring error caused by limited measuring time. Recordings of even higher states would be included by increasing the measuring time. Evidently in the computation of M_n, this error has an ever increasing influence the greater n is.

In Chapter 10 criteria have been presented giving necessary and sufficient conditions for presentation of the moments in the form that follows from the theory of intensity variations. When investigating the extent to which these conditions are satisfied for the moments determined by measurements, *the relative moments* μ_n are now introduced, determined by

$$\mu_n = \frac{M_n}{M_1{}^n} = \frac{M_n}{A^n} \tag{153 a}$$

We then have $\mu_0 = \mu_1 = 1$. For zero variation, i.e. for pure random traffic, all the relative moments are equal to unity. For random traffic modified by intensity variations, all $\mu_n > 1$ for $n > 1$.

According to *Stieltjes*' criterion, the conditions (148 b) of *Hankel* determinants defined by (149 a) shall be valid for every positive n. Because of the relation (149 b) we can now substitute the relative moment μ_r for M_r everywhere in these conditions. In tests with moments determined by measurements, it is of advantage to use the determinants based on the differences according to relation (149 d) instead of the determinants (149 a). By forming the differences of the series of numbers derived from the moments we in fact immediately observe – in the same way as described regarding the distribution function of the inter-arrival times – a limit caused by the uncertainty of the measuring results, also constituting a limit for the necessary satisfying of the conditions (148 b).

The condition (149 e) is also interesting in the study of moments determined by measurements. It is necessary, but not sufficient, and can be supplemented by a series of analogous conditions that successively reinforce one another. By applying the same method as used in Chapter 6 in deriving (60 b), the general condition is obtained without difficulty as follows:

$$\frac{\mu_{r+1}}{\mu_r} < \left(\frac{\mu_{r+2}}{\mu_r}\right)^{\frac{1}{2}} < \left(\frac{\mu_{r+3}}{\mu_r}\right)^{\frac{1}{3}} < \dots$$

$$\dots < \left(\frac{\mu_{r+n}}{\mu_r}\right)^{\frac{1}{n}} \tag{153 b}$$

which is valid for every integer $r \geqq 0$.

The simple expression (134 a) in Chapter 9 is valid for the moments by traffic in normal form 1, giving for the relative moments

$$\mu_n = \left(1 + \frac{\eta_0}{y}\right)\left(1 + 2\frac{\eta_0}{y}\right)\dots\dots$$

$$\dots\dots\left(1 + (n-1)\frac{\eta_0}{y}\right) \tag{154 a}$$

From this relation a number of conditions for the moments can be derived, for instance

$$\frac{\mu_n}{\mu_{n-1}} = 1 + (n-1)\frac{\eta_0}{y} \tag{154 b}$$

For fulfilling this relation for all n-values, the ratio μ_n/μ_{n-1} must be a linear function of n, from which a mean value of η_0 can be calculated. For its computation we can also use the second degree equation in η_0, which is obtained from the relation

$$\frac{\mu_n}{\mu_{n-2}} = \left(1 + (n-2)\frac{\eta_0}{y}\right)\left(1 + (n-1)\frac{\eta_0}{y}\right) \tag{154 c}$$

If we have determined by measurements the successive moments μ_2, μ_3 ... and wish to study in a more versatile manner to what extent they satisfy the expressions for normal form 1, it is feasible to form a difference scheme in the following way: From (154 a) we obtain

$$\frac{\mu_n - \mu_{n-1}}{n-1} = \frac{\eta_0}{y}\mu_{n-1} \tag{154 d}$$

First we form the differences defined in this way and go on in the same manner with the results obtained. Naturally we shall then have

$$\frac{\frac{\eta_0}{y}\mu_{n-1} - \frac{\eta_0}{y}\mu_{n-2}}{n-2} = \left(\frac{\eta_0}{y}\right)^2 \mu_{n-2}$$

By continuing in the same way we shall obtain a difference scheme

$$\begin{array}{llllll}
1 & & & & & \\
\frac{\eta_0}{y} & & & & & \\
\mu_2 & \left(\frac{\eta_0}{y}\right)^2 & & & & \\
\frac{\eta_0}{y}\mu_2 & & \left(\frac{\eta_0}{y}\right)^3 & & & \\
\mu_3 & \left(\frac{\eta_0}{y}\right)^2\mu_2 & & \left(\frac{\eta_0}{y}\right)^4 & & \\
\frac{\eta_0}{y}\mu_3 & & \left(\frac{\eta_0}{y}\right)^3\mu_2 & & \left(\frac{\eta_0}{y}\right)^5 & \\
\mu_4 & \left(\frac{\eta_0}{y}\right)^2\mu_3 & & \left(\frac{\eta_0}{y}\right)^4\mu_2 & & \\
\frac{\eta_0}{y}\mu_4 & & \left(\frac{\eta_0}{y}\right)^3\mu_3 & & & \\
\mu_5 & \left(\frac{\eta_0}{y}\right)^2\mu_4 & & & & \\
\frac{\eta_0}{y}\mu_5 & & & & & \\
\mu_6 & & & & &
\end{array} \tag{154 e}$$

It can be expected that a scheme of this kind will give a clear picture of the extent to which the moments satisfy the formula (154 a), and will allow for various possibilities of determining the mean of η_0. It turns out in practice, however, that the higher differences decrease rapidly and hence make a more accurate computation of η_0 impossible. The formula (154 c) and similar expressions for quotients betweeen moments more distant from one another offer a greater advantage. The computations in Chapter 14 illustrate this condition in somewhat greater detail.

If it now turns out that we cannot satisfactorily represent the values of the measurement-based moments by means of normal form 1, it remains to be studied whether a better representation is possible using normal form 2. The expressions of the moments in this case will be more complicated. It has not been possible to develop a similar table to the one for normal form 1, to guide the appropriate choice. For lower moments there are explicit expressions at the end of this work, in conjunction with the calculations for table 2, in Chapter 9. From these we obtain, for instance, by means of moments μ_2, μ_3, μ_4, the equations applicable for direct computations of η_0 and κ

$$\left.\begin{aligned} &\frac{\eta_0}{y} + \varkappa = \mu_2 - 1 \\ &\frac{\eta_0}{y} \cdot \varkappa = 2\mu_2^2 - \mu_2 - \mu_3 \\ &\left(\frac{\eta_0}{y}\right)^2 \varkappa = 3\mu_2^2 - 6\mu_2^3 + \mu_3(6\mu_2 - 2) - \mu_4 \end{aligned}\right\} \quad (155\text{ a})$$

Already from these equations we shall have guidelines as to whether it will be rewarding to apply normal form 2. In order to carry out a more systematic study, the following method can be applied. We introduce

$$\vartheta(t) = \int_0^\infty e^{-xt}\, dG(x) = \sum_{n=0}^\infty \int_0^\infty \frac{(-xt)^n}{n!}\, dG(x)$$

Now according to the definition

$$\mu_n = \frac{1}{y^n}\int_0^\infty x^n\, dG(x)$$

and we obtain the series

$$\vartheta(t) = \sum_{n=0}^\infty \mu_n \frac{(-yt)^n}{n!}$$

Differentiation results in

$$\varphi(t) = \sum_{n=1}^\infty \mu_n \frac{(-yt)^{n-1}}{(n-1)!}$$

and further

$$\varphi'(t) = -y\sum_{n=2}^\infty \mu_n \frac{(-yt)^{n-2}}{(n-2)!}$$

For normal form 2 the formula (152 e) just derived is valid, the left-hand member being

$$\frac{-\varphi'(t)\,\vartheta(t)}{\{\varphi(t)\}^2}$$

By means of the series expansions above we obtain the relation

$$\frac{\displaystyle\sum_{n=0}^\infty \mu_n \frac{(-yt)^n}{n!} \sum_{n=2}^\infty \mu_n \frac{(-yt)^{n-2}}{(n-2)!}}{\displaystyle\left\{\sum_{n=1}^\infty \mu_n \frac{(-yt)^{n-1}}{(n-1)!}\right\}^2} =$$

$$= 1 + \frac{\eta_0}{y} + \varkappa + \varkappa \ln(1 + \eta_0 t) \quad (155\text{ b})$$

This relation can be used in such a way that the left-hand member is calculated by means of the values of the measurement-based moments for a number of t-values. The values of η_0 and κ are then determined by comparison with the right-hand member. Another possibility is derived from a series expansion of both members with increasing powers of t, and a comparison between the respective coefficients. We obtain in this way the quantities

$$\left(\frac{\eta_0}{y}\right)^n \varkappa$$

directly expressed in the μ_2, μ_3, ... μ_{n+2} moments. However, the expressions, clearly identical with those in table 2, for instance (155 a), become very complicated for higher powers. It might then pay to carry out the series expansion of the left-hand member of (155 b) only after the introduction of the numerical quantities.

The methods for the study of moments shown above imply – to the extent the moments are calculated from measured states – also methods of indirect studies of the state quantities. It is most desirable to have access to more direct procedures for the study of the consistency between the results of a state measurement and the conditions expected according to the theories. The states are in fact the quantities that have the greatest importance for the dimensioning of groups and for this reason it is particularly interesting to have possibilities of directly comparing these with the theoretical results.

The state n, i.e. the mean of the part of the time, when n, and only n, of the devices in the group are occupied at the same time, has for intensity variations the form (138 a), Chapter 9. From this it is apparent that the expression

$$\frac{n!}{s^n} P_n$$

can be understood as the nth moment of the function $e^{-sx}\, dG(x)$. We can thus apply *Stieltjes'* moment criterion to the states to investigate the extent to which they can be expressed in the proposed form. The states, however, will seldom be determined with such accuracy by measurements that a check of this kind has any practical significance. We have then, as a rule, to be content with representing the moments of the function $dG(x)$ as previously shown, and applying the mentioned criterion to these quantities. In that way they are usually determined with considerably greater accuracy than the separate states.

The possibilities of a direct check are much more favourable with regard to the representation of the states in normal form 1. From the expression (140 a), valid in this case for the state P_n, we obtain the simple relation

$$\frac{n}{s} \frac{P_n}{P_{n-1}} = \frac{y + (n-1)\,\eta_0}{1 + s\eta_0} \qquad (156\text{ a})$$

which might be suitable for determining η_0. Unfortunately an extensive measurement material is required, and consequently longer measurement times, so that the ratio P_n/P_{n-1} is so accurately determined as to permit a reasonably safe estimation of η_0 by means of this relation. However, it always offers a possibility of determining a mean value of η_0 using several different n-values, and this can be used as a first approximation. (See further below.)

A determination of η_0, less dependent on random variations of the measuring results, can be obtained from the relation

$$\frac{n\,(n-1)}{s^2} \frac{P_n}{P_{n-2}} = \frac{(y + (n-1)\,\eta_0)\,(y + (n-2)\,\eta_0)}{(1 + s\eta_0)^2} \qquad (156\text{ b})$$

We can, of course, also form the ratios between every third or every fourth, etc. state, but we then obtain equations which are rather awkward for numerical calculations.

For every given P_n it is desirable to determine the corresponding η_0 from (140 a) to allow for a more systematic investigation of the possibility of representing measured state quantities in normal form 1. For lack of tables of the expressions, which – due to the presence of several parameters – will be rather extensive, we shall be constrained to resort to a trial-and-error method. This may be rather cumbersome, especially for higher n-values. The calculations can, however, be considerably simplified by transforming the expression (140 a) by means of *Stirling*'s series expansion of the gamma function. We then first introduce a new parameter $\tau_0 = y/\eta_0$ instead of η_0. The expression (140 a) can then be written

$$P_n = \frac{A^n}{n!} \frac{\tau_0\,(\tau_0 + 1)\,(\tau_0 + 2) \ldots (\tau_0 + n - 1)}{(A + \tau_0)^{n+\tau_0}} \,\tau_0^{\tau_0}$$

where $A = sy$. By means of the gamma function this can be written

$$P_n = \frac{A^n}{n!} \frac{\tau_0^{\tau_0}}{(A+\tau_0)^{n+\tau_0}} \frac{\Gamma(n + \tau_0)}{\Gamma(\tau_0)}$$

Now, according to *Stirling*[20] the following asymptotic series expansion of the gamma function is valid

$$\Gamma(x) = \sqrt{2\pi}\, e^{-x}\, x^{x-\frac{1}{2}} \left\{1 + \frac{1}{12x} + \frac{1}{288\,x^2} - \frac{139}{51840\,x^3} - \ldots\right\}$$

In the present applications, the parameter τ_0 has the order of magnitude $10-100$. The desired accuracy allows terms containing $1/x$ in second or higher power to be neglected. The expression for P_n will then, after some transformations, be

$$P_n = \frac{A^n}{n!} e^{-n} \left(1 + \frac{n - A}{\tau_0 + A}\right)^{\tau_0 + n}$$

$$\sqrt{\frac{\tau_0}{\tau_0 + n}} \frac{1 + \frac{1}{12(\tau_0 + n)}}{1 + \frac{1}{12\tau_0}} \quad (157\text{ a})$$

As a comparison quantity we then suitably introduce

$$P_{n,0} = \frac{A^n}{n!} e^{-A}$$

which expresses the corresponding state for random traffic. It can also be obtained from (157 a) by letting $\tau_0 \to \infty$. We then find

$$\frac{P_n}{P_{n,0}} = e^{-(n-A)} \left(1 + \frac{n - A}{\tau_0 + A}\right)^{n + \tau_0} \cdot$$

$$\cdot \sqrt{\frac{\tau_0}{n + \tau_0}} \frac{1 + \frac{1}{12(n + \tau_0)}}{1 + \frac{1}{12\tau_0}} \quad (157\text{ b})$$

If we again introduce the parameter $\eta_0 = y/\tau_0$ and put $y = 1$ for the sake of simplicity, we obtain

$$\frac{P_n}{P_{n,0}} = \frac{e^{-(n-A)}}{\sqrt{1 + n\eta_0}} \left(1 + \frac{n - A}{1 + A\eta_0}\eta_0\right)^{n + \frac{1}{\eta_0}}$$

$$\left(1 - \frac{n\eta_0^2}{(12 + \eta_0)(1 + n\eta_0)}\right) \quad (157\text{ c})$$

The expressions of P_n presented here, especially for greater n, are considerably better adapted to numerical computations than the basic formula (140 a). Only in the case of a wanted tabulation of P_n, for all n-values with given A and η_0, might the last formula be preferable. If we now have a value determined by measurements of a certain P_n, and wish to calculate the corresponding η_0-value, we have to use a trial-and-error method by repeated introduction in (157 c) of different η_0-values. As an initial value we can use a mean of η_0 determined from one of the formulas (156). Another possibility of getting a first approximate value of η_0 is to use the approximate formula (71), Chapter 6, viz.

$$\frac{P_n}{P_{n,0}} \approx 1 + \frac{\eta_0}{2}(n(n-1) - 2nA + A^2) \quad (158)$$

However, within certain areas this formula gives a rather poor approximation of η_0.

If now an approximate value η_0' for η_0 has been determined and the corresponding $P_n(\eta_0')$ is calculated by (157 c), a more accurate approximation η_0'' can be computed by means of the formula

$$\frac{1}{\eta_0'' - \eta_0'} \ln \frac{P_n(\eta_0'')}{P_n(\eta_0')} =$$

$$= -\frac{n}{2(1 + n\eta_0')} - \frac{n\eta_0'}{6} + \frac{1}{2}\left(\frac{n - A}{1 + A\eta_0'}\right)^2 \cdot$$

$$\cdot \left\{1 - \frac{2}{3}\frac{n - A}{1 + A\eta_0'}\eta_0' + \frac{1}{2}\left(\frac{n - A}{1 + A\eta_0'}\eta_0'\right)^2 - \right.$$

$$\left. - \frac{2}{5}\left(\frac{n - A}{1 + A\eta_0'}\eta_0'\right)^3\right\} \quad (159)$$

which can be derived by series expansion of the natural logarithm of (157 c). This formula is rather complicated and it hardly pays to use it in numerical computations. Besides, with further simplification, it will as a rule be too unreliable. It might therefore be more convenient to use a direct trial-and-error method by iteration of (157 c) to find the desired value of η_0.

If we then have a series of state quantities P_n, obtained from measurements, we can determine the parameter η_0 for every value of n. The task is now to determine a suitable mean value of the different parameter values and study whether a number of P_n-quantities, based on this mean value, satisfactorily agree with the P_n-quantities determined by the measurements.

To obtain a suitable mean, it should not always be necessary to carry out the rather complicated calculations described above. A mean value derived from one of the formulas (156) might be useful. Another possibility is to determine η_0 from (158) for different n-values and from it compute a mean. It should be pointed out that there exists a very convenient method of determining a relatively accurate separate η_0-value. For $n = A$, (157 c) has a particularly simple form, allowing direct computation of η_0. Of course it is quite seldom that A is so close to an integer that this condition can be made use of. We can, however, draw a continuous curve through the different P_n points and endeavour to apply the relation (157 c) to a non-integer $n = A$.

If it now turns out that the formulas valid for normal form 1 cannot be adapted with adequate accuracy to the state quantity values obtained from measurements, the question arises whether a better adaptation can be reached by means of formulas of normal form 2. Unfortunately, the expressions for the P_n quantities for normal form 2 are so inconvenient for numerical calculations that it is hardly possible to carry out a study of this kind. This fact, however, is of minor practical importance. Due to the complications of normal form 2, it will be out of the question to use it for practical traffic problems. Normal form 2 is introduced mainly for theoretical reasons to describe superposing conditions. Numerically it differs insignificantly from normal form 1. In judging the measurement results it must be kept in mind that the purpose of the theories put forward in this work is the development of simple, convenient formulas which agree better with real traffic conditions than those based on random traffic. Therefore it is desirable, but not absolutely necessary, that there is complete agreement between the measurement results and the formulas; it suffices that they agree considerably better than the formulas used until now. We should not, however, when occasion arises, refrain from checking the feasibility of normal form 2, as for the moments and the distribution function of the inter-arrival times. However, normal form 2 is interesting mainly for measuring superposition phenomena.

As to the evaluation of the measurement results, some words should be devoted to the presentation of the frequency function $g(x)$ and the distribution function $G(x)$ of the intensity distribution. The behaviour of these functions has no direct importance for questions about the agreement of the measuring results with the theories, but they are interesting all the same as they give a clear picture of the variation of the traffic intensity. The frequency function has the form (118 c), Chapter 8, for normal form 1. It is easy to calculate numerically. The *Stirling* formula is advantageously used for calculating the gamma function. To determine the distribution function

$$G(x) = \int_{z=0}^{x} g(z)\, dz$$

numerical or graphical integration should be used. We can also use existing tables of the incomplete gamma function, but they are not as a rule suitably arranged for this purpose.

For normal form 2 the frequency function of the intensity acquires a considerably more complicated form which makes numerical calculations cumbersome. It differs little, however, from the frequency function of normal form 1 if in the one case the parameters are η_0 and κ and in the other $\eta_0 + y\kappa$. The two frequency functions then have the same variance as has been verified previously. This is η_0/y for normal form 1, and $\eta_0/y + \kappa$ for normal form 2.

In judging the intensity distribution we often find it useful to compare it with the distribution of the mean traffic values for measured partial times of the total measuring time. If, for instance, the measured traffic comprises the traffic during 100 busy hours, a certain idea of the intensity distribution can be obtained already from the distribution of the mean value of every hour. This applies for the traffic volume, the call volume and the mean occupation time during each of these hours. However, this distribution gives no real measure of the intensity distribution, since during each separate hour a separate intensity variation can occur and also since the means of the different hours may differ from one another due to random variations. The influence of these two components on the results can in principle only be separated by means of the methods for intensity analysis. It is evident, however, that the shorter the measured partial times are, the greater the influence of the random variations on the distribution of the mean values, and the longer the partial times are, the greater the influence of the intensity variations. It is of interest to be able to compare with the distribution of mean values that would occur if only random varations were present. As this is relatively easy to determine, we shall touch shortly upon it. Strangely enough, this does not previously seem to have been evaluated.

With regard to the traffic volume, the problem can be formulated in the following way: We assume a traffic with the call volume y and the mean occupation time s. What is the probability that the sum of the occupation times during a time span T is t, i.e. that it is between t and $t + dt$?

The solution to this problem obviously makes it possible to determine the probability that the total traffic volume measured during the time T deviates to a certain extent from the theoretical mean value syT of the traffic volume during this time. Now it turns out to be advantageous for the mathematical treatment to modify the problem somewhat. Instead, we consider therefore the following formulation of the problem:

What is the probability that the occupations starting during a time span T have a total duration t?

The difference between the two formulations lies in the fact that in the latter case we have decreased the traffic volume considered by the part that is derived from the occupations in progress at the start of the time span T. Further, we have increased it by the remaining part of the occupations in progress at the termination of the time span T. A treatment of the start and termination states would certainly involve mathematical complications. The latter formulation, on the other hand, gives a simple treatment. As it seems probable that the two problem formulations lead to results very close to one another, we can without risk apply computations based upon the latter formulation, especially as we can be satisfied with approximate results.

In order to arrive at a simple mathematical treatment, we must make a further assumption, namely that the occupation times have an exponential distribution so that

$$e^{-\frac{x}{s}}$$

expresses the probability that an occupation has at least the duration x. This would not, however, be a necessary condition for the validity of the result.

As is well known, and follows, for instance, from the derivation of formula (69), Chapter 6, the probability with random traffic, that during the time T, exactly n calls arrive, is

$$\frac{(yT)^n}{n!} e^{-yT} \tag{160}$$

If now the occupation times follow the exponential distribution mentioned, the probability that n occupations have a total duration of exactly the time t is further

$$\frac{1}{s} \frac{\left(\frac{t}{s}\right)^{n-1}}{(n-1)!} e^{-\frac{t}{s}} dt \tag{161}$$

The validity of this formula can be shown by induction. Its validity for $n = 1$ follows directly from the form of the distribution function. If now the formula is supposed to be valid up to the value n, the probability that $n + 1$ occupations will have a total duration t is

$$\int_{x=0}^{t} \frac{1}{s} \frac{\left(\frac{x}{s}\right)^{n-1}}{(n-1)!} e^{-\frac{x}{s}} \cdot \frac{1}{s} e^{-\frac{t-x}{s}} \, dx \, dt$$

as the probabilities of the durations of the different occupations are always assumed to be independent of each other. In carrying out the integration we obtain

$$\frac{1}{s} \frac{\left(\frac{t}{s}\right)^{n}}{n!} e^{-\frac{t}{s}} dt$$

which is the same as formula (161), only with $n + 1$ inserted instead of n. Thereby the general validity of (161) is proved.

The probability that precisely n occupations occur during the time T, and that their total duration is precisely t, will then be the product of the expressions (160) and (161), since with random traffic the durations of the occupations are assumed to be independent of the number of calls and their placing. This probability is therefore

$$\frac{1}{s} \frac{(yT)^n}{n!} e^{-yT} \frac{\left(\frac{t}{s}\right)^{n-1}}{(n-1)!} e^{-\frac{t}{s}} dt$$

Now a total occupation time t may occur for $n = 1, 2, \ldots$ and we find that the probability of a total occupation time t, independent of how many calls arrive during the time T, is

$$f_T(t)\, dt = \frac{1}{s} e^{-yT-\frac{t}{s}} \sum_{n=1}^{\infty} \frac{(yT)^n}{n!} \frac{\left(\frac{t}{s}\right)^{n-1}}{(n-1)!} dt \quad (162)$$

This function $f_T(t)$ obviously constitutes the frequency function of the total occupation time and, multiplied by dt, gives the wanted probability that the occupations starting during the time T will have the total duration t. The formula is valid only for $t > 0$. The case $t = 0$ occurs only if there has been no occupation during the time T and has the probability e^{-yT}, which is not a differential quantity. The distribution function of the total occupation time thus at the origin has a discontinuity of the mentioned magnitude. The sum of the probabilities for all the t-values must be equal to 1. We also easily verify that

$$e^{-yT} + \int_0^{\infty} f_T(t)\, dt = 1$$

The mean of the total traffic volume of the incoming calls during the time T now becomes, as expected

$$\int_0^{\infty} t\, f_T(t)\, dt = syT$$

where the discontinuity at the point $t = 0$ has no influence. Of interest also is the variance, which by simple calculations will be

$$\int_0^{\infty} (t - syT)^2 f_T(t)\, dt = 2 s^2 yT \quad (163)$$

It would now seem useful to introduce a new variable x, which expresses the ratio between the total occupation time t and the mean syT of the total occupation time, i.e.

$$x = \frac{t}{syT}$$

We then have $dt = syT\, dx$. Further we introduce the constant $R = yT$, which clearly expresses the mean value of the number of calls during the time T. If, finally, we put $h(x) = f_T(syTx)$, we obtain from (162)

$$h(x)\, dx = R^2 e^{-R(1+x)} \sum_{n=1}^{\infty} \frac{(R^2 x)^{n-1}}{n!\,(n-1)!} dx \quad (164\text{ a})$$

This then expresses the probability that the ratio between the traffic volume during a time T and the mean value of the traffic volume during the same time has a value between x and $x + dx$. In (164 a) there is only one constant, i.e. R.

The function $h(x)$ appears to be rather awkward for numerical calculations. Indeed it can be transformed to a known function, a *Bessel* function, with a pure imaginary argument. We have, as is well known[16]

$$J_1(2i\sqrt{t}) = i\sqrt{t}\sum_{\mu=0}^{\infty}\frac{t^{\mu}}{\mu!\,(1+\mu)!}$$

and we can then write

$$h(x) = \frac{R\,e^{-R(1+x)}}{i\sqrt{x}}\,J_1(2iR\sqrt{x})$$

For the calculation of J_1 there are tables, for instance in the work by *Jahnke-Emde* quoted above. These are, however, of little interest here as the argument for the applications in this case is very great. On the other hand the known asymptotic expansions of the *Bessel* function can be used with advantage. From these (see *Jahnke-Emde*) we can, after some transformations, obtain

$$h(x) = \sqrt{\frac{R}{4\pi}}\,e^{-R(1+x-2\sqrt{x})}\,x^{-\frac{3}{4}}$$
$$\left\{1-\frac{0{,}1875}{R\sqrt{x}}-\frac{0{,}029297}{R^2 x}+\ldots\right\} \quad (164\text{ b})$$

The series is not convergent. The error is, however, always less than the last included term. Only values of x in the close vicinity of 1 have importance for the current applications, as the exponential function, containing the usually great constant R, causes $h(x)$ to be very small for x-values differing much from 1.

To get an idea of the general behaviour of the function, it is sufficient to consider the approximate expression

$$h(x) \approx \sqrt{\frac{R}{4\pi}}\,e^{-R(1+x-2\sqrt{x})} \quad (164\text{ c})$$

Writing the exponent in the following way

$$1+x-2\sqrt{x} = 2+(x-1)-2\sqrt{1+(x-1)}$$

we can expand the square root for $|x-1|<1$ and then obtain

$$1+x-2\sqrt{x} = +\frac{(x-1)^2}{4}-\frac{(x-1)^3}{8}+\ldots$$

If we here neglect the higher powers of $(x-1)$ we obtain from (164 c)

$$h(x) \approx \sqrt{\frac{R}{4\pi}}\,e^{-\frac{R}{4}(x-1)^2} \quad (164\text{ d})$$

The random variable x is thus approximately normally distributed with the standard deviation

$$\sigma = \sqrt{\frac{2}{R}}$$

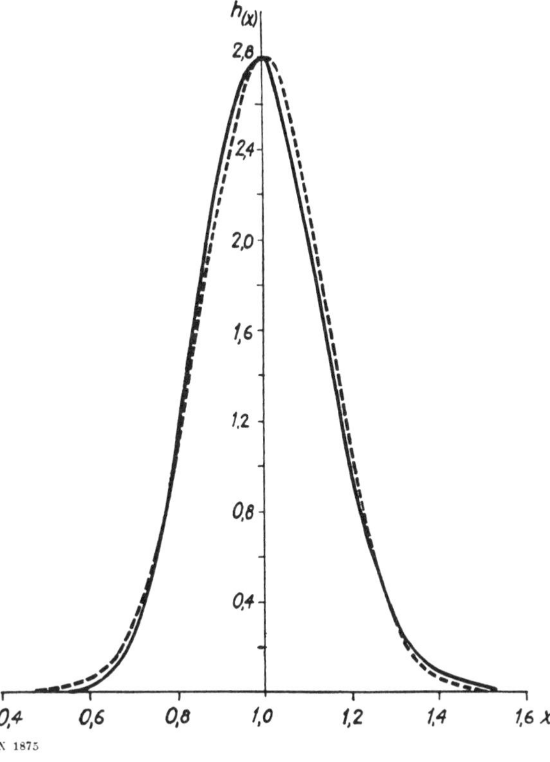

Fig. 23. The frequency function $h(x)$ of the traffic volume at $R = 100$. The continuous curve shows the results computed from (164 b) and the dashed curve the result according to the approximate formula (164 d).

an expression which is clear already from (163). As the normal or *Gaussian* distribution is tabled in most statistical manuals and collections of statistical tables, we have then an easy method of calculating $h(x)$. In order to study the influence of the approximations made, we have carried out a computation for $R = 100$. This corresponds for a mean occupation time of 2 minutes to a traffic of 3.3 traffic units, which is a rather small traffic. The standard deviation is then 0.1414. Fig. 23 shows the result of the computation of $h(x)$ from (164 b) and from the approximate formula (164 d). As is apparent from the figure, the insignificant difference present between the curves is in the form of a certain skewness compared with the symmetrical normal frequency function and is mainly due to the influence of the factor

$$x^{-\frac{3}{4}}$$

in (164 b). The difference between the curves decreases very rapidly by increasing R. We may then use the normal distribution for all practical applications.

In addition to the variations of the traffic volume, also the variations of the call arrivals during a time span T are of interest, i.e. the distribution represented by (160). The mean of the total number of calls during the time T will then be

$$\sum_{n=0}^{\infty} n \frac{(yT)^n}{n!} e^{-yT} = yT$$

which is self-evident. Further the variance

$$\sum_{n=0}^{\infty} (n - yT)^2 \frac{(yT)^n}{n!} e^{-yT} = yT$$

thus becomes equal to the previously introduced constant R. If in (160) instead of n we introduce the variable $x = n/yT$, then x will be approximately normally distributed for great n and T values, which can easily be proved by means of the *Stirling* formula for the gamma function. The mean value of x is evidently 1 and the standard deviation

$$1 / \sqrt{R}$$

Finally, also the variation of the mean of a number of measured occupation times is of interest. If the separate occupation times follow an exponential distribution, (161) gives the probability that n occupations will have a total duration t. The mean length of these n occupations is then t/n and the mean of these means evidently is s. If we in (161) introduce a new variable $x = t/ns$, we obtain

$$n \frac{(nx)^{n-1}}{(n-1)!} e^{-nx} \, dx$$

as an expression of the probability that the ratio between the mean of n occupation times and the mean value s will have a value between x and $x + dx$. The mean of x is obviously 1 and the standard deviation will be

$$1 / \sqrt{n}.$$

The distribution also in this case approaches the normal distribution for great n. This follows immediately from the central limit theorem for addition of independent random variables.

Chapter 13

Measurement accuracy

Before we can draw any conclusion from measurement data, it is necessary to have a fairly definite idea of the tolerances of the measured values. The smallest deviation that can be read from the measuring instrument is called the threshold value of the accuracy of the measurement. The threshold value thus forms a natural limit for the measurement equipment. In the measurements reported on in this work, all reading instruments have been counters. The smallest increment of the recording is a step on the counter, which makes it very easy to determine in every single case the threshold value in question. The threshold, however, is only a lower limit of the tolerance. In most cases we should reckon on a considerably wider tolerance due to the presence of measuring errors. These can, as a rule, be of two different kinds, systematic errors and random errors, although a division such as this might be difficult to make consistently. The systematic errors may be said to depend on whether for any reason the measuring equipment measures a quantity other than the intended one. We can also count among the systematic errors those arising from the possibility that for some reason the measured object is not identical with the object whose properties we wish to determine. Obviously, if systematic errors appear, for the measurements to be of any value, we should either be able to make corrections for them or also prove that their influence is negligible. Certainly the best thing would be for the measurements to be performed in such a way that systematic errors are avoided. Sometimes, however, it is necessary, in order not to complicate the measurement conditions too much, to permit certain systematic errors and to correct the measurements for their influence.

Based upon experience from earlier measurements, different mechanisms resulting in systematic errors have been studied in the planning of the measurements reported on in this work. It has therefore been possible in most cases to avoid them. In other cases we can prove that their influence can be disregarded, as will be shown below. Exceptions are errors caused by occasional faults in the measuring equipment that should be counted among the systematic errors. These errors do not generally derive from the equipment itself – which has worked surprisingly reliably – but from faults in switches and other devices belonging to the part of the telephone exchange handling the traffic studied. These faults, which appear as blockings and other phenomena not belonging to the normal switching process, have as a rule an insignificant influence on the normal operation; for this it is sufficient that they are localized and repaired at periodic checks. Such otherwise rather harmless disturbances can, however, cause very serious errors in measurements of the type in question. It has been necessary therefore to have a continuous monitoring of the entire switching equipment handling the traffic investigated. Thanks to monitoring the faults appearing have always been identified and the equipment repaired within a short time. The measurement results from the periods when faults occurred have of course been omitted in the treatment. The principle has been followed that, when minor faults influencing only a small part of all simultaneously recorded values have occurred, only the values directly concerned have been omitted. In the case of some faults with greater influence, all measurements during that time span have been omitted in the treatment.

All errors that have a tendency to neutralize one another belong as a rule to the other kind of errors, the random errors. The mean of the random errors in the measurements should therefore approach zero when the number of observations increases. Thus no systematic displacement of the mean of the measured values will appear, as compared with the true values. The random errors are generally due to phenomena which cannot be controlled by the measurement equipment used. There are thus random errors in the interval measurements due to a certain variation of the time determination. Another reason for the appearance of random errors is that the functioning of the equipment can in certain respects be characterized as a statistical sampling method. With random errors of this kind, as shown below, we can estimate with fair accuracy the probability of a resulting error of a certain magnitude.

Systematic and random errors may appear in measurements of all physical quantities. In measurements determining statistical quantities we can readily distinguish errors of a third kind that can be termed statistical errors, although they may be considered to be random. The tolerances which appear due to such statistical errors can be determined in advance only if we know the theoretical value of the probability empirically searched for. Thus, when casting a die n times to empirically determine the probability that a certain side of the die will come up, we can compute the probability that the tolerance will have a certain value. We then neglect the influence of systematic errors due to imperfections of the die or the performance of the cast. A similar tolerance determination should in principle be possible in traffic measurements if we could rely on the assumption of random traffic. The computations will, however, be very complicated for most of the quantities. Also in the case of slow intensity variations it should be possible, at least in principle, to make prior tolerance determinations. By means of these it should be possible to present a quantitative norm for judging whether a measurement result is consistent with the assumption of intensity variations. Every effort to determine such tolerances seems to involve extremely great difficulties. We have to consider that the traffic intensity according to the assumptions may vary almost arbitrarily. Hence the measured values can show considerable variations even if no statistical variations appeared. Though it is quite conceivable that the criteria in Chapter 10 could be supplemented by tolerance indications, we seem to be restricted to drawing conclusions from the measurement results only as regards the order of magnitude of the statistical errors. We could, however, as with all statistical phenomena, assume that the tolerances decrease with the inverse square root of the number of observations. We thereby obtain guidance for estimation of the influence of the measuring time on the accuracy of measurement.

In this chapter no attempt will be made to further investigate the tolerances due to statistical errors. The following account will be devoted to possible systematic error mechanisms and how to avoid them. An investigation will also be made of the tolerances of the random errors associated with the function of the measuring equipment. Some of the possibilities of systematic errors observed in earlier measurements (and their removal) have been dealt with earlier. Consequently, only a short recapitulation will be given here.

The measurements carried out in conjunction with this work are aimed at discovering the properties of "fresh" traffic, i.e. traffic that has not been deformed by switching in establishing the connections. In order to avoid systematic errors due to the fact that the object of measurement is not identical with the intended one, special means have been introduced to avoid the deformation in question. Thus the 500-group in which the measurements took place has been so abundantly dimensioned that not only did blocking not occur, but also there was never any fear that the operation times of the switches would cause systematic displacement of the internal positions of the occupations. It has been found necessary to measure the inter-arrival times between the terminations of the occupations, thus also avoiding the disturbances of parallel tests and other phenomena such as the repeated on-off hook signalling by the subscribers. In order to obtain the necessary symmetry for the combinations of occupations, the starting allotter of the line finders has been redesigned. Finally, special attention has been devoted to the connection between the measuring equipment and the devices of the group studied. Hence no corrections regarding homing times, etc. have been needed.

Faults have in some instances appeared in the measurement equipment and in the devices belonging to the part of the telephone exchange handling the traffic studied. As a consequence, the total results from two days of measurement have not been included in the final material, nor have certain isolated results of measurements of combinations of occupation. Further, at the end of the measurement period, a fault in the interval meter occurred which was of such a kind that it was not observed until the preliminary examination of the results. By special analysis of the meter the fault in question has been exactly localized. For this reason the usable results of the interval meter do not refer to quite the same time span as the other results. To some extent this has influenced the subdivision of the measuring material, to which further reference is made in the next chapter.

The fact that traffic originates from a number of subscribers, great but not infinitely great, can have the result that the studied traffic in some respects presents undesirable properties. It can therefore be considered as a source of systematic errors. We have no reason to expect, even at constant intensity, that the traffic would be random in the sense that the *Erlang* formulas would be exactly valid, but they are in such cases replaced by considerably more complicated expressions. There is, however, no hindrance in principle to applying the theory of slow intensity variations also to these expressions. We obtain expressions, however, which are so complicated that analysis of the intensity variations will be much more difficult. It would be of advantage to disregard entirely the influence of the number of subscribers being in reality limited or at least to make do with some simple corrections.

The influence of the limited number of subscribers on the properties of random traffic arises from the fact that the probability of a new call will depend on the number of subscribers already conversing. The result will appear as a certain smoothing of the traffic peaks compared with the conditions of random traffic. There are no difficulties in developing expressions for these conditions. For a detailed study of the conditions, reference is made to an earlier paper by the author.[8] It may be mentioned that two fairly plausible assumptions can be made which lead to somewhat different results. In the first case we assume that the calls of every single subscriber are distributed at random over the part of the time when he is not conversing. It can be stated, however, that it is equally justified to assume that the need to call is distributed at random over the entire time regardless of whether the subscriber is conversing or not. During conversation there is an accumulation of the need to call, appearing as an increase of the probability of a new call immediately after the termination of the ongoing one. Both alternatives have been studied in the paper just referred to. The results show that the conditions deviate considerably from pure random traffic when the number of subscribers is small. As soon as the number of subscribers is fairly substantial the difference becomes insignificant. Some numerical calculations have shown that, for traffic on which the reported measurements were made, no appreciable influence of the limited number of subscribers can be expected. No less than 428 subscribers were connected to the 500-group in question. Table 3 at the end of this work gives a survey of the call conditions of these subscribers.

Due to the mode of operation of the measuring equipment, random errors may also occur. The task of all measuring apparatus except the interval meter is to record the total duration of different kinds of states of the traffic. This is accomplished by scanning the devices in the group and marking on counters to what extent the states in question are present. This mode of operation can undoubtedly be characterized in principle as a sampling method giving certain tolerances in the measurement results. It is now easy to estimate the tolerances with adequate accuracy. Besides, this has been the condition for venturing to use these measuring principles. The results by which the tolerances can be estimated were presented in an earlier work by the author.[10] A short account of the results will be given below.

To begin with we will deal with n occupations of a device. The device is scanned at fixed time intervals. For each scanning a counter is moved one step if the device is found busy. The product of the scanning interval and the total number of steps will then give an approximate value of the total length of the n occupations. To determine the probability that these two values will show a certain deviation, we assume that no correlation exists between the starting points of time of occupations and the scanning points of time. Further we assume that the occupation times follow an exponential distribution. Now let x be the ratio between the product of the scanning interval and the total number of counter steps, and the total length of the n occupations. According to the quoted work, the probability that this ratio has a value between x and $x + dx$ is

$$\frac{1}{\sigma\sqrt{2\pi}}\, e^{-\frac{(x-1)^2}{2\sigma^2}}\, dx \tag{165}$$

provided that n is not too small. We have then a normal distribution around the mean value 1. The standard deviation σ is dependent on the number of occupations n and the length of the scanning interval in such a way that we have

$$\sigma = \frac{\sigma_0}{\sqrt{n}} \tag{166 a}$$

and

$$\sigma_0 = \sqrt{\frac{1+e^{-a}}{1-e^{-a}}\,a-2} \tag{166 b}$$

where a is the ratio between the length of the scanning interval and the mean of the occupation times.

The detailed studies in the quoted work have shown that we should have no apprehension about limiting the usefulness of the results due to the assumption of exponential distribution of the occupation times. Further has been shown that the possible correlation between the placing of the occupations resulting in clusters in certain time spans corresponds to a decrease in the standard deviation. We would then get less, but never greater, measuring errors than those appearing in (165). Finally the result in unchanged form will be applicable also when we do not have a single device but an entire group of devices with simultaneously appearing occupations.

Naturally the results presented cannot be applied only to the occupation of devices but are also valid for measurements of all kinds of time states. Thus they can be directly applied to measurements of states and combinations of device occupations described in Chapter 11. Instead of the mean occupation time, we shall use the mean duration of the time states in question in determining α. It is of minor importance that the mean duration is only approximately estimated, as shown in the quoted work. The fact that the scanner cannot scan all devices in the group exactly at the same time does not cause any systematic errors in the measurements, as was also shown.

In order to show the application of the theorem put forward, some values will be presented from the mesurements reported in detail in the next chapter.

The total measuring time comprised 838,788 s. The traffic volume meter recorded during this time an absolute traffic volume of 12,277,353 s, giving a mean of 14.63702 traffic units. There were altogether 98,686 occupations, giving a mean occupation time of 124.41 s. The scanning interval of the traffic volume meter used was exactly 1.6 s, giving $\alpha = 0.01286$ and $\sigma_0 = 0.00525$. We obtain $\sigma = 0.0000167$ according to (166 a), signifying an extraordinary accuracy of measurement. This corresponds to a standard deviation of the absolute traffic volume of 205 s and of the mean traffic value of 0.00024. If we put the mean traffic volume at 14.637 the threefold standard deviation is less than one unit in the last figure.

In measurements with the state meter we cannot compute directly from the counter readings the mean duration of the various states. They can, however, be estimated by a comparison with random traffic conditions. It has been shown earlier[8] that for random traffic the mean of the individual lengths of the state n, i.e. the mean of the continuous time spans, where n and only n occupations occur, is

$$\frac{s}{n + sy} \qquad (167)$$

where s, as before, represents the mean occupation time and y the mean number of calls per time unit. Due to the presence of intensity variations or for other reasons the mean value in real traffic might deviate somewhat from (167). As mentioned above, the influence of such a deviation is of no practical significance for the determination of σ. We can thus, without any risk, use the value determined from (167).

As an example, we take the recording of the state 15 during the total measurement period. By means of the information above regarding mean occupation time and traffic volume we obtain the expression

$$\frac{124.41}{15 + 14.637} = 4.198$$

for the mean duration of the state according to (167). The scanning interval was exactly 2.0 s in the state measurements, giving $\alpha =$ $= 2/4.198 = 0.476$ and $\sigma_0 = 0.194$. The recording of state 15 now gave a total duration time of 71,260 s. As the mean duration of the separate state has been calculated to 4.198, the total number of states with totally 15 occupations will be 16,975. We then obtain

$$\sigma = \frac{0.194}{\sqrt{16,882}} = 0.00149$$

or a considerably poorer accuracy than in the traffic volume measurements.

Among other things, in the treatment of the measurement results, additions are made of type (139 b) for the state quantities. To determine the tolerances of values of the moments M_n obtained in this way, we might start from the errors in the different state quantities, determined in the above manner. These errors are not, however, independent of one another owing to the fact that during the measurement period the states succeed one another directly and without interruption. Starting from the errors in measurement of the individual state quantities and applying the rules of independent random variables, we therefore obtain values, much too great, of the errors of the moments M_n. Unfortunately it seems difficult as a rule to estimate the real errors of the moments. Regarding the first moment M_1, which is identical with the traffic intensity, it is easy to make a computation. When we determine the state quantities P_n, with the state meter and thereafter compute M_1 from formula (139 b), we perform in reality traffic volume measurements of exactly the same kind as the one that the traffic volume meter itself performs. The difference is only formal, since the counter recording works in different ways. Besides, the scanning period of the meter was 2.0 s as against 1.6 s for the traffic volume meter. The standard deviation of the error of M_1 will therefore be 25 % greater than in the direct traffic volume measurement. As a matter of fact the relative accuracy in the determination of M_1 becomes in spite of this more than 100 times greater than in the determination of the state quantities.

An important consequence of the smoothing of the measurement errors arising in the calculation of the moments of the state quantities is that the latter should be computed from the counter readings with a much higher accuracy than is motivated by their own tolerances.

The errors of the higher moments can probably not be determined by any method similar to that for the first moment. It is clear, however, that the relative accuracy of the higher moments should be less than that of the first moment, although it may be supposed that it is still of the same order of magnitude.

Also regarding the measurement of the combinations of occupations we can apply, finally, the formulas mentioned above for measuring errors. In these measurements, however, the devices concerned are not scanned the whole time but only when all are busy. The recordings take place at points of time which are fixed by the time source and not connected with the start and termination points of time of the combinations of occupations. We have thus in principle a scanning count where the intervals (as indicated in Chapter 11) are 0.2 s for the occupations of 3, 4, 6 and 9 devices and 0.1 s for occupations of 12 devices. The mean duration between the point of time when n specific devices become busy and the point of time when any of the occupations terminates is s/n. That can easily be proved, assuming that the occupations follow an exponential distribution with the mean value s. This assumption, however, is not necessary. We can derive the same result from formulas valid for an arbitrary distribution function of the occupation times, which were presented by the author in an earlier work.[9]

As an example of the use, the results obtained for six devices during the total measurement period can be chosen. In this case we get, using the previously adopted value,

$$\alpha = \frac{0{,}2 \cdot 6}{124{,}41} = 0{,}00965$$

which gives $\sigma_0 = 0.00394$. During the measurement period the total recording time for such combinations was 67,096 s. Since the mean duration for each combination will be 20.74 s the total number of combinations of occupations will be 3,235. From this we obtain

$$\sigma = \frac{0{,}00394}{\sqrt{3\,235}} = 0{,}000069$$

The relative accuracy is thus somewhat lower than in the traffic volume measurement.

Also the errors appearing in the interval measurement depend on the uncertainty of the time determination, but are of a quite dif-

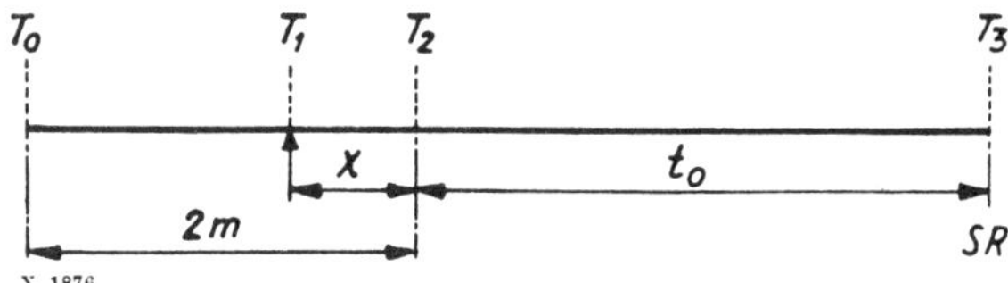

Fig. 24. Time diagram explaining the influence of the uncertainty domain in the time determination of the interval meter.

ferent kind from those studied above. According to the description in Chapter 11 of the interval meter and the time source, there is always a period of uncertainty of 0.040 s, such that the interval meter does not take into consideration at what moment during this period that an occupation impulse terminates. This is illustrated in Fig. 24. At the point of time T_1 an impulse terminates but the switch AV starts only after a time x at the point of time T_2. (The switch AV performs the interval measurements.) If no new impulse arrives during the entire time $x + t_0$, a recording is made on the counter SR at the point of time T_3. The sequence described can now take place for all values of x between 0 and $2m$. If $x > 2m$ the switch AV starts at an earlier point of time. The time $2m$ between T_0 and T_2 indicates the length of the period of uncertainty. Let now $f(x)\,dx$ be the probability that the distance T_1 to T_2 has the value x. If $\varphi(t)$ is the distribution function of the interval lengths, then $\varphi(x + t_0)$ gives the probability that no new impulse appears during the time T_1 to T_3. Then

$$\varphi(x + t_0)\, f(x)\, dx$$

is the part of the total number of intervals when SR makes a step and the distance between T_1 and T_2 thereby has the value x. It becomes clear that

$$\int_{x=0}^{2m} \varphi(x + t_0)\, f(x)\, dx \tag{168}$$

gives the part of the total number of intervals when SR moves a step regardless of the occurrence of T_1 within this uncertainty period. Obviously (168) is the distribution function determined by the measurements. If it is not identical with $\varphi(t)$, a systematic error appears, due to the fact that the measurement apparatus measures a quantity other than the desired one.

As the impulses arrive independently of the points of time T_1 and T_2, which are fixed by the time source, the probability must be equal for all x-values between 0 and $2m$. Therefore $f(x)$ must be a constant determined by

$$\int_0^{2m} f(x)\, dx = 1$$

from which we get $f(x) = 1/2m$. The mean of the variable x then becomes

$$\int_0^{2m} x\, f(x)\, dx = m$$

In order to study (168), we suitably introduce a series expansion

$$\varphi(t_0 + x) = \sum_{\sigma=0}^{\infty} \varphi^{(\sigma)}(t_0 + m) \frac{(x-m)^\sigma}{\sigma!}$$

and (168) becomes

$$\sum_{\sigma=0}^{\infty} \varphi^{(\sigma)}(t_0 + m) \int_0^{2m} \frac{(x-m)^\sigma}{\sigma!} f(x)\, dx$$

or, since $f(x) = 1/2m$,

$$\sum_{\varrho=0}^{\infty} \varphi^{(2\varrho)}(t_0 + m) \frac{m^{2\varrho}}{(2\varrho + 1)!}$$

or

$$\varphi(t_0 + m)\left\{1 + \frac{m^2}{6} \frac{\varphi''(t_0 + m)}{\varphi(t_0 + m)} + \right.$$

$$\left. + \frac{m^4}{120} \frac{\varphi''''(t_0 + m)}{\varphi(t_0 + m)} + \ldots \right\}$$

The distribution function from the interval meter is equal to $\varphi(t_0 + m)$ times a correction factor. To estimate its magnitude we can assume that $\varphi(t) = e^{-yt}$, which is true with good approximation. We then obtain from (168)

$$\varphi(t_0 + m)\left\{1 + \frac{(y\,m)^2}{6} + \frac{(y\,m)^4}{120} + \ldots\right\}$$

In these measurements the uncertainty period was 0.040 s, and thus $m = 0.020$. Furthermore in these measurements y was always less than 0.2 so that the maximum value of the correction factor will be 1.000003. This means that the systematic error caused by the uncertainty of the time determination is so small that it cannot influence the final results. Regarding the change of the argument from t_0 to $t_0 + m$ this has already been taken into consideration in Chapter 11 in the determination of the recording times of the counters.

Furthermore the magnitude is of interest of the random errors that may arise in measuring a limited number of intervals due to irregular distribution of the placing of the impulses within the uncertainty period. It will be sufficient to study the error of the mean of the variable x. The variance around the mean m is

$$\int_0^{2m} (x-m)^2 f(x)\,dx = \frac{m^2}{3}$$

and the standard deviation is

$$m / \sqrt{3}.$$

In a measurement comprising p intervals, the standard deviation of the mean becomes

$$m / \sqrt{3\,p}.$$

As for other reasons we cannot determine the recording times with greater accuracy than some milliseconds, it is clear that the influence of the random errors can be disregarded at least in all more extensive measurements.

The values calculated numerically above refer to a length of the uncertainty period of 40 ms. As has been remarked in Chapter 11, we should in reality count on a somewhat longer uncertainty period, around 55 ms, whereby allowance has been made also for the fact that the impulse times of the sequence switches exhibit certain variations. As can easily be seen, with a somewhat longer uncertainty period we also obtain entirely reliable values of the standard deviation.

Chapter 14

Measurement results and conclusions

The line finder group whose traffic conditions were subject to study in the reported measurements had, as mentioned, a capacity of 500 subscriber lines. During the measurements 428 subscriber lines were connected. The other 72 connection points formed a spare group (greater than usual). As a close relation probably exists between the general statistical character of the traffic and the structure of the corresponding subscriber population, as regards subscriber class and call volume, information of this kind concerning subscriber population has been compiled in Table 3 as guidance for comparisons. (All tables and figures quoted in this chapter will be found in the Appendix). It should be remarked in connection with the table that subscribers with at most 10,000 calls/year are divided into seven tariff classes whose boundaries are presented in the table. We obtain on each subscriber meter a recording of each connected and answered outgoing call. "Busy" and "no answer" are not counted. Calls, however, answered by a PABX operator, or an operator of a manual suburban exchange, are recorded even if the wanted connection has not been obtained. It should further be remarked that, when setting the tariffs, only a certain percentage of the recorded calls will be counted.

It is clear from the table that we may consider the traffic studied as normal traffic from residential subscribers in a metropolitan area with a certain element of business traffic. The mean number, 1971, of recordings per year and subscriber line can serve as an overall characterization of the subscriber population.

After having rejected the measurement material that could not be used due to faults in the measuring equipment or in the switching equipment in the relevant part of the exchange, the usable material comprised a total measurement period of 233.5 hours from measurements during 33 days in July and August 1941. Measurements were made on week-days, though not on days before Sundays and public holidays. During the first 14 days, measurements were made in the period 10 a.m. to 4.30 p.m. During the other 19 days from 9 a.m. to 4.30 p.m. It might perhaps be questioned why measurements were not made exclusively in the busy hours, as it is generally only the traffic during this time that is assigned decisive importance in computation of the quantities of equipment required. This condition is of no importance, however, in the present studies, which have only the character of the traffic as their final goal. There is no reason to presume that this character – disregarding the magnitude of the traffic volume – during the different hours of the day would be so fundamentally different that, during certain periods, it would show to a higher degree the basic properties of traffic with intensity variations. By extension of the measurements over several hours of the day, some advantages were obtained. We could thus, during a rather short period of time, obtain a rather extensive and, in respect of the variation of the traffic volume, fairly inhomogeneous material. The fact that the measurements were made partly during the low traffic month of July and partly during the more intense traffic month of August made it possible to obtain a further variation of the traffic volumes without reconnection of the measurement equipment to other groups.

As mentioned above, the readings of the counters were made by photographing the counter rack, as a rule at 10 and 11 a.m. and at 4.30 p.m. each day. In this way specific readings for the generally busy hours 10 to 11 a.m. were obtained. After having transferred the values obtained to tables, a preliminary treatment of the material was undertaken by calculating the recordings during the intervals between subsequent photographing times. The numerical material thus obtained was called *the primary material.* Thus this indicates for every day of measurement the total number of recordings for every counter having occurred partly during the period 9 – 10 a.m. (for the days when the equipment was connected during this time), and partly during the period 10 – 11 a.m., and partly during the period 11 a.m. – 4.30 p.m. As the primary material is very extensive, it has not been considered necessary to publish it in extenso in the present work.

In order to get guidance for evaluation of the necessary grouping of the primary material, certain mean values were calculated for different *traffic samples*. By traffic samples in what follows is meant the units into which the primary material is divided. The traffic 9 to 10 a.m., 10 to 11 a.m., and 11 a.m. to 4.30 p.m. of every day of measurement thus each constitutes one traffic sample. The results of these calculations are shown in Table 4, giving the total measurement period, the traffic volume, the number of calls, and the mean occupation time for every traffic sample. The table gives an overall idea of the general magnitude of the different traffic values during the measurement period. The table also shows the numbering of the traffic samples, i.e. they have been given three digit numbers. The first digit represents the time of measurement during the day, the second digit the week, and the third the day of the week.

The digits are as follows:

Time of day	First digit
9 a.m. – 10 a.m.	1
10 a.m. – 11 a.m.	2
11 a.m. – 4.30 p.m.	3

Week		Second digit
1	7 – 11 July	1
2	14 – 18 July	2
3	21 – 25 July	3
4	4 – 8 August	4
5	11 – 15 August	5
6	18 – 22 August	6
7	25 – 29 August	7

Day of week	Third digit
Monday	1
Tuesday	2
Wednesday	3
Thursday	4
Friday	5

For instance, traffic sample 364 means the traffic during the time 11 a.m. – 4.30 p.m. on Thursday, August 21, 1941.

Principles for the grouping of the material

Theoretical traffic, in the sense in which the expression has been used in the foregoing, always corresponds to an ongoing process during an infinite period. By measurements we can only determine the traffic conditions in a device group during a limited period. Measurements are not, as a rule, carried out continuously over a lengthy period, but we let each measurement comprise only the traffic conditions during a short period, e.g. a busy hour. If a number of such measurements are made during short periods in the same group of devices, we usually say that the measurements refer to one and the same traffic. It is difficult, however, to define clearly the meaning of this traffic concept. In order to avoid misunderstandings, we shall in this work use the expression *measured traffic* to describe the measured traffic conditions during a number of periods when measurements have been made in one and the same group of devices. The aforementioned traffic samples are obviously special cases of measured traffic, although the expression traffic sample is used here only to represent the traffic conditions during a continuous measurement period. The measured traffic studied in the following is obtained by compounding a great number of traffic samples. The expression *compounding of traffic samples* used here means that the results of measurements of the different traffic samples are treated as if they belong to one and the same theoretical traffic. It should be observed that the term compounding in this sense has nothing to do with superposing of traffic flows. By superposing we mean a kind of combination of several traffic flows occurring at the same time in different groups. By compounding of traffic samples in the sense mentioned, we can therefore, for the sake of clarity, speak of *compounding in time-sequence,* whereby is clearly understood that it is not a question of simultaneous traffic flows.

As the measurements comprise a considerable number of traffic samples, the compounding can be performed in many different ways. A question of great importance is how to select the measured traffic that should be compounded and studied, having regard to the intention of the measurements. It might be sufficient in a case like this, where all measurements have been made in the same unchanged device group, to consider only the measured traffic obtained from all traffic samples, i.e. the total measurement result. In this way the measured traffic is also obtained which contains the largest possible measurement material. Considering that the measurements undertaken have been intended to permit a rather extensive examination of the extent to which the theoretical results could be applied to real telephone traffic, it is desirable to extend the study to several cases of measured traffic which should as far as possible have differing characters. As guidance for determining the compounding of the selected cases of measured traffic, arguments used in the foregoing, to justify the introduction of the theoretical assumptions valid for traffic with slow variations and the different normal forms, have been taken into account.

In the compounding of measured traffic from the traffic samples, we can follow two different principles which might be called *compounding in regard to the place in time* and *compounding in regard to the magnitude of the mean traffic values.* In the first case we select the traffic samples that are placed in a certain way, i.e. occur within certain times of the day or on certain week-days. In the other case we select traffic with certain mean values. As an example of compounding according to placing in time, the measured traffic can be considered which is obtained by compounding all traffic samples during the time 10 – 11 a.m. all days of measurement. As an example of compounding according to the magnitude of the mean traffic values, the measured traffic can be considered which is obtained by compounding all the traffic samples whose mean traffic value lies between the limits 12.0 and 12.2 traffic units.

We have now selected from the total measurement material in all 14 cases of measured traffic. Their constituent samples are presened below. The motivations given at the same time show also the application of the principles of compounding according to placing in time or to the magnitude of the mean traffic value. In the selection special consideration has been given to the fact that the selected cases of measured traffic as a matter of course should preferably have differing characters and not contain too many traffic samples in common. It should be remarked that every measured traffic should be compounded of a great many traffic samples, thus containing a rather voluminous measurement material, otherwise it will be difficult to obtain unequivocal results of the influence of the intensity variations. Thus it has turned out that it will be of little use to study measured traffic which does not contain at least around 10,000 occupations.

Measured traffic 1 is compounded of all traffic samples, thus containing the total measurement material. The total measurement time is 233.5 hours with a total of 98,686 calls. As mentioned previously, at the end of the measurement period a fault in the interval meter occurred which was not discovered until the treatment of the results. Because of this fault the results from five days, i.e. August 21, 22, 25, 26, and 27, could not be used to determine the distribution of the inter-arrival times. We have therefore been forced in the treatment to consider a special measured traffic somewhat different from measured traffic 1, namely

Measured traffic 2. This is compounded of all traffic samples except those occurring during the five days mentioned, comprising in all 196 hours of measurement and 79,276 calls. From Table 4 it is apparent that the traffic volumes of the traffic samples as a rule have a tendency to increase from low values at the beginning of July to high values at the end of August. Weeks 2, 3, 4, and 5 show, however, almost equally large traffic volumes. As it was found to be of interest to study also a rather homogeneous material, a

Measured traffic 3 was formed, comprising all traffic samples occurring during weeks 2, 3, 4, and 5. The measured traffic comprises 140 hours and 55,677 calls. A less homogeneous, but from certain points of view interesting material has been compiled in

Measured traffic 4, comprising all traffic samples of one hour's length, except those occurring on August 21, 22, 25, 26, and 27. The traffic comprises 42 hours of measurement and 19,771 calls.

As it proved desirable – without drawing upon the compounding based on the magnitude of the mean traffic values – to obtain a still more homogeneous measured traffic than the above-mentioned number 3,

Measured traffic 5 has further been studied. This traffic comprises the traffic samples 11 a.m. – 4.30 p.m. during weeks 2, 3, 4, and 5, in all 110 hours of measurement and 41,674 calls.

The five cases of measured traffic mentioned above have been compounded of traffic samples chosen with regard to their occurrence in time (i.e. their placing in time). Such measured traffic can be considered as representative of the type of material that is at our disposal as a basis for practical calculations. However, there is also a great interest in measured traffic obtained from a material arranged according to the magnitude of the mean traffic values. For selection of such measured traffic, all traffic samples have been sorted according to increasing values of the traffic intensity and the call intensity, i.e. according to the magnitude partly of A, partly of y in Table 4. A similar division according to increasing values of the mean occupation time s has not been considered worth while. By studying the diagrams obtained in this way, the following cases of measured traffic have been chosen:

Measured traffic 6. This comprises 13 traffic samples of one hour each, the traffic intensities of which lie between 16.1 and 17.0. This measured traffic is thus considerably more homogeneous than the measured traffic 4, having traffic samples also of one hour's length but whose A-values vary between 12 and 25. By contrast, measured traffic 6 comprises a rather small amount of material, only 6,285 calls, which would seem to be on the low side considering the purpose of the treatment. It was, however, deemed suitable to include such a small measured traffic to indicate the conditions in such a case.

Measured traffic 7 comprises 26 traffic samples of one hour each, the traffic values A being between 16 and 18, thus showing a greater variation than in measured traffic 6. Measured traffic 7 comprises 12,335 calls. Among the traffic samples also those during which the interval meter was defective have been included. An investigation of the inter-arrival times has therefore not been possible. In spite of this fact measured traffic 7 has been considered to be of sufficient interest to justify its inclusion.

Measured traffic 8 has been selected only among traffic samples with a length of 5.5 hours, comprising 5 such traffic samples. The traffic intensities A lie between 12.01 and 12.21, showing a very strong concentration. Measured traffic 8 comprises 10,022 calls.

Measured traffic 9 was also selected among only traffic samples of 5.5 hours' length, the A-values of which, however, have a greater dispersion than in the foregoing case, as they lie between 11.16 and 12.95. The measured traffic 9 comprises 82.5 hours and 29,888 calls.

Measured traffic 10 is of the same size and dispersion as measured traffic 9, but comprises traffic samples with a somewhat higher average of A-values. It contains in all 13 traffic samples of 5.5 hours and with A-values between 12.01 and 13.44. The number of calls is 26,832.

Measured traffic 11 has, in contrast to the foregoing, been selected according to the mean number of calls y of the included traffic samples. It comprises 17 traffic samples of 5.5 hours each, and with y-values between 0.0990 and 0.1095. The number of calls is 35,084.

Measured traffic 12 was also selected according to the magnitude of the y-values and is part of measured traffic 11. It comprises 11 traffic samples of 5.5 hours each, with y-values between 0.0990 and 0.1053. The number of calls is 22,216.

Measured traffic 13 has been included to permit comparisons together with measured traffic 1. The latter namely includes relatively few traffic samples with very high A-values. These have been excluded in the forming of measured traffic 13, which then comprises all traffic samples with A-values smaller than 18.12 (the lowest being 11.16). Measured traffic 13 contains 218.5 hours of measurement and 90,093 calls. Like measured traffic 1 and 7, it also includes traffic samples during which the interval meter was faulty. The inter-arrival times have not, therefore, been studied.

Measured traffic 14 is also formed with regard to the magnitude of the A-values but in this case another principle has been applied, as a number of traffic samples with very small and with very great A-values have been brought together. Traffic samples with one hour's duration, as well as 5.5 hours, have been included. The A-values are partly between 11.16 and 11.84, partly between 16.50 and 25.15. Measured traffic 14 comprises 62 hours of measurement and 25,837 calls.

Table 5 contains particulars of the traffic samples of the different cases of measured traffic. Table 6 shows the main data of the different cases of measured traffic, i.e. A, y, and s. It also comprises the standard deviations, computed from the corresponding traffic samples, of the three quantities mentioned. It should be observed that the computed standard deviations concern the dispersion of the means of traffic samples, in certain cases with one hour's length and in other cases with 5.5 hours' length, depending on the kind of traffic samples included in the measured traffic in question. In the measured traffic 1, 2, 3, 13, and 14, traffic samples of one hour's length as well as such of 5.5 hours' are included. In these cases the standard deviation has been computed as follows. Assume that in a measured traffic the following traffic samples are included: n_1 traffic samples, each of one hour's length and with the A-values $A_{1,1}$, $A_{1,2}$, ... A_{1,n_1} and n_2 traffic samples of 5.5 hours' length each with the A-values $A_{2,1}$, $A_{2,2}$, ... A_{2,n_2}. The mean traffic intensity is A_m. The standard deviation has then been calculated from the square root of the following expression:

$$\frac{\sum_{\sigma=1}^{n_1} A_{1,\sigma}^2 + 5{,}5 \sum_{\sigma=1}^{n_2} A_{2,\sigma}^2}{(n_1 + 5{,}5\, n_2)\, A_m^2} - 1 \qquad (169)$$

As is apparent from this formula, the standard deviations in the table are relative, i.e. they express the dispersion as a fraction of the mean. In the table we find also the corresponding values of the standard deviations valid theoretically for random traffic. They are obtained from the formulas at the end of Chapter 12. According to these, the relative standard deviation of the traffic volume becomes

$$\sqrt{2/R}$$

and of the number of calls and the mean occupation time

$$\sqrt{1/R}.$$

Here $R = yT$ where T represents the length of every included traffic sample. In the calculations we have put y equal to the mean number of calls of the measured traffic. The length of the traffic samples has been 1, and 5.5 hours respectively. In measured traffic including traffic samples of 1 as well as 5.5 hours' duration, the variance has been calculated from the following expression analogous to (169)

$$\frac{1}{3\,600\, y} \frac{n_1 + n_2}{n_1 + 5{,}5\, n_2} \qquad (170)$$

where the measure of y is calls/s. This expression is valid for the computation of the standard deviation of the call volumes and of the mean occupation times. In a computation of the traffic volumes, (170) must be multiplied by 2.

Table 6 now shows that in all cases of measured traffic which have been selected according to the placing in time of the traffic samples, the dispersions computed from the mean values of the traffic samples are much greater than those computed from the formulas for random traffic. This undoubtedly shows that the traffic in question has intensity variations. For cases of measured traffic which have been selected according to the mean values of the traffic samples, the conditions are somewhat different. In those cases of measured traffic that show a strong concentration of the A-values of the included traffic samples, the measured dispersion is only slightly greater than the theoretical or, as for measured traffic 6, 7, and 8, less. This is quite natural, as the random dispersion as well as that caused by the intensity variation must be cancelled in part by the concentration arising from the "unnatural" selection. It is interesting that a certain, but less striking influence also shows up in the dispersion of y and s. For measured traffic 11 and 12, selected with reference to the y-values of the traffic samples, a strong influence is apparent on the dispersion of the y-value and a weaker influence on the dispersion of the A-value.

The condition mentioned, characteristic of all cases of measured traffic composed in an "unnatural" way, will be further elucidated in the discussion of the measurement results.

Evaluation of states and moments

In Chapter 12 several methods were presented for the study of the measurement material in different respects. As a rule it was not possible, however, to form an opinion with any certainty as to the extent to which the proposed evaluation methods could be used, i.e. to which they give sufficiently accurate results. To get an idea of these conditions, the application of the methods to the measured traffic formed by the total measurement material will be discussed in detail below. Regarding other cases of measured traffic, the treatment was concentrated on determination of the parameter η_0 of normal form 1 and to examination of the feasibility of this normal form in various cases. This is also most important from a practical point of view.

To determine the states P_n of measured traffic 1, the recordings of the respective state counters were added up for all traffic samples. For each value of $n = 0, 1, 2, \ldots 36$, we then obtain a number indicating how many times the state in question has been recorded during the measurement period. The total number of recordings of all states was 419,408. If we divide the recordings for the different n-values by the total number of recordings, we obtain an expression of the fraction of the total measurement period in which the different states occurred, i.e. we obtain directly P_n. As a check we have the condition that the sum of all P_n equals 1. The first columns of Table 12 show the values obtained in this way for measured traffic 1.

The next step concerned the calculations of the moments from the resulting values of P_n according to the formula (139 b) (a rather laborious computation). We then obtain $M_1 = 14.6420$ indicating also the mean value A of the traffic intensity in accordance with the discussion above. From the traffic volume meter we obtain, however, $A = 14.6365$. The difference between these values is greater than might be expected from an estimate of the accuracy based on the formulas in Chapter 13. The reason was found to be that on one day of measurement, July 22, the traffic volume meter had been faulty. After excluding the values from that day the difference between the traffic volumes from the state meter and the traffic volume meter was only 0.0001. Due to the fault in the traffic volume meter, we accepted the above-mentioned value of M_1 as a correct value of A.

From the values of the moments M_n, thus obtained, the relative moments μ_n were calculated by means of formula (153 a). In Table 7 we find the values of the relative moments up to $n = 12$. As expected, they form an increasing series of numbers, indicating immediately that the measured traffic in question has no obvious random character. For random traffic all relative moments equal 1. In order to establish whether the divergence is consistent with the prerequisites for random traffic, modified by slow intensity variations, we might endeavour to find out to what extent the criteria in Chapters 10 and 12 are met for the moments. First we may test the condition (153 b). That is equivalent to the expressions

$$\frac{1}{n-r}\{\log \mu_n - \log \mu_r\} \tag{171}$$

forming an increasing series of numbers for every r with increasing n. The expressions (171) have, therefore, been calculated in Table 7. We find that the conditions for all moments at $r = 0, 1, 2$, and 3 have been met. For $r = 4$ a decrease in the 12th moment occured, as also for $r = 5$. For higher r-values the decrease starts already in the 11th moment. Considering that the conditions (153 b) very narrowly limit the permitted variations in the moment values, the results in Table 7 may be regarded as surprisingly satisfactory. The tendency at higher n and r to a decrease of the values of (171) is not unexpected (meaning that the moments are smaller than they should be according to the assumptions of the theory of slow intensity variations). According to the discussion in Chapter 12, p. 139, a decrease of this kind in the values of the moments would be an unavoidable consequence of a limited measurement period. This does not explain, though, the entire decrease in the moment values. We shall show below that very likely other causes exist as well. This is also why it has not been regarded as worth while to compute higher moments than those presented in the table.

The studied relations (153 b) constitute necessary, though not sufficient, conditions for the moments. We should therefore also study the conditions (148 b) of the *Hankel* determinants. As mentioned in Chapter 12, the absolute moments might be replaced by the relative ones. Further, according to the remarks in Chapter 10, p. 111, the expression

$$(\mu_r, \mu_{r+1}, \ldots \mu_{r+2n}) > 0 \qquad (172\text{ a})$$

must be met for all values of r and n. For $n = 1$ this condition becomes

$$\mu_r \mu_{r+2} > \mu_{r+1}^2 \qquad (172\text{ b}),$$

being, however, the same inequality as the one in (153 b). It can be shown from Table 7 that this condition has been met in respect of the measured moments.

It would seem, however, that for higher n-values, the expression (172 a) gives considerably narrower limits than (153 b). On numerical computation of the *Hankel* determinants, formed from the moments, it now turns out that for higher n-values these become very small. To arrive at results forming a basis for an estimate, we must determine the moments to at least six decimals. With regard to the estimated accuracy of measurement, it might now seem questionable whether there is any firm justification for the last figures. The computations that have been made seem to indicate, however, that that is the case, at least in respect of the lower moments, i.e. we find the following values for the left-hand member in (172 a):

$+ 0.000062$ for $r = 0$ and $n = 2$
$+ 0.000037$ for $r = 1$ and $n = 2$
$- 0.000008$ for $r = 2$ and $n = 2$
$- 0.000001$ for $r = 0$ and $n = 3$

Larger r- and n-values result as a rule in negative values, which in most cases are numerically considerably greater and vary irregularly in magnitude. The fact that negative values are obtained is obviously due to the systematic errors mentioned, causing too low values of the moments for higher n-values. The aforesaid values in themselves, however, suggest a considerable accuracy of the lower moments and may also be regarded as providing some support for the occurrence of properties of the traffic to be expected according to the theory of slow intensity variations.

As the properties of the *Hankel* determinants are closely connected to the difference values of the elements, it has been considered justified to include a table of the differences of the relative moments, Table 8. From the fifth difference on, values with irregular steps occur throughout.

Obviously, an investigation of the moments of the above kind can give results of value only if the measured traffic comprises very extensive measuring material. This has therefore been carried out only in respect of measured traffic 1. The other cases of measured traffic of similar magnitude differ relatively little from measured traffic 1, apparently giving no differing results of interest. However, the results obtained above for measured traffic 1 are considered to be of great interest. They have shown that the properties, which should generally characterize random traffic modified by slow intensity variations, also appear in reality insofar as the accuracy of measurement allows a verification.

The next phase of the treatment consists in the adaptation to normal form 1. The task is then to investigate whether the relative moments can be represented in the form (154 a) and, if so, what value η_0 obtains. In the following the time unit is chosen so that $y = 1$. Hereby η_0 becomes a dimensionless quantity, as is also κ. To compute η_0 from the relative moments, the expressions (154 b) and (154 c) are at our disposal. Table 9 shows the values obtained in this way. Apparently, the higher the moments that have been used in the computations, the lower the values of η_0

become. Up to the fifth moment the differences are, however, very small and we realize from their magnitude that it is justified to state η_0 to four decimals. The differences for the higher moments become considerable. The η_0 obtained from μ_{12}/μ_{11} is only about half of the first η_0-value. This decrease in the η_0-values is a consequence of the above-mentioned condition, viz. that the moment values of higher n are less than they should be according to the theory of slow variations. Regarding the difference between η_0-values obtained from μ_n/μ_{n-r} for $r = 1$ and $r = 2$, it turns out that the latter as expected represents a mean value of the former. Thus the η_0 obtained from μ_n/μ_{n-2} is, with very good accuracy, the geometrical mean of the η_0 obtained from μ_n/μ_{n-1} and μ_{n-1}/μ_{n-2}.

We can also calculate η_0 directly from the higher moments according to (154 a), involving unfortunately an awkward trial-and-error procedure. In some preliminary computations it has turned out that, from μ_{10}, $\eta_0 = 0.028$ is obtained, from μ_{11}, $\eta_0 = 0.027$ and from μ_{12}, finally, $\eta_0 = 0.025$.

In Chapter 12 a special kind of a difference scheme has also been shown, (154 e), valid for normal form 1 and indicating several possibilities for computing η_0. Table 10 shows the result of this scheme applied to the measured moments. It is obvious that in this case already the third differences become negative (except the first one). The computation of η_0 from the first differences evidently gives the same values as the computation presented above from μ_n/μ_{n-1}. The η_0-values which come from the second differences will be found in the last column of Table 9. They are throughout lower and decrease more rapidly than those previously obtained. Moreover, as is clear from the values in Table 10, we cannot expect equally good accuracy from the latter determination as from the former. The difference scheme (154 e), then, seems to be of little practical use.

Fig. 25 shows graphically the results of the η_0-calculations in Table 9 according to the three different methods.

In an endeavour to adapt the measurement results to normal form 1 a systematic displacement of the η_0-values appears owing to the fact that the moments decrease more and more in relation to the values they should have according to normal form 1. There is reason to believe that this decrease can be explained by the systematic measurement errors due to the limitation of the material. It is not certain, however, that this is the cause of the total observed decrease. There is every reason to study how an adaptation to normal form 2 will turn out. It seems likely that in normal form 2 a phenomenon of exactly the observed kind would appear. To form an idea of the numerical conditions, the equations (155 a) in Chapter 12 are considered first. These equations give for the measured values of μ_2, μ_3 and μ_4.

$$\eta_0 + \varkappa = 0{,}034314$$
$$\eta_0 \varkappa = 0{,}000416$$
$$\eta_0^2 \varkappa = 0{,}000600$$

Obviously the equations for the sum and the product of η_0 and κ have no real roots. The maximum value which can be obtained for the product $\eta_0\kappa$ is 0.000294 occurring for $\eta_0 = \kappa = 0.017157$. Hereby $\eta_0^2\kappa = 0.000005$. Already this should prove that it is of no avail to determine η_0 and κ by a deeper analysis, for instance by applying the formula (155 b). It is interesting, however, to form an idea as to the extent of the observed decrease of the moment values that at the best can be explained by normal form 2. As has been shown previously, the greatest difference between normal forms 1 and 2 at constant $\eta_0 + \kappa$ seems to occur around $\eta_0 = \kappa$. For $\eta_0 = \kappa = 0.017157$ some moment values according to normal form 2 were calculated. The results will be found in Table 11, comprising also a comparison between the measured moments and those calculated from normal form 1 at

$\eta_0 = 0.034314$. Clearly, by assuming that normal form 2 is valid, we can explain a not insignificant part of the observed deviation from normal form 1.

The next phase in the evaluation of the findings concerned the study regarding the extent to which the measured state quantities P_n (Table 12) can be adapted directly to normal form 1. The state quantities $P_{n,0}$, theoretically applicable with random traffic and the same traffic intensity as for measured traffic 1, were therefore first calculated. Table 12 contains the values of the resulting quantities $P_{n,0}$, $P_n/P_{n,0}$ and $\log P_n/P_{n,0}$. Clearly $P_n/P_{n,0}$ has exactly the behaviour that would be expected from the discussion in Chapter 6 of random traffic with slow intensity variations. For small n-values the ratio is very large, decreasing for n-values in the neighbourhood of the mean traffic value to somewhat below 1. For large n-values the ratio $P_n/P_{n,0}$ increases again to around 17 for $n = 32$. For still higher n-values up to $n = 36$ (the number of devices in the group), the ratio decreases to zero. Evidently for sufficiently high n-values P_n must be zero, as arbitrarily high states cannot appear during a limited measurement period. The fact that the state 36 did not appear at all during the measurement period further shows that the aims as set out in the planning of the measurements have been fulfilled, i.e. that the group should comprise so many devices that blocking would not occur during the measurement period. Of special interest, however, is that the P_n-values for $n = 33$, 34, and 35 are noticeably smaller than would be expected. We can here assume an influence of the limited number of devices in the group together with the operating times of the line finders. If only a small number of the line finders are free, the hunting for the calling subscriber line will take longer than normally (see the description of the functioning of the switching machine in Chapter 11). This will reduce the appearance of states with only a minority of free line finders. Strong support for the soundness of this explanation of the observed phenomenon is obtained by comparing the conditions of different cases of measured traffic. A corresponding decrease of P_n at high n-values appears for measured traffic with A-values of the size that recordings of n appear in the vicinity of 36. On the contrary, this phenomenon is hardly noticeable with measured traffic for which no recordings of states higher than about 30 occur. The condition mentioned regarding the measured values of the higher states is obviously to be regarded as a systematic error caused by too few devices in the group. Due to the strongly varying traffic intensity, it is always difficult to form a correct opinion in advance as to the number of devices needed to avoid such faults. Evidently, in this systematic error, we have one of the reasons for the decrease in the moment values observed earlier.

In connection with the further discussion of Table 12 we should say something about the principles applied for determining the number of digits included in the summary tables. The numerical calculations have always been carried out with a higher accuracy than seems to be warranted by the estimated accuracy of measurement. This ensures that no unnecessary errors are introduced by the arithmetical treatment itself. The values obtained in this way have then been inserted in the tables in an abbreviated form. Consideration has been paid to the approximate accuracy of measurement estimated in Chapter 13 and also to the mutual variations of the values in question. The latter is necessary especially for quantities whose calculation from the primary material is so complicated that it becomes difficult to understand the connection between the accuracy of the starting value and the final value. Furthermore, as a principle, in doubtful cases one digit too many has been given instead of one too few.

For each of the state values P_n derived from the measurement material, we have computed the η_0 corresponding to normal form 1. The computations have generally been carried out by a trial-and-error method using the relation (157 c), a rather tedious work. For the lowest n-values it turned out to be easiest to start from the explicit expression (140 a) for P_n. The η_0-values obtained are presented in Table 12. For some isolated values of n, no acceptable values of η_0 were obtained; in the table the corresponding places have been left out. The η_0-values for different n-values of course present mutual deviations, but as a rule these seem to be less than might be expected. An idea of the nature of the deviations can be obtained from Fig. 26, showing η_0 as a function of n (the continuous curve). The curve seems to indicate that as a rule systematic deviations occur, the influence of random measurement errors being rather small. The strange behaviour of the curve around $n = 11$ and $n = 20$ is of special interest. The reason for this is easily realized. In the vicinity of these points the ratio $P_n/P_{n,0}$ passes the value 1. For normal form 1 very small deviations of the ratio from the value 1 near such points correspond to considerable changes of η_0. Also a very small difference between the measured P_n-curve and the curve computed from normal form 1, within these areas therefore causes considerable changes in the computed η_0-values. The curve of these values can be expected to behave as if there were asymptotes in both points mentioned. By putting the right-hand member of (158) equal to 1 we can evidently obtain an approximate value for the position of these two *limit points*. We then obtain

$$\left.\begin{matrix} A_1 \\ A_2 \end{matrix}\right\} = A + \frac{1}{2} \mp \sqrt{A + \frac{1}{4}} \tag{173}$$

where A_1 is the low and A_2 the high limit point. In measured traffic 1, A_1 becomes 11.3, and A_2 19.0. In Fig. 26 the limit points and the traffic intensity A are indicated by arrows.

If we wish to study the accuracy with which the P_n-curve obtained from the measurements can be produced from normal form 1, using a common η_0-value, it is appropriate to calculate a mean of the different η_0-values shown in Fig. 26. It would be desirable to use a weighted mean, whereby consideration might be paid in the determination of η_0 to the condition that normal form 1 gives different sensitivities for different n-values, especially significant in the vicinity of the limit points. By means of the derivative of (157 c) in respect of η_0 we can form an idea of the weights that might be allotted to the different η_0-values. In determining the weights we should also take into account the accuracy of the measured P_n-values, but this is rather difficult. As pointed out in Chapter 13, we do not have any definite notion of the tolerances of the statistical variations. Finally, in determining the weights, we should consider the intended use of the required mean value. For computation of quantities of equipment, conditions for higher n-values are the most important. Clearly it would seem to be difficult to issue recommendations as to how the weighting for forming the mean value should be arrived at. In the present case, therefore, no attempts have been made in this direction; in some other cases, elaborated on below, a kind of weighting was arrived at by leaving out certain extreme values.

The mean of the η_0-values presented in Table 12, calculated from (157 c), was 0.0369. It is somewhat higher than the previous value obtained from the second moment, 0.0343. The explanation of this otherwise insignificant difference can be found in the fact that the moment value is influenced by the often mentioned decrease in P_n-values for the highest n-values. The η_0-value, formed as a mean value, is not influenced by this fact, as no η_0-values are included for $n > 33$.

As mentioned, the determination of η_0 from (157 c) by a trial-and-error method results in extremely cumbersome calculations. It is of interest to examine to what extent the convenient approximate formula (158) can be used to calculate η_0. In Table 12 the η_0-values obtained in this way are presented for the sake of comparison, and in Fig. 26 they are summarized in the curve marked by dashes. In this case we obtain a still stronger influence by the limit points A_1 and A_2. Furthermore a significant systematic variation of η_0 is obtained due to the fact that (158) is only an approximation and within certain areas a very poor one. The right-hand member in (158) is, as a matter of fact, too small for large and for small n-values, and likewise for n-values around A. Only around the limit points is the right-hand member in (158) greater than the left-hand member. The η_0-curve, computed from (158) in Fig. 26, gives a good picture of the general character of the deviations caused by these conditions. For very small and very large n-values it is obviously meaningless to make computations from (158). If we exclude the extreme values for larger and smaller n, and around the limit points, a good mean value for η_0 can be obtained from (158). Hence, from 23 of the points in Fig. 26, the mean value 0.0357 has been obtained.

In Chapter 12 some further methods have been proposed for determining η_0 from P_n, viz. those that appear from the formulas (156 a) and (156 b). An idea of the usefulness of this method is obtained from Table 12, containing η_0 computed from these two formulas. The results are summarized in the form of curves in Fig. 27. Clearly, and as would be expected, in this case we obtain a considerably stronger influence of random errors. As expected, the variations of η_0, determined from (156 b) also seem to be somewhat less than those obtained from (156 a). The tendency to systematic displacement of the η_0-values that can be traced in the curves is further of great interest. The limit points do not seem to affect this case, which as a matter of fact cannot be expected. On the other hand remarkable irregularities appear in the area round $n = A$. This can apparently be explained by the fact that (156 a) and (156 b) nearly equal 1 in this area, where, as a consequence, η_0 can be determined only with very poor accuracy.

If we exclude the unreliable values around $n = A$, a fair mean is obtained also from the points in Fig. 27. The full-line curve gives 0.0352 and the dashed curve 0.0354. The relations (156), convenient for numerical computations, would thus be applicable in determining suitable mean values of η_0.

Starting from the mean value of $\eta_0 = 0.0369$, obtained by means of (157 c), P_n has been theoretically calculated from normal form 1 to allow a comparison with the measurement result. It should be mentioned that in such computations it is appropriate to start from (157 c) and compute a P_n-value for an n in the neighbourhood of A. The other P_n-values can thereafter be conveniently computed recurrently from the ratio P_n/P_{n-1}. This method is advantageously applied also in computations of $P_{n,0}$. As a check we know that the sum of all state quantities is 1. In Table 12 the values of P_n, $P_n/P_{n,0}$, and $\log P_n/P_{n,0}$ are shown for $\eta_0 = 0.0369$. An idea of the agreement between measured and computed values is obtained from Fig. 28, showing P_n computed and measured as well as $P_{n,0}$ in graphic form. It is clear that the difference between P_n obtained by measurements and $P_{n,0}$ valid for random traffic is very considerable and, moreover, precisely of the expected form. The figure also contains the theoretically computed P_n from $\eta_0 = 0.0369$ as a dot and dash line. This coincides very well with the measured P_n in the applied scale of the figure. The only exception is in the area $n = 12$ to $n = 16$, where a minor deviation appears.

Fig. 28 gives a very fine picture of the possibility of expressing the state values obtained from the measurement by means of normal form 1. Such a presentation, however, does not give the desired accurate idea about the means of adaptation at very small and very great n-values, the P_n-values being very small within these areas. In spite of that the conditions within these areas, especially the higher one, are of great practical importance as the state values of high n-values are decisive for computation of quantities of equipment. The interest, therefore, is not so much in the difference $P_n - P_{n,0}$ as in the ratio $P_n/P_{n,0}$. In order to obtain a suitable graphic representation, it has seemed appropriate to look at $\log P_n/P_{n,0}$. Figs. 29 a and 29 b show this quantity in different scales, partly according to the measurements and partly computed from normal form 1 with $\eta_0 = 0.0369$. In judging the goodness of the fit we shall keep in mind that in this case the abscissa gives the corresponding curve of random traffic. The figure must be considered to show that, disregarding the anomaly at $n \geqq 33$, normal form 1 gives an extremely good representation of the measurement results.

In Figs. 29 a and 29 b, a dot and dash curve is also included, showing values obtained with $\eta_0 = 0.0369$ by computation using the approximate formula (158) (see also Table 12). As will be seen, this approximate formula should be applied with some caution.

Table 12 also shows computations with the η_0-value 0.0343, obtained from the second moment. Figs. 29 c and 29 d show the fit to the measurement result. It is obvious that a somewhat better agreement is obtained than for $\eta_0 = 0.0369$ in the area round $n = A$, which is compensated by a slightly poorer agreement for the higher n-values.

Finally the results obtained for measured traffic 1 from the measurements of the combinations of device occupations should be mentioned. Table 13 shows the results and relative moments μ_p computed from them according to (151 b). As a comparison the moments computed from the states in the manner previously shown are presented. It will be seen that we obtain quite good agreement. As expected, the moments computed from the device occupations do not show the decrease with increasing p, which was mentioned in the discussion of the moment values determined from the states. If we attempt to compute η_0 from the moment values obtained by the device occupations, we find, e.g. from μ_3 the value $\eta_0 = 0.0332$ and from μ_4/μ_3 the value 0.0344, thus in good agreement with the values previously obtained.

Evaluation of the distribution of inter-arrival times

A rather detailed study has been made regarding the inter-arrival times – as was the case in respect of states and moments – for the measured traffic comprising the greatest material, i.e. in this case measured traffic 2. For the other cases of measured traffic, the treatment has concerned mainly the adaptation to normal form 1. In the following a detailed account is given of the treatment of measured traffic 2.

The total number of recorded inter-arrival times (or, more correctly, inter-termination times) pertaining to measured traffic 2 was 79,276. From that and from the total measurement period, we obtain $y = 0.11264$ call/s, giving 8.878 s as the mean value of the inter-arrival times. Table 14 shows the total number of recordings during the measurement periods on the different counters of the interval meter and the corresponding recording times. The table further shows the distribution function $\varphi(t)$ of the inter-arrival times, computed from the recordings. In order to determine the difference curve, the exponential distribution function e^{-yt} has also been computed, giving thereafter the difference function $D(t) = \varphi(t) - e^{-yt}$. Clearly, $D(t)$ is numerically rather small, the maximum being only about 0.01.

To investigate whether the inter-arrival times follow a distribution compatible with the theory of intensity variations, we shall study, in accordance with the investigations in Chapters 10 and 12, the extent to which the difference quotients form an entirely monotone series of numbers. For this task, a number of difference schemes have been formed from $\varphi(t)$ based on different time intervals. The scheme shown in Table 15 starts from the values of $\varphi(t)$ obtained from measurements at about every other measuring point, i.e. for every second counter. In the computations of the difference quotients it appears difficult to form any general view of the influence of the tolerances on the measured values. Taking into account the purpose of the formation of the difference scheme, it has therefore been considered appropriate always to use as many places as correspond, by a good margin, to the threshold value for measurement of the distribution function. Furthermore, in order to carry out a continued formation of differences, all differences of the same order should be computed to the same number of decimals. In considering the values of Table 15 it should be observed that the various differences cannot be expected to have the same absolute accuracy. On the contrary, an attempt to estimate the relative accuracy has indicated it to be of the same order of magnitude for all differences of the same order. This is due to the choice of the various time differences. It is clear from Table 15 that up to and including the third difference we obtain everywhere the signs which should signify an entirely monotone function. By the fourth difference a reversed sign appears and by the fifth difference the signs vary so much that the monotone character of the series of numbers has almost disappeared. We can show by tests, however, that these anomalies do not generally correspond to greater errors than about 0.0001 in determining $\varphi(t)$. If, for instance, we perform a corresponding numerical difference computation for the tabled values of the exponential function, we find regular signs for as many differences as the given number of decimals. We can claim, therefore, that the measured distribution function meets the fully monotone requirement within the margin of the expected accuracy.

In Chapter 12 the possibility of performing the difference study on a smoothed curve of measured values has also been discussed. The result of such a study is shown in Table 16. In a graphic presentation of the difference function $D(t)$, a curve subjectively smoothed has been inserted, after which $\varphi(t)$ has been computed by addition of corresponding values of the exponential function. In Table 16 the values of the distribution function so obtained is given for the points $t = 1, 2, \dots 12$ with the time unit given as $1/y$. Table 16 therefore gives a series of numbers with greater time intervals than those in Table 15. By the use of equidistant points we can further apply the usual differences defined by (146 a) for the difference study instead of the difference quotients. Also in this context a sign irregularity appears already in the fourth difference. This, however, is of less importance and not until the seventh and eighth differences can the monotone property be considered as being seriously disturbed.

The criteria in Chapter 10, regarding the check on measuring functions in traffic with slow intensity variations, could not, as already mentioned in Chapter 13, be supplemented by criteria of the permitted tolerances. Due to this fact, it is obviously not possible to draw any binding conclusions from the aforesaid difference studies regarding the extent to which the measurment results fulfil the criteria laid down. We shall have to be content with more or less subjective assessments of the conditions and thereby come to the conclusion that there is no reason to abandon the assumptions set up for the theories.

The next phase in the treatment of the inter-arrival times is the study of the extent to which the difference curve can be represented in normal form 1. For this task the η_0-values corresponding to the different measured points on the difference curve have been calculated. Preliminarily the approximate formula (128) has been applied. The result is clear from Table 14 and Fig. 30, in which the abscissa is graduated according to the numbers of the counters from which the recordings have been obtained of the points used on the difference curve. The figure gives clear indications as to the appearance of systematic deviations between the measured difference curve and normal form 1. The conditions around counter 28 are of interest; this corresponds to the time around twice the mean inter-arrival time, i.e. 17.16 s. As is shown in Chapter 9, the difference curve for normal form 1 always has a zero point around this value. Small changes in the curve correspond to great changes of η_0. As a consequence, we can expect a similar condition for the η_0-values derived from the measured difference curve in the vicinity of $t = 2/y$, as was the case in respect of limit points of the P_n-curve. This is also the case, as can be seen from the figure. Further, a noticeable increase of η_0 appears at very large t-values and also for the very smallest t-values. We should not attach much importance to the latter, as in this area the uncertainty of the time determination may have a considerable influence.

In determining η_0-values suitable for designing a comparison curve, mean values have been computed from the η_0-values in the areas marked by arrows in Fig. 30. For the mean of the η_0-values from the negative branch of the difference curve we obtain 0.0557 and from the positive branch 0.0494. From all the points included we get the mean value 0.0522.

The values presented are not quite correct, as they have been calculated by means of the approximate formula (128) which for this order of magnitude of η_0 has a noticeable error. For more accurate estimations we have determined from the measured curve of $D(t)$ the values of the difference curve for the t-values, corresponding to the curves of η_0 for normal form 1 given in Diagram 2. We have then determined the corresponding η_0. The result is given in Table 17. The formation of a mean value of the η_0-values so obtained, excluding some extreme values, has given $\eta_0 = 0.0550$. In Table 17 $D_{\eta_0}(t)$ is computed for this η_0-value according to normal form 1, as also the remaining difference between the measured $D(t)$ and the computed $D_{\eta_0}(t)$. As is seen, the remaining difference is very small. Fig. 31 shows in different scales $D(t)$ and $D_{\eta_0}(t)$ computed as above.

Although the difference between the measured curve and the curve computed from normal form 1 is mostly highly insignificant with regard to the expected accuracy of measurement, investigation is justified of the extent to which an improved adaptation can be obtained by means of normal form 2. For this purpose an attempt was made to use the calculation methods given in Chapter 12. In accordance with formula (152 a) the distribution function of the next call arrival $\vartheta(t)$ has been determined first by graphic integration of the difference curve. An idea regarding the extent to which the relation (152 b), valid for normal form 1, is fulfilled is obtained from Fig. 32 showing the ratio $\vartheta(t)/\varphi(t)$ (computed from the measurements) as a function of t. In Figs. 32 a and 32 b, showing the behaviour in different areas and to different scales, the points (partly interconnected by a broken line) indicate the values obtained from the measurements. Up to $t = 12$ s the points fall very close to a straight line, extended in the figures. From the slope of this line we obtain, under the assumption of normal form 1, the value $\eta_0 = 0.0574$, and under the assumption of normal form 2, the value $\eta_0 + \kappa = 0.0574$. It should be observed in this connection that –

as previously mentioned – it is assumed that η_0 is a dimensionless constant corresponding to the fact that $1/y$ has been assigned as time unit. If we wish to use another time unit, e.g. the second, we only have to substitute η_0 by $y\,\eta_0$.

The η_0 so obtained is about 4 % higher than the value previously calculated as a mean using the various points of the difference curve. The rather insignificant difference might be due to the fact that the area used above, i.e. 0 – 12 s, falls entirely within the negative branch of the difference curve which, according to Fig. 30, gives on an average a somewhat higher η_0-value than the positive branch.

The question is now to what extent the noticeable unilateral deviation from the straight line for higher t-values can be explained by adaptation to normal form 2. Following (152 d) we can with normal form 2 expect a deviation of the kind indicated. Here, however, the same phenomenon appears as in the former study of the moments, viz. the observed deviation from normal form 1 goes in the right direction for adaptation to normal form 2 but is considerably greater than can be explained by the latter normal form. As in this case we shall have $\eta_0 + \kappa = 0.0574$, the square term in (152 d) has its maximum value for $\eta_0 = \kappa = 0.0287$. In such a case we obtain for $\vartheta(t)/\varphi(t)$ at $t = 40$ s the value 1.266 instead of 1.258 from normal form 1 and 1.38 from the measurements.

Another method for investigation of the possibilities of adaptation to normal form 2 is by formula (152 e), as shown in Chapter 12. The value of the left-hand member of this expression, computed from the measuring material of measured traffic 2, is clear from Fig. 33, in which the different measured points are connected by a broken line. An endeavour to adapt it to the right-hand member in (152 e) seems to be all too hazardous; it has, however, on average, a slope of the kind to be expected from normal form 2. The undulating form of the measured curve is due to the influence of the derivative of the difference curve.

For the study of the maximum improvement of the adaptation of the difference curve which can be attained by means of normal form 2, $D_{\eta_0,\kappa}(t)$ has been computed for $\eta_0 = \kappa = 0.0287$. The result is given in Fig. 34, showing on a larger scale $D(t) - D_{\eta_0}(t)$ for $\eta_0 = 0.0550$ and $D_{\eta_0,\kappa}(t) - D_{\eta_0}(t)$ for $\eta_0 = \kappa = 0.0287$, $D(t)$ being the measured difference curve. As is shown, normal form 2 gives a highest difference from normal form 1 of 0.00011, a value which can be disregarded in measurements of the magnitude of measured traffic 2.

As a further contribution to the question of the adaptation of the measured distribution function to any of the normal forms, the integral (152 f) has finally also been computed. For normal form 1 and $\eta_0 = 0.0550$, it must have the value 83.40 and for $\eta_0 = 0.0574$ the value 83.61. In measured traffic 2 the value 83.45 was obtained, corresponding to $\eta_0 = 0.0556$ in normal form 1. As the integral in question should theoretically have an infinitely great value in normal form 2, the measurement result would strongly support the preference for the adaptation to normal form 1.

Survey and discussion of the measurement results

An account has been given above of the rather detailed treatment of the states of measured traffic 1 and the inter-arrival times of measured traffic 2. As mentioned previously, the treatment of the remaining material has been concentrated to the possibilities of adaptation to normal form 1. The following quantities have thus been treated:

– The states P_n from the readings of the state meter

– The states $P_{n,0}$ computed for random traffic
– The relative moments μ_n up to $n = 6$, computed from the P_n quantities
– η_0 computed from μ_n/μ_{n-1}
– The relative moments, computed from the results of measurements of combinations of occupations
– The distribution function of the inter-arrival times, $\varphi(t)$, from the measuring results of the interval meter
– The distribution function at random traffic e^{-yt}
– The difference curve $D(t)$
– η_0 computed as a mean value from different points of $D(t)$.

For the cases of measured traffic in which an adaptation to normal form 1 has been possible, the theoretical values of P_n and $D(t)$ have been computed from the respective η_0-values.

The result of the computation of moments from the states is apparent from Table 18, showing the 6 first moments for all 14 cases of measured traffic. Generally the moments (except, obviously, the first) are greater than 1, increasing with increased n-value, indicating the presence of intensity variations. A reversed condition appears, however, in measured traffic 6, 7, and 8, where the moments are less than 1 and decrease with increased n-value. This is because these cases of measured values are composed solely of traffic samples with A-values very close to one another. It has been remarked above, in conjunction with Table 6, that the dispersion of the A-values of these traffic samples is less than that theoretically valid for random traffic. Through the selection of traffic samples with A-values close to one another, obviously an unnatural concentration of states to the mean arises which abolishes or outweighs the influence of any intensity variations. The conditions will be further elucidated in connection with the study of the P_n-curves. It should be remarked that also measured traffic 10 shows a similar phenomenon, as the moments at the beginning are insignificantly greater than 1, decreasing below 1 for higher n-values.

A study of Table 18 shows that the moments, here too with the exception of measured traffic 6, 7, 8 and 10, give a good picture of the character of measured traffic such as it will appear in the ratio of the real to the theoretical dispersion of the traffic samples according to Table 6. A still clearer picture appears from Table 19, showing η_0, computed from μ_n/μ_{n-1} for different n-values and different cases of measured traffic. In this table measured traffic 6, 7, and 8 are not included, as they do not give any positive values of η_0. First it should be remarked that for all measured traffic except measured traffic 4, the η_0-values exhibit a similar decrease with increasing order of the moments from which they are determined, as has been observed previously for measured traffic 1. The decrease, however, is as a rule insignificant for the measured traffic that has not been obtained from traffic samples with almost similar A-values. For measured traffic 4 even a small increase is obtained. It is further of interest to compare the η_0-values for different cases of measured traffic starting from η_0 obtained from μ_2, as they seem to give the most reliable estimation. If we compare with the values in Table 6, we observe that η_0 has a noticeable connection with the homogeneity of the measured traffic, insofar as it can be estimated from the variations of the A-values of the traffic samples in question. This connection will be discussed in some detail later. Here it should only be remarked that, if it were possible by making more extensive measurements to determine such a connection unambiguously, it would be possible in practice to estimate η_0 solely from the result of common traffic measurements.

The result of the computations of the state quantities P_n of the different cases of measured traffic and their theoretical equivalents is apparent from Figs. 35–47, all presenting $\log P_n/P_{n,0}$, computed partly from the measurement results, partly according to normal form 1. In the latter case the η_0-values from Table 19 for $n = 2$ have been used. As a comparison, a value has been computed for all cases of measured traffic, except measured traffic 6, 7 and 8, by means of (157 c) from a P_n with n in the neighbourhood of A. The results are presented in the last column of Table 19. For cases of measured traffic with large η_0-values a fairly good agreement is obtained with the η_0-value computed from μ_2, but in other cases the agreement is poorer. We cannot, however, expect any great accuracy from a single state value.

Fig. 35 shows the conditions for measured traffic 2. It appears that very good agreement is obtained between measured and theoretically computed values from normal form 1. In measured traffic 3, Fig. 36, the agreement is somewhat poorer. As a consequence of the fact that this measured traffic is considerably smaller than measured traffic 1 and 2, the points determined by the measurements are considerably more irregular than previously. It has been considered suitable, therefore, to present the measurement-based curve as a broken line connecting the measured points, so as to give the best idea of their dispersion. The same principle has been applied for all cases of measured traffic in which the measured points do not lie so regularly that a smooth curve can be drawn to better advantage. Measured traffic 4 and 5, Figs. 37 and 38, show an adaptation to the theoretical curve which is considerably poorer than for measured traffic 1 and 2, but it must still be considered as being good. On the other hand cases of measured traffic composed of traffic samples with almost equal A-values show no possibilities of adaptation. For these we obtain, as is clear from Figs. 39 to 43, a P_n-curve of an entirely deviating type and virtually the direct opposite of the theoretical one according to normal form 1, which has not therefore been inserted. This condition is quite natural. As can be seen from Fig. 28, the influence of the intensity variations is evident from the fact that the P_n-curve is as a rule much flatter than the corresponding $P_{n,0}$-curve, so that the maximum becomes lower and the lateral branches higher. If we now had a theoretically random traffic and from it formed a measured traffic by selecting traffic samples with almost equal A-values, we would clearly get a P_n-curve with an entirely different tendency than the one caused by intensity variations, i.e. a steeper curve with higher and sharper maximum and lower values in the lateral areas. The present cases of measured traffic obtained from traffic samples with nearly equal A-values should therefore show P_n-curves resulting from the influence of both these mutually counteracting factors. In this way the appearance of the strange curves in Figs. 39 to 43 can be explained. It is interesting to observe that we do not generally obtain such a compensation that the result is similar to that for random traffic.

Of interest are also the conditions for measured traffic 11 and 12, Figs. 44 and 45, composed of traffic samples with almost equal y-values. Due to the strong correlation between y and A, we obtain here as well a certain influence, though weaker, of the concentration of A. For measured traffic 11 quite good adaptation to the theoretical curve is obtained.

Measured traffic 13, Fig. 46, has been obtained from measured traffic 1 by omitting some traffic samples with very high A-values. As will be seen, good adaptation to the theoretical curve is obtained.

Measured traffic 14, Fig. 47, is very interesting as it is obtained by mixing traffic samples with very low and very high A-

values. Its η_0-value has therefore become greater than for any of the other cases of measured traffic studied. The agreement with the theoretical curve must be considered good, showing the great elasticity of normal form 1. Due to the composition of measured traffic 14, we can expect that its intensity distribution will present two maxima, rather far distant from one another, while normal form 1 assumes an intensity distribution with only one maximum. The fact that we obtain, none the less, such a satisfactory agreement between measured and theoretical values indicates that normal form 1 has a very wide domain of application.

To sum up, we can venture the conclusion from the P_n-curves of the 14 cases of measured traffic studied that normal form 1 at "normal" traffic offers good possibilities for computing the state quantities. "Normal" traffic is a traffic which has not been obtained from measurement material by selection according to the magnitude of the mean traffic values. In judging the accuracy of the theoretical presentation we should keep in mind that the formulas used up to now, valid for random traffic, correspond to the abscissa on the graphs for log $P_n/P_{n,0}$. The main concern is to present a correction factor, which may be fairly great in certain cases, for random traffic which has adequate accuracy for practical needs. Taking this into account, the claim seems justified that the reported measurement results have shown the usefulness of the theories presented.

In connection with the study of the results of measurements of the states, it will also be reported on the moment values obtained from the measurements of the combinations of occupations. Table 20 presents the relative moments thereby obtained; for μ_3, μ_4, and μ_6 also compared with the previously from the state quantities computed relative moments. Quite obviously the combinations of occupations generally give higher moment values than the states. This is evident especially for measured traffic composed of traffic samples with nearly the same mean traffic values. It would seem difficult to present any plausible reason for this effect. The μ_n-values, however, obtained from the combinations of occupations, agree quite well, at least for the lower n-values, with the η_n obtained previously for the different cases of measured traffic. For the higher moments, especially μ_{12}, there is a tendency towards a decrease, which can possibly be explained by the fact that, as mentioned previously, (151 b) implies a poorer accuracy the greater n is. It should be remarked further that the values of the combinations of occupations seem to exhibit strong random variations, with the result that the relative moments of cases of smaller measured traffic cannot be considered quite accurate. This is the case especially for the higher moments μ_9 and μ_{12}. To sum up, it should be stated that the measurements have shown that the state quantities offer possibilities for a more reliable analysis than the combinations of occupations. This is of interest in conjunction with future measurements.

The result of the computations for the different cases of measured traffic of the difference curve $D(t)$ and its theoretical counterpart for normal form 1 is apparent from Figs. 48 – 57. In the cases where an adaptation to normal form 1 has been possible, the used η_0-value has been computed as the mean of a number of η_0-values, normally around 20, which has been obtained from different points within the negative as well as the positive branch of the difference curve obtained from the measurements.

The inter-arrival times for measured traffic 2 have been discussed above. In Fig. 48 the conditions for measured traffic 3 are shown. The full-line curve shows $D(t)$ and the dashed curve $D_{\eta_0}(t)$, in which η_0 has been found to be 0.0470. As will be seen the agreement is fairly accurate. The deviations seem to be of a character similar to that of measured traffic 2.

The measured difference curve for measured traffic 4 is shown in Fig. 49. The measuring points are here so irregularly placed that it would seem best to show the curve by a broken line through them. The behaviour – which is irregular compared with measured traffic 2 and 3 – should be ascribed to the fact that measured traffic 4 contains a considerably smaller number of calls. It must be considered that the behaviour has turned out so irregular that it has been impossible to arrive at a fair adaptation to a difference curve according to normal form 1. For the same reason it has seemed meaningless to try to determine any mean value of η_0.

Measured traffic 5 also contains less calls than measured traffic 3, but has anyway double the number of calls as compared with measured traffic 4. As seen in Fig. 50, we have obtained so regular a behaviour between the measured points as to permit the representation of $D(t)$ as a smooth curve. A fairly satisfactory adaptation to normal form 1 has also been obtained. η_0 has been found to be 0.0455.

Fig. 51 shows the case of measured traffic 6. We have here a measured traffic with a very small number of calls, only 6285, which furthermore is made up of traffic samples with A-values very close to one another. The latter condition implies also a concentration of y-values, causing a decrease in the number of very small, and very long, inter-arrival times as compared with normal traffic. This should be compensated by an increase of the number of inter-arrival times round the mean value. It will thereby be seen that the difference function for small t-values should have greater values than normal, and smaller values for very great t-values. This is excellently illustrated by the difference curve in Fig. 51. It is obvious that a curve such as this – even disregarding the considerable irregularities caused by the limited volume of the measured traffic – does not allow adaptation according to normal form 1.

For measured traffic 7, as for 1 and 13, no study has been made of the inter-arrival times, as it comprises traffic samples during which the interval meter was out of order. Measured traffic 8, Fig. 52, contains, as does measured traffic 6, a rather small number of calls, showing also clear indications of an influence of the concentration of A-values of traffic samples. This applies likewise to measured traffic 9 and 10, in which the influence is noticeable for small t-values. Notwithstanding this it has been possible to draw up rather well-adapted $D_{\eta_0}(t)$-curves, Figs. 53 and 54. In measured traffic 11 and 12, composed of traffic samples with y-values quite close to one another, a particularly strong influence of the concentration of the y-values could be expected. Nevertheless it has been possible to adapt the curves quite closely to normal form 1, Figs. 55 and 56.

Finally, Fig. 57 displays the condition for measured traffic 14. The adaptation to normal form 1, here giving an, easily explained, exceptionally high η_0-value, i.e. 0.0800, should be characterized as very good.

It should be remarked with regard to the figures of the various difference curves that, for the sake of comparison, it has seemed to be most suitable to graduate the abscissa in yt, instead of in seconds, whereby the same scale has been obtained everywhere, regardless of the y-values. Further, the behaviour for $yt > 7.5$ has not been included, as in this area the absolute values of the difference curves become very small and it would be difficult, therefore, to form an opinion of the possible deviations from the theoretical curves.

A survey is given in Table 21 of the different values of η_0 which have been obtained from the states as well as from the inter-arrival times. The values which have been obtained from the relative moment μ_2 have been denoted η_{0Z} and the values from the difference curve η_{0A}. As can be seen, η_{0A} is throughout greater than η_{0Z}. In certain instances the difference is quite considerable, which will be evident from the column for η_{0A}/η_{0Z}. Regarding measured traffic composed of traffic samples with concentrated mean traffic values, it is quite natural for η_{0A} and η_{0Z} to deviate from one another. As has been repeatedly pointed out, there are otherwise at least two causes for the occurrence of different values of η_{0A} and η_{0Z}. One is to be found in the possible presence of intensity variations of the mean occupation times. The values in Table 6 of the measured and theoretical dispersion of the traffic samples seem to indicate that such variations are quite common. It would seem, however, as stated previously, to be very difficult to find methods of determining the intensity variations of the occupation times and their correlation with the intensity variations of the call and traffic volume. Hopefully, further research will offer opportunities to tackle this important problem; for the time being we shall have to be content with establishing the fact that the influence of the intensity variations of the mean occupation times does not seem to make impossible the practical application of the formulas developed hitherto.

The other possible cause of the deviations between η_{0A} and η_{0Z} consists in the varying influence which the speed of the intensity variations exercises on the state values and the distribution of the inter-arrival times (shown in Chapter 7). This influence is unique in that as soon as s is considerably greater than $1/y$, then $\eta_{0A} > \eta_{0Z}$ as is the case for the values in Table 21. Under the assumption that this dissimilarity is due solely to the influence of the speed of the intensity variations, the different values of η_{0A} and η_{0Z} also provide, for one and the same measured traffic, a possibility of estimating, very approximately, the average order of magnitude of the speed of the intensity variations. We must hereby assume that the call intensity has a simple, periodic form, e.g. (83) in Chapter 7. According to the investigations in Chapter 7 we obtain different equilibrium traffic intensities for the state values and for the inter-arrival distances. The latter intensity can be considered to be equal to the real traffic intensity (83). By means of the formulas in Chapter 8 of normal form 1, we obtain, after some computations,

$$\eta_{0A} = \frac{\omega}{\pi}\left(\frac{y_1}{y_0}\right)^2 \int_0^{\frac{\pi}{\omega}} \cos^2 \omega T \, dT$$

and from that

$$\eta_{0A} = \frac{1}{2}\left(\frac{y_1}{y_0}\right)^2$$

The equilibrium traffic intensity for the states has the form (85 a), giving

$$\eta_{0Z} = \frac{1}{2}\left(\frac{y_1}{y_0}\right)^2 \frac{1}{1+s^2\omega^2}$$

If, for example, we use the values of measured traffic 2, we obtain a period length of 775 s and $y_1 = 0.33$ and $y_0 = 0.037$. It can, of course, be assumed that entirely different values will be obtained if we presuppose a more complicated form of $y(T)$. The example stated seems, however, in all its simplicity, to indicate that the intensity variations of the real traffic might be unexpectedly rapid and extensive.

The measurements reported in this chapter have had as a goal to find out whether, in real telephone traffic we can find the properties which were the prerequisites in the theoretical studies carried out in this work. The measurement results may in this respect be considered to have given a clearly positive answer, so that making further studies on the same basis can be regarded as justified. Certainly such investigations require considerably more extensive measurements than those performed here. It is to be hoped that they will be carried out with improved and at the same time simplified equipment. With due regard to the applications for practical use, an endeavour should be made to determine the connection between η_0 and the dispersion of the mean traffic values obtained from the traffic samples (e.g. from measurements during busy hours). The division into different cases of measured traffic in the measurements mentioned was not intended to be a study of that kind. Also the measurement material was too small for this purpose. Furthermore the intention of combining and investigating measured traffic obtained from traffic samples with almost similar mean traffic values – which do not directly influence the applications in practice – has been to examine whether, by artificial interference, similar or contrary effects to the intensity variations could be caused. It is tempting, in spite of this and in spite of the limited volume of measured traffic studied, to try to obtain some approximate values regarding the connection that presumably exists between η_0 and the dispersion of the mean values of the traffic samples. In order to obtain a quantity characteristic of this dispersion and at the same time not too dependent on the lengths of the various traffic samples, we have chosen the square of the ratio between the relative standard deviations computed from the traffic samples (given in Table 6) and the corresponding quantities computed from the formulas for random traffic. These deviation quotients are shown in Table 21 with the notations k_A, k_y and k_s, in which the index indicates the quantity to which the quotient belongs. A study of the values in Table 21 seems to suggest the possibility of unique connections between the respective η_0- and k-values.

In connection with the discussion of the dispersions it may be of interest to look also at an example of the intensity distributions. To compute numerically the frequency function $g(x)$ of the intensity, according to (118 c), the *Stirling* formula for the development of the gamma function may be used (Chapter 12). We easily find then from (118 c), in which, for the sake of convenience, y is set to 1,

$$g(x) \approx \frac{1}{\sqrt{2\pi\eta_0}}\, x^{\frac{1}{\eta_0}-1}\, e^{-\frac{x-1}{\eta_0}} \qquad (174\text{ a}),$$

an expression valid with very good accuracy for the present values of η_0. If, further, we write

$$x = e^{\ln x} \approx e^{x-1-\frac{(x-1)^2}{2}+\dots}$$

we obtain

$$g(x) \approx \frac{1}{x\sqrt{2\pi\eta_0}}\, e^{-\frac{(x-1)^2}{2\eta_0}} \qquad (174\text{ b}).$$

As only x-values close to 1 are of any importance, we can finally set

$$g(x) \approx \frac{1}{\sqrt{2\pi\eta_0}}\, e^{-\frac{(x-1)^2}{2\eta_0}} \qquad (174\text{ c})$$

The intensity x is therefore approximately normally distributed. The factor x in the denominator in (174 b) causes some skewness in relation to the normal distribution. As an example, Fig. 58 shows the frequency functions of the intensity variations in A and y for measured traffic 2 and computed from the respective η_0-values by means of the formulas indicated.

Concluding remarks

The measurements reported in the foregoing should show that it is highly feasible to use the theories presented in this work for computations regarding real traffic conditions. Before we venture, with confidence, to base calculations of quantities of equipment on computations of this kind, a considerable amount of additional measurements will, of course, be required under varying traffic conditions. In particular it will be necessary for the applications to carry out measurements regarding superposing conditions, which up to the present time it has not been possible to study. There is also the possibility, by measurements combined with further theoretical investigations, to study the question of intensity variations of the mean occupation time and their influence on the traffic conditions. It would seem that there are also other conceivable means to analyse by measurements the character of the traffic in greater detail than reported here. Thus it would appear to be possible to obtain a more accurate idea of the speed of intensity variations by measurements of the behaviour of successive interarrival times. These times, whose distribution function appears from (69) in Chapter 6, are as a matter of fact sensitive in different degrees to the speed of the intensity variations, depending on the magnitude of n. It would seem possible, therefore, by a detailed study of the distribution functions, to develop a "frequency spectrum" of the intensity variations. As we can never in principle determine the momentary value of the intensity, a frequency spectrum of this kind seems to represent the last resort for analysing intensity variations.

As is seen, there exist many possibilities of supplementing by measurements the results obtained in the present work, which should be regarded merely as a modest beginning. The same can be said of the purely theoretical parts of the work. Many interesting aspects of development which have come to light during the progress of the work have had to be left aside. As an example may be mentioned the applications to graded groups of the results in Chapter 4, and further the influence of rapid intensity variations on the overload capacity of delay systems treated in Chapter 7. It may be observed, finally, that a great part of the theoretical studies, due to the necessity of avoiding too great mathematical complications, has been based on assumptions which are probably unnecessarily restricted. This is the case especially for the assumption – stated throughout – that the occupation times are distributed as a simple exponential function. It is possible that this assumption can in many cases be avoided by a deeper mathematical analysis.

Appendix

A property of the call congestion

It will be shown in this appendix that the second derivative of the kernel function of the call congestion, presented in Chapter 6,

$$D_n(x) = x\,E_{1,n}(x) = \frac{\dfrac{x^{n+1}}{n!}}{1 + \dfrac{x}{1!} + \dfrac{x^2}{2!} + \ldots + \dfrac{x^n}{n!}} \qquad \text{(A 1)}$$

is always positive. First some auxiliary theorems will be shown. If we form the first derivative of $D_n(x)$, we obtain

$$D_n'(x) = \frac{\dfrac{x^n}{n!}}{\left(1 + \dfrac{x}{1!} + \ldots + \dfrac{x^n}{n!}\right)^2} \left\{ \left(n+1\right)\left(1 + \frac{x}{1!} + \ldots + \frac{x^n}{n!}\right) - \right.$$

$$\left. - x\left(1 + \frac{x}{1!} + \ldots + \frac{x^{n-1}}{(n-1)!}\right) \right\}$$

After some reduction we see that the term in the numerator containing the highest power in x is $(x^n/n!)^2$. The same is valid for the denominator. From this it follows that $D_n'(x) \to 1$ when $x \to \infty$.

Further we write $D_n(x)$ according to (A 1) in the form

$$D_n(x) = x - n + \frac{C_n}{1 + \dfrac{x}{1!} + \ldots + \dfrac{x^n}{n!}} \qquad \text{(A 2)}$$

whereby

$$C_n = n\left(1 + \frac{x}{1!} + \ldots + \frac{x^n}{n!}\right) -$$

$$- x\left(1 + \frac{x}{1!} + \ldots + \frac{x^{n-1}}{(n-1)!}\right)$$

which can be transformed to

$$C_n = \left(n-1\right)\left(1 + \frac{x}{1!} + \ldots + \frac{x^{n-1}}{(n-1)!}\right) -$$

$$- x\left(1 + \frac{x}{1!} + \ldots + \frac{x^{n-2}}{(n-2)!}\right) +$$

$$+ \left(1 + \frac{x}{1!} + \ldots + \frac{x^{n-1}}{(n-1)!}\right)$$

giving the recursion formula

$$C_n = C_{n-1} + \left(1 + \frac{x}{1!} + \ldots + \frac{x^{n-1}}{(n-1)!}\right)$$

We now find easily – as an example – $C_1 = 1$. We must then always have $C_n > 0$. From (A 2) we thus obtain

$$D_n(x) > x - n \qquad \text{(A 3)}$$

In order to investigate the second derivative of $D_n(x)$, we now start from the recursion formula (12) in Chapter 3, which is written in the form

$$D_n = \frac{x\,D_{n-1}}{n + D_{n-1}}$$

If this expression is differentiated twice in respect of x, we obtain, after some simplifications

$$\frac{(n + D_{n-1})^3}{n} D''_n = x\,(n + D_{n-1})\, D''_{n-1} + \\ + 2\, D'_{n-1}\,(n - x\, D'_{n-1} + D_{n-1})$$

Suppose now that D''_{n-1} is always positive for $x > 0$. Then D'_{n-1} is continuously increasing with x towards a maximum value for $x \to \infty$, which, according to the previous proof, is 1. If, therefore, in the last bracket in the right-hand member of the expression above, we substitute 1 for D'_{n-1} the entire right-hand member must decrease, giving

$$\frac{(n + D_{n-1})^3}{n} D''_n > x\,(n + D_{n-1})\, D''_{n-1} + \\ + 2\, D'_{n-1}\,(n - x + D_{n-1})$$

The last bracket in the right-hand member is now with certainty positive due to (A 3). From the assumption now that D''_{n-1} is always positive it follows that the entire right-hand member above is always positive, D''_n is therefore also always positive. It is now easily proved that, as an example, D''_1 is positive for $x > 0$. Then follows recursively that all D''_n are positive for $x > 0$.

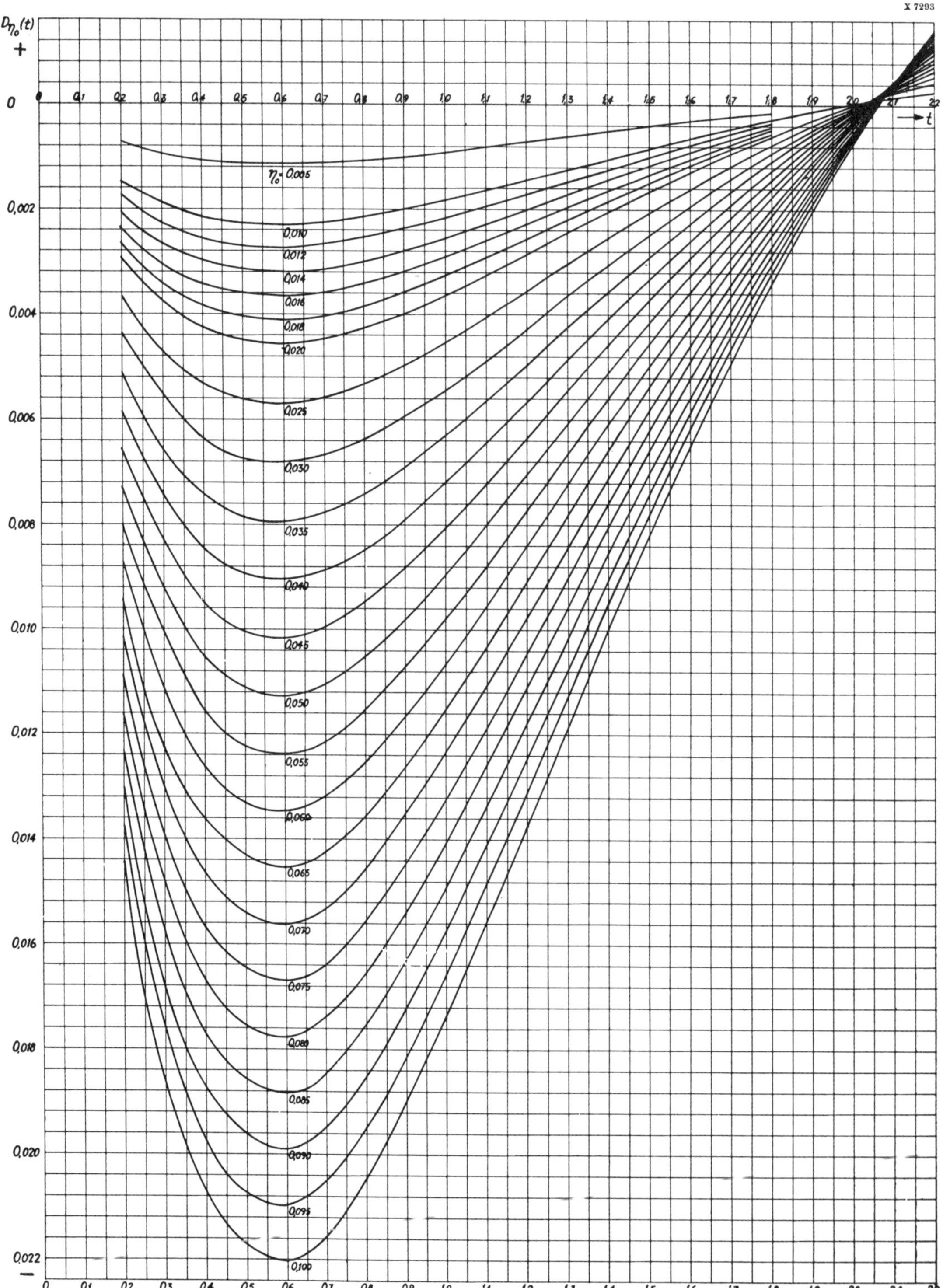

Diagram 1 A. The difference curve $D_{\eta_0}(t)$ of normal form 1. The family of curves shows the course of the negative branch of the difference curve as a function of t at different values of η_0.

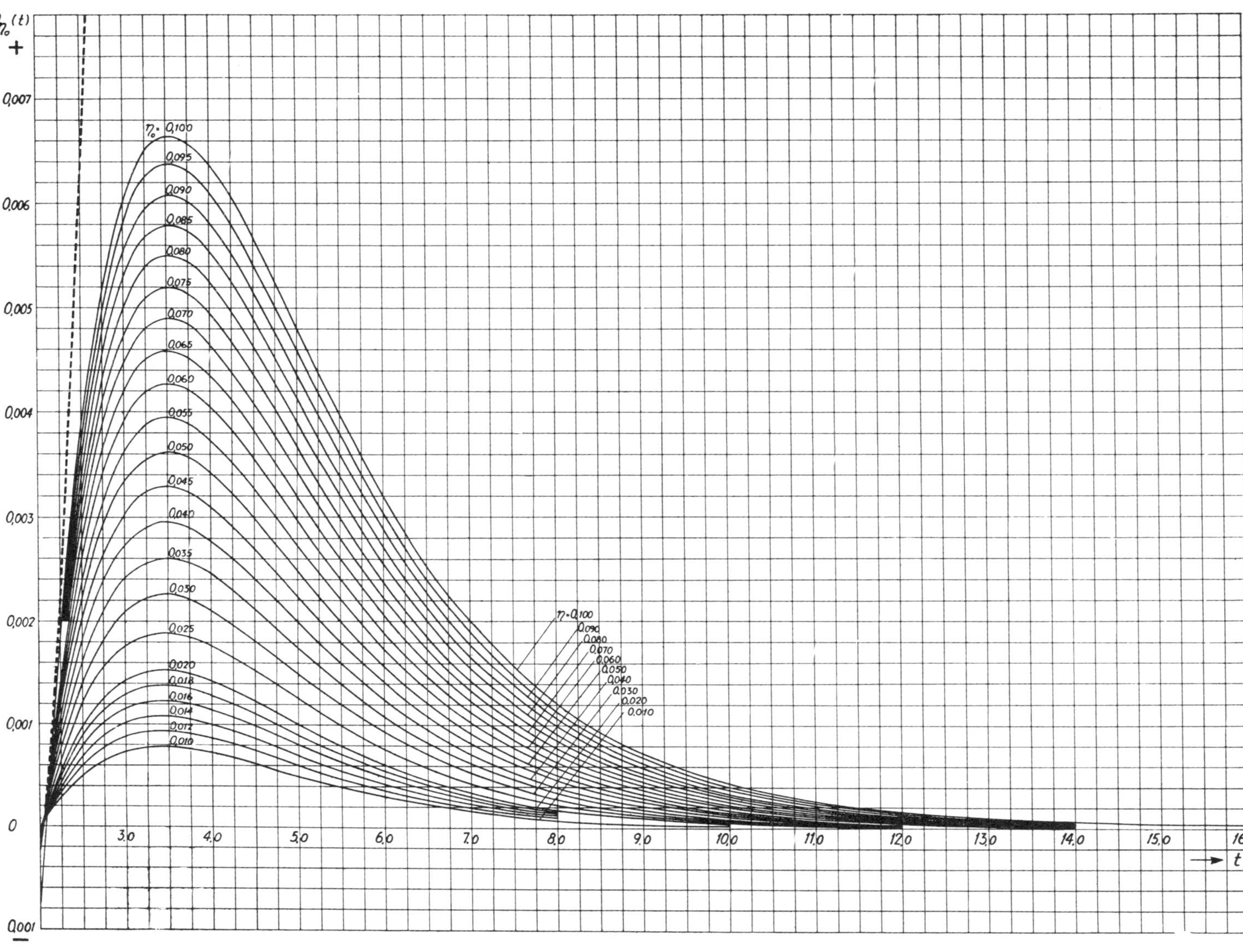

Diagram 1 B. The difference curve $D_{\eta_0}(t)$ of normal form 1. The family of curves shows the course of the positive branch of the difference curve as a function of t at different values of η_0.

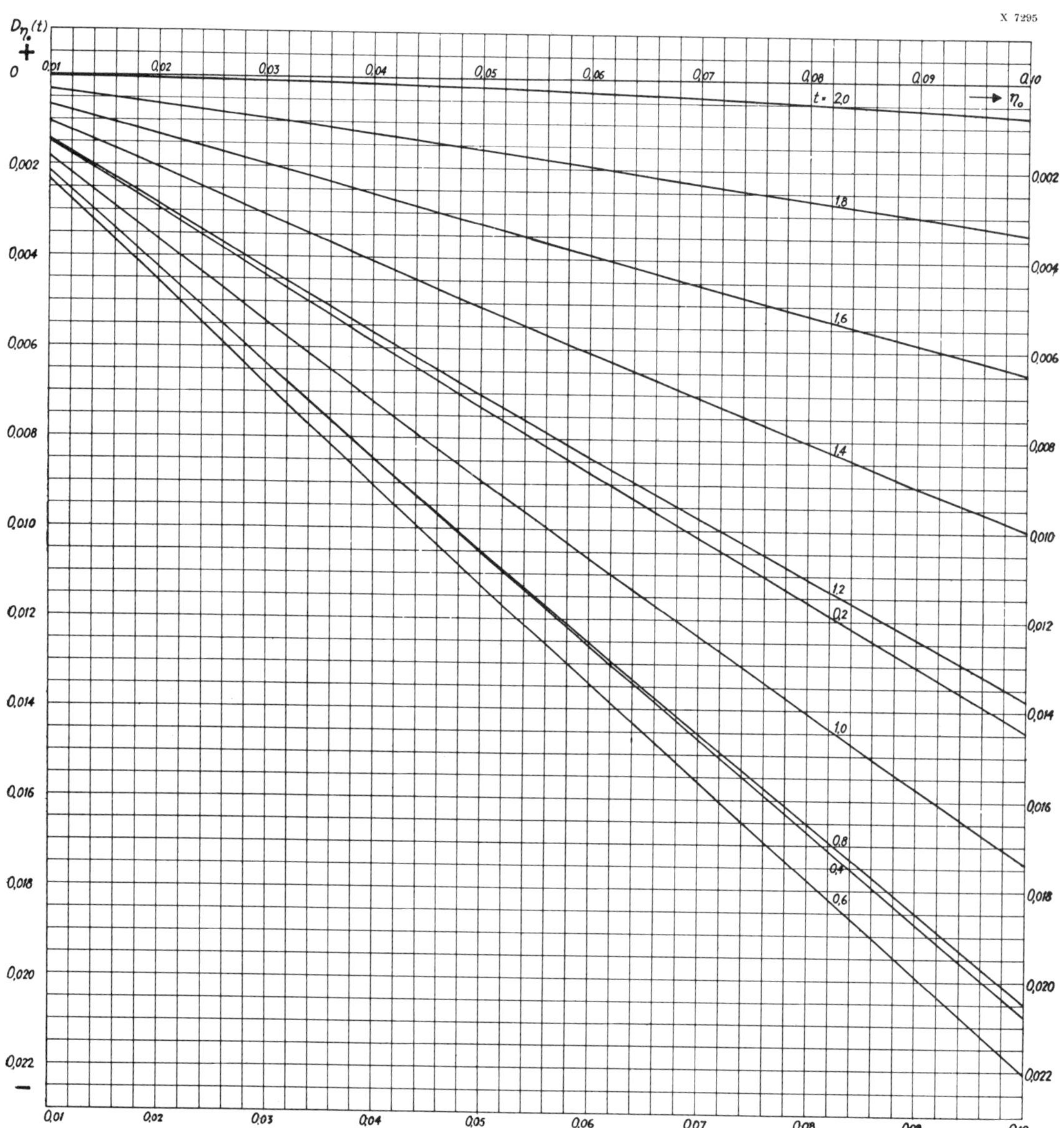

Diagram 2 A. The difference curve $D_{\eta_0}(t)$ of normal form 1. The family of curves shows the course of the negative branch of the difference curve as a function of η_0 at different values of t. η_0: 0.01 − 0.10

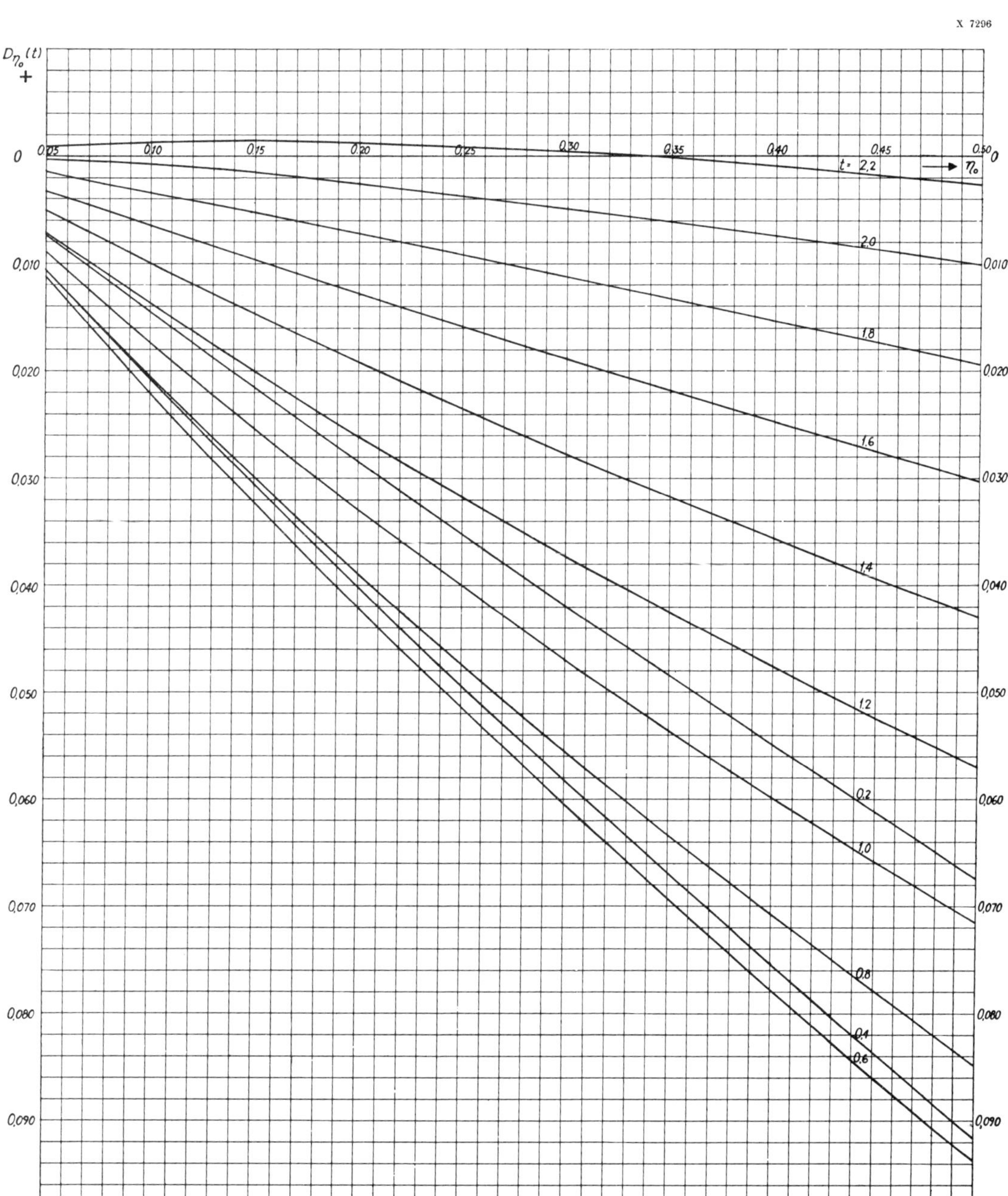

Diagram 2 B. The difference curve $D_{\eta_0}(t)$ of normal form 1. The family of curves shows the course of the negative branch of the difference curve as a function of η_0 at different values of t. η_0: 0.05 − 0.50

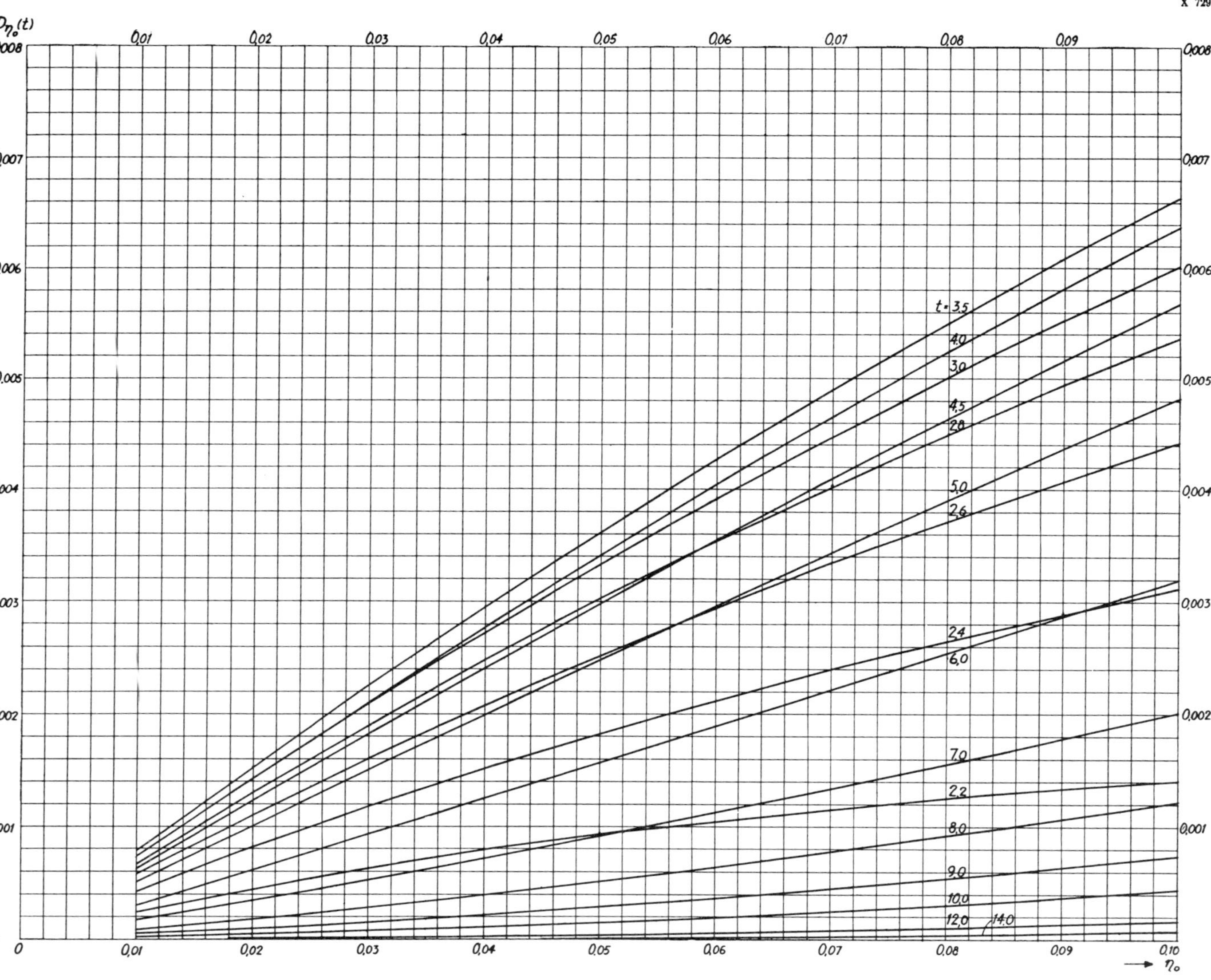

Diagram 2 C. The difference curve $D_{\eta_0}(t)$ of normal form 1. The family of curves shows the course of the positive branch of the difference curve as a function of η_0 at different values of t. η_0: 0.01 − 0.10

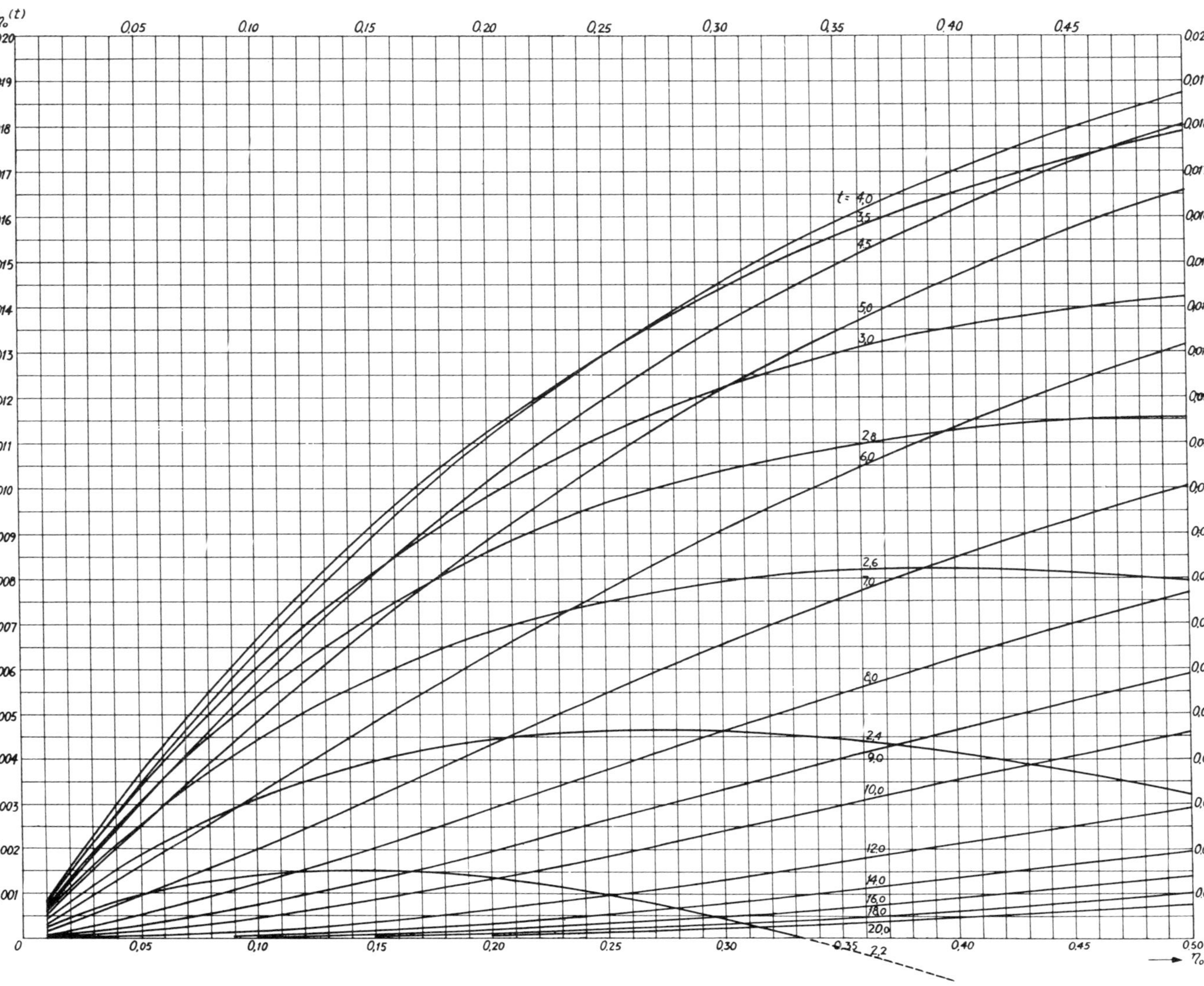

Diagram 2 D. The difference curve $D_{\eta_0}(t)$ of normal form 1. The family of curves shows the course of the positive branch of the difference curve as a function of η_0 at different values of t. η_0: 0.05 – 0.50

Table 1. Auxiliary table for the approximate computation of $D_{\eta_0}(t)$. For lower values of η_0, $D_{\eta_0}(t)$ can be expressed in the form $k(t)\,\eta_0$. $k(t)$ is then given by the table. It shows also the approximate value of η_0. The error of $D_{\eta_0}(t)$ computed from $k(t)\,\eta_0$ can be maximum 0.0001. For $t > 12$ $D_{\eta_0}(t) < 0.0001$ for $\eta_0 < 0.07$.

t	$k(t)$	η_0	t	$k(t)$	η_0
0,2	— 0,1474	0,05	2,6	+ 0,0579	0,03
0,4	— 0,2145	0,04	2,8	+ 0,0681	0,03
0,6	— 0,2305	0,03	3,0	+ 0,0747	0,03
0,8	— 0,2157	0,03	3,5	+ 0,0793	0,03
1,0	— 0,1839	0,03	4,0	+ 0,0733	0,04
1,2	— 0,1446	0,04	4,5	+ 0,0625	0,05
1,4	— 0,1036	0,07	5,0	+ 0,0505	0,07
1,6	— 0,0646	0,20	6,0	+ 0,0298	0,07
1,8	— 0,0298	0,05	7,0	+ 0,0160	0,05
2,0	0,000	0,03	8,0	+ 0,0080	0,05
2,2	+ 0,0244	0,03	9,0	+ 0,0039	0,04
2,4	+ 0,0435	0,03	10,0	+ 0,0018	0,06

Table 3. Classification of the subscriber population of the investigated traffic in number of calls per year and traffic class. The number printed after the class is the maximum number of calls per year in the class.

Subscriber type	Number of subscribers in % of total	Mean number of measured calls per subscriber and year	Number of measured calls in % of all measured calls
Class 1 1,200	64,7	829	27,2
Class 2 1,800	11,4	1 623	9,4
Class 3 2,500	7,2	2 430	8,9
Class 4 3,500	2,8	3 261	4,6
Class 5 5,000	2,1	4 164	4,4
Class 6 7,000	2,1	6 622	7,1
Class 7 10,000	0,7	9 577	3,4
Subscriber with nominal calls	3,0	11 471	17,7
Subscriber with group number	4,4	6 083	13,7
Other subscribers with operator(s) for incoming calls	0,5	10410	2,5
Other subscribers with "line selector" equipment	0,7	2 676	1,0
Subscribers free of charge	0,2	648	0,1

Table 2. Auxiliary table for computations according to normal form 2. The table gives $L_{n,\mu}$ according to (137 b) up to $n = 12$ as well as the relative moments μ_n explicit up to $n = 6$.

n \ μ	0	1	2	3	4	5	6	7	8	9	10	11
1	1											
2	1	1										
3	1	3	2									
4	1	6	11	6								
5	1	10	35	50	24							
6	1	15	85	225	274	120						
7	1	21	175	735	1 624	1 764	720					
8	1	28	322	1 960	6 769	13 132	13 068	5 040				
9	1	36	546	4 536	22 449	67 284	118 124	109 584	40 320			
10	1	45	870	9 450	63 273	269 325	723 680	1 172 700	1 026 576	362 880		
11	1	55	1 320	18 150	157 773	902 055	3 416 930	8 409 500	12 753 576	10 628 640	3 628 800	
12	1	66	1 925	32 670	357 423	2 637 558	13 339 535	45 995 730	105 258 076	150 917 976	120 543 840	39 916 800

$$\mu_1 = 1$$
$$\mu_2 = 1 + \varkappa + \eta_0$$
$$\mu_3 = (1+\varkappa)(1+2\varkappa) + 3\eta_0(1+\varkappa) + 2\eta_0^2$$
$$\mu_4 = (1+\varkappa)(1+2\varkappa)(1+3\varkappa) + 6\eta_0(1+\varkappa)(1+2\varkappa) + 11\eta_0^2(1+\varkappa) + 6\eta_0^3$$
$$\mu_5 = (1+\varkappa)(1+2\varkappa)(1+3\varkappa)(1+4\varkappa) + 10\eta_0(1+\varkappa)(1+2\varkappa)(1+3\varkappa) + 35\eta_0^2(1+\varkappa)(1+2\varkappa) + 50\eta_0^3(1+\varkappa) + 24\eta_0^4$$
$$\mu_6 = (1+\varkappa)(1+2\varkappa)(1+3\varkappa)(1+4\varkappa)(1+5\varkappa) + 15\eta_0(1+\varkappa)(1+2\varkappa)(1+3\varkappa)(1+4\varkappa) + 85\eta_0^2(1+\varkappa)(1+2\varkappa)(1+3\varkappa) + 225\eta_0^3(1+\varkappa)(1+2\varkappa) + 274\eta_0^4(1+\varkappa) + 120\eta_0^5$$

Table 4. Survey of the traffic samples of the original measurement material.

Date 1941	Traffic sample No.	Time of the day	Meas-ure-ment time in s	*A* Traffic in Erlangs	*y* Mean call value per s	*s* Mean occupa-tion time in s
July 7	211	10—11	3 600	14,11	0,1211	116,5
	311	11—16,30	19 800	11,60	0,0943	123,0
8	212	10—11	3 416	13,03	0,1133	115,0
	312	11—16,30	19 782	11,16	0,0897	124,5
9	213	10—11	3 600	15,00	0,1125	133,3
	313	11—16,30	19 800	12,09	0,0949	127,3
11	215	10—11	3 578	16,52	0,1316	125,5
	315	11—16,30	19 800	12,60	0,1026	122,9
14	221	10—11	3 600	15,91	0,1394	114,1
	321	11—16,30	19 800	12,05	0,1047	115,1
15	222	10—11	3 600	19,43	0,1414	137,4
	322	11—16,30	19 800	12,91	0,1081	119,4
16	223	10—11	3 600	15,01	0,1256	119,5
	323	11—16,30	19 800	12,06	0,1015	118,8
17	224	10—11	3 600	15,50	0,1150	134,7
	324	11—16,30	19 800	11,41	0,0948	120,4
18	225	10—11	3 600	14,15	0,1142	123,9
	325	11—16,30	19 800	11,27	0,0993	113,5
21	231	10—11	3 600	16,28	0,1275	127,7
	331	11—16,30	19 800	12,10	0,1040	116,3
22	232	10—11	3 600	16,58	0,1361	121,8
	332	11—16,30	19 800	12,95	0,1111	116,5
23	233	10—11	3 600	16,17	0,1425	113,5
	333	11—16,30	19 800	11,17	0,0990	112,7
24	234	10—11	3 600	15,75	0,1333	118,1
	334	11—16,30	19 800	12,21	0,1010	119,0
25	235	10—11	3 600	18,31	0,1469	124,6
	335	11—16.30	19 800	11,83	0,1004	117,9
Aug. 4	141	9—10	3 600	13,84	0,1058	130,7
	241	10—11	3 600	17,92	0,1339	133,7
	341	11—16,30	19 800	14,38	0,1092	131,6
5	142	9—10	3 600	13,16	0,1150	114,4
	242	10—11	3 620	19,96	0,1552	128,6
	342	11—16,30	19 780	13,22	0,1076	122,8
6	143	9—10	3 600	12,22	0,0936	130,6
	243	10—11	3 600	16,25	0,1347	120,6
	343	11—16,30	19 800	12,40	0,1041	119,1
7	144	9—10	3 600	14,36	0,1164	123,4
	244	10—11	3 600	17,70	0,1252	141,3
	344	11—16,30	19 800	13,44	0,1082	124,1
8	145	9—10	3 600	12,45	0,1183	105,2
	245	10—11	3 600	16,13	0,1347	119,7
	345	11—16,30	19 810	14,25	0,1053	135,4

Date 1941	Traffic sample No.	Time of the day	Meas-ure-ment time in s	*A* Traffic in Erlangs	*y* Mean call value per s	*s* Mean occupa-tion time in s
Aug. 11	151	9—10	3 606	14,80	0,1215	121,9
	251	10—11	3 594	16,90	0,1338	126,3
	351	11—16,30	19 800	14,44	0,1126	128,2
12	152	9—10	3 600	16,35	0,1239	132,0
	252	10—11	3 600	18,12	0,1450	125,0
	352	11—16.30	19 800	13,24	0,1001	132,4
13	153	9—10	3 600	16,52	0,1342	123,1
	253	10—11	3 602	22,08	0,1416	155,9
	353	11—16,30	19 798	13,33	0,1073	124,2
14	154	9—10	3 600	13,76	0,1250	110,1
	254	10—11	3 600	17,52	0,1469	119,2
	354	11—16,30	19 800	15,13	0,1095	138,2
15	155	9—10	3 600	16,41	0,1244	131,8
	255	10—11	3 600	16,28	0,1375	118,3
	355	11—16,30	19 800	15,42	0,1169	131,9
18	161	9—10	3 600	16,90	0,1256	134,6
	261	10—11	3 600	25,15	0,1728	145,5
	361	11—16,30	19 800	17,15	0,1288	133,1
19	162	9—10	3 600	17,64	0,1258	140,2
	262	10—11	3 600	20,48	0,1728	118,5
	362	11—16,30	19 802	15,17	0,1179	128,7
20	163	9—10	3 600	15,19	0,1183	128,4
	263	10—11	3 600	16,98	0,1603	105,9
	363	11—16,30	19 804	16,52	0,1361	121,4
21	164	9—10	3 600	15,36	0,1267	121,3
	264	10—11	3 600	24,57	0,1778	138,2
	364	11—16,30	19 800	16,61	0,1261	133,2
22	165	9—10	3 596	15,45	0,1126	137,2
	265	10—11	3 606	22,23	0,1661	133,8
	365	11—16,30	19 804	17,20	0,1363	126,2
25	171	9—10	3 600	18,32	0,1572	116,5
	271	10—11	3 600	20,91	0,1858	112,5
	371	11—16,30	19 800	16,72	0,1410	118,6
26	172	9—10	3 600	18,02	0,1653	109,0
	272	10—11	3 600	22,75	0,1789	127,2
	372	11—16,30	19 800	18,01	0,1474	122,2
27	173	9—10	3 600	18,66	0.1478	126,3
	273	10—11	3 600	23,66	0,1869	126,6
	373	11—16,30	19 800	17,73	0,1396	127,0
29	175	9—10	2 000	19,91	0,1565	127,2
	275	10—11	3 720	20,34	0,1624	125,3
	375	11—16,30	19 680	17,12	0,1353	126,5

Table 5. Cases of measured traffic formed from the traffic samples

Measured traffic 1	All traffic samples
Measured traffic 2	All traffic samples except 164, 264, 364, 165, 265, 365, 171, 271, 371, 172, 272, 372, 173, 273 and 373
Measured traffic 3	All traffic samples with 2, 3, 4 and 5 as second digit
Measured traffic 4	All traffic samples with 1 or 2 as first digit except 164, 264, 165, 265, 171, 271, 172, 272, 173 and 273
Measured traffic 5	All traffic samples with 3 as first digit and 2, 3, 4 and 5 as second digit
Measured traffic 6	Traffic samples 245, 233, 243, 255, 231, 152, 155, 215, 153, 232, 161, 251 and 263
Measured traffic 7	All traffic samples in measured traffic 6 as well as 213, 223, 163, 164, 165, 224, 234, 221, 254, 162, 244, 241 and 172
Measured traffic 8	Traffic samples 334, 321, 323, 313 and 331
Measured traffic 9	All traffic samples in measured traffic 8 as well as 312, 333, 325, 324, 311, 335, 343, 315, 322 and 332
Measured traffic 10	All traffic samples in measured traffic 8 as well as 343, 315, 322, 332, 342, 352, 353 and 344
Measured traffic 11	Traffic samples 333, 325, 352, 335, 334, 323, 315, 331, 343, 321, 345, 353, 342, 322, 344, 341 and 354
Measured traffic 12	Traffic samples 333, 325, 352, 335, 334, 323, 315, 331, 343, 321 and 345
Measured traffic 13	All traffic samples but 235, 171, 173, 222, 175, 242, 275, 262, 271, 253, 265, 272, 273, 264 and 261
Measured traffic 14	Traffic samples 312, 333, 325, 324, 311, 335, 153, 232, 161, 251, 263, 375, 361, 254, 162, 244, 241, 252, 235, 222, 175, 242, 275, 262, 253 and 261

Table 6. Survey of data for the 14 selected cases of measured traffic

Measured traffic No.		1	2	3	4	5	6	7
Total measuring time in s		838 816	703 786	504 137	149 319	395 978	46 778	93 574
Total number of calls		98 686	79 276	55 677	19 771	41 674	6 285	12 335
A in traffic units		14,642	13,998	13,650	16,49	12,986	16,53	16,44
y calls/s		0,11765	0,11264	0,11044	0,1324	0,1052	0,1344	0,1318
s mean occupation time in s		124,46	124,28	123,57	124,74	123,39	123,0	124,7
Relative standard deviation of A	for random traffic	0,0414	0,0420	0,0424	0,0648	0,0310	0,0642	0,0649
	from traffic samples	0,1857	0,1225	0,1453	0,1556	0,0949	0,0167	0,0525
Relative standard deviation of y	for random traffic	0,0293	0,0297	0,0300	0,0458	0,0219	0,0454	0,0459
	from traffic samples	0,1703	0,1449	0,1145	0,1289	0,0501	0,0685	0,0934
Relative standard deviation of s	for random traffic	0,0293	0,0297	0,0300	0,0458	0,0219	0,0454	0,0459
	from traffic samples	0,0598	0,0612	0,0652	0,0811	0,0602	0,0619	0,0735

Measured traffic No.		8	9	10	11	12	13	14
Total measuring time in s		99 000	296 992	257 378	336 588	217 820	786 268	221 616
Total number of calls		10 022	29 888	26 832	35 084	22 216	90 093	25 837
A in traffic units		12,12	11,99	12,73	12,84	12,29	14,222	14,48
y calls/s		0,1012	0,1006	0,1043	0,1042	0,1020	0,11458	0,1166
s mean occupation time in s		119,7	119,2	122,1	123,2	120,5	124,12	124,2
Relative standard deviation of A	for random traffic	0,0315	0,0317	0,0311	0,0311	0,0315	0,0658	0,0447
	from traffic samples	0,0052	0,0470	0,0411	0,0831	0,0675	0,1643	0,2484
Relative standard deviation of y	for random traffic	0,0223	0,0224	0,0220	0,0220	0,0222	0,0465	0,0316
	from traffic samples	0,0339	0,0539	0,0398	0,0337	0,0209	0,1411	0,2052
Relative standard deviation of s	for random traffic	0,0223	0,0224	0,0220	0,0220	0,0222	0,0465	0,0316
	from traffic samples	0,0361	0,0330	0,0386	0,0618	0,0603	0,0565	0,0720

Table 7. Measured traffic 1. The relative moments μ_n. The expressions $\frac{1}{n-r}\{\log \mu_n - \log \mu_r\}$ provide a check of the conditions (153 b).

n	μ_n	$\log \mu_n$	$\frac{1}{n-r}\{\log \mu_n - \log \mu_r\}$										
			$r=0$	$r=1$	$r=2$	$r=3$	$r=4$	$r=5$	$r=6$	$r=7$	$r=8$	$r=9$	$r=10$
1	1,0000	0,00000	0,00000										
2	1,0343	0,01465	0,00733	0,01465									
3	1,1049	0,04332	0,01444	0,02166	0,02867								
4	1,2167	0,08518	0,02130	0,02839	0,03527	0,04186							
5	1,3776	0,13913	0,02783	0,03478	0,04149	0,04791	0,05395						
6	1,5976	0,20347	0,03391	0,04069	0,04721	0,05338	0,05915	0,06434					
7	1,8876	0,27591	0,03942	0,04599	0.05225	0,05820	0,06358	0,06839	0,07244				
8	2,2595	0,35402	0,04425	0,05057	0,05656	0,06214	0,06721	0,07163	0,07528	0,07811			
9	2,7221	0,43490	0,04832	0,05436	0,06004	0,06526	0,06994	0,07394	0,07714	0,07950	0,08088		
10	3,2799	0,51586	0,05159	0,05732	0,06265	0,06751	0,07178	0,07535	0,07810	0,07998	0,08092	0,08096	
11	3,9105	0,59223	0,05384	0,05922	0,06418	0,06861	0,07244	0,07562	0,07775	0,07908	0,07940	0,07867	0,07637
12	4,6097	0,66367	0,05531	0,06033	0,06490	0,06893	0,07231	0,07493	0,07670	0,07755	0,07741	0,07626	0,07391

Table 8. Measured traffic 1. The differences of the relative moments.

n	μ_n	$\Delta\mu$	$\Delta^2\mu$	$\Delta^3\mu$	$\Delta^4\mu$	$\Delta^5\mu$
1	1,000000					
		0,034314				
2	1,034314		0,036252			
		0,070566		0,005027		
3	1,104880		0,041279		0,002771	
		0,111845		0,007798		— 0,000646
4	1,216725		0,049077		0,002125	
		0,160922		0,009923		— 0,000980
5	1,377647		0,059000		0,001145	
		0,219922		0,011068		— 0,000306
6	1,597569		0,070068		0,000839	
		0,289990		0,011907		— 0,004152
7	1,887559		0,081975		— 0,003313	
		0,371965		0,008594		— 0,000579
8	2,259524		0,090569		— 0,003892	
		0,462534		0,004702		— 0,023293
9	2,722058		0,095271		— 0,027185	
		0,557805		— 0,022483		+ 0,045668
10	3,279863		0,072788		+ 0,018483	
		0,630593		— 0,004000		
11	3,910456		0,068628			
		0,699221				
12	4,609677					

Table 9. Measured traffic 1. η_0 computed from μ_n/μ_{n-r} as well as from the second difference of the difference scheme (154 e)

n	η_0 from $\mu_n : \mu_{n-r}$		η_0 from (154 e)
	$r = 1$	$r = 2$	
2	0,0343		
3	0,0341	0,0342	0,0311
4	0,0337	0,0339	0,0311
5	0,0331	0,0337	0,0298
6	0,0319	0,0324	0,0277
7	0,0303	0,0310	0,0251
8	0,0282	0,0291	0,0224
9	0,0256	0,0268	0,0191
10	0,0228	0,0241	0,0152
11	0,0192	0,0209	0,0066
12	0,0163	0,0177	0,0039

Table 10. Measured traffic 1. Difference scheme for μ_n according to (154 e)

n	μ_n	1. Difference	2. Difference	3. Difference
1	1,000000			
		0,034314		
2	1,034314		0,000969	
		0,035283		0,000030
3	1,104880		0,000999	
		0,037281		—
4	1,216725		0,000983	
		0,040231		—
5	1,377647		0,000938	
		0,043984		—
6	1,597569		0,000870	
		0,048332		—
7	1,887559		0,000801	
		0,053138		—
8	2,259524		0,000688	
		0,057817		—
9	2,722058		0,000520	
		0,061978		—
10	3,279863		0,000120	
		0,063059		—
11	3,910456		0,000051	
		0,063566		
12	4,609677			

Table 11. Measured traffic 1. The relative moments μ_n according to the measurements and computed from normal form 1 for $\eta_0 = 0.034314$ and from normal form 2 for $\eta_0 = \kappa = 0.017157$

n	μ_n Normal form 1	μ_n Normal form 2	μ_n Measured
1	1,000000	1,000000	1,000000
2	1,034314	1,034314	1,034314
3	1,105297	1,105003	1,104880
4	1,219078	1,218101	1,216725
5	1,386404	1,383006	1,377647
6	1,624269	1,616661	1,597569

Table 13. Measured traffic 1. $\alpha_{p,n}$ and the relative moments computed according to the measurements of combinations of occupied devices

p	$\alpha_{p,n}$	$\frac{A^n}{n(n-1)\ldots(n-p+1)}$	μ_p computed from $\alpha_{p,n}$	μ_p computed from the states
3	0,080741	0,073274	1,1019	1,1049
4	0,039520	0,032512	1,2156	1,2167
6	0,012611	0,007026	1,7948	1,5976
9	0,0028368	0,0009054	3,133	2,722
12	0,0007903	0,0001619	4,880	4,610

Table 12. Measured traffic 1. The states P_n and their adaptation to normal form 1

n	Number of recordings	P_n	$P_{n,0}$	$\frac{P_n}{P_{n,0}}$	$\log \frac{P_n}{P_{n,0}}$	η_0 from (157 c)	η_0 from (158)	η_0 from (156 a)	η_0 from (156 b)	Computation for $\eta_0 = 0,0369$				$\log P_n:P_{n,0}$ from (157 c) for $\eta_0=0,0343$	n
										P_n from (157 c)	$P_n:P_{n,0}$ from (157 c)	$\log P_n:P_{n,0}$ from (157 c)	$\log P_n:P_{n,0}$ from (158)		
0	2	0,00001	0,00000	—	—	0,030	—	—	—	0,00001	18,8	+ 1,274	+ 0,696	+ 1,206	0
1	64	0,00015	0,00001	25	+ 1,406	0,0516	0,265	—	—	0,00008	12,2	+ 1,086	+ 0,645	+ 1,030	1
2	192	0,00046	0,00005	9,7	+ 0,989	0,0411	0,111	0,135	—	0,00039	8,23	+ 0,915	+ 0,592	+ 0,868	2
3	626	0,00149	0,00023	6,52	+ 0,814	0,0408	0,0833	0,0424	0,0836	0,00131	5,74	+ 0,759	+ 0,537	+ 0,720	3
4	1 449	0,00346	0,00084	4,12	+ 0,615	0,0368	0,0572	0,0587	0,0497	0,00347	4,14	+ 0,617	+ 0,479	+ 0,585	4
5	2 860	0,00682	0,00245	2,78	+ 0,444	0,0322	0,0396	0,0555	0,0572	0,00756	3,08	+ 0,489	+ 0,425	+ 0,464	5
6	5 577	0,01330	0,00599	2,221	+ 0,3465	0,0329	0,0355	0,0300	0,0426	0,01494	2,370	+ 0,3748	+ 0,3555	+ 0,3565	6
7	9 643	0,02299	0,01253	1,836	+ 0,2638	0,0348	0,0325	0,0284	0,0292	0,02355	1,880	+ 0,2742	+ 0,2885	+ 0,2610	7
8	14 252	0,03398	0,02293	1,482	+ 0,1709	0,0323	0,0267	0,0460	0,0335	0,03520	1,536	+ 0,1864	+ 0,2209	+ 0,1777	8
9	19 672	0,04690	0,03730	1,2575	+ 0,0995	0,0308	0,0226	0,0343	0,0372	0,04816	1,2912	+ 0,1110	+ 0,1520	+ 0,1063	9
10	25 362	0,06047	0,05461	1,1073	+ 0,0443	0,0309	0,0186	0,0307	0,0325	0,06099	1,1167	+ 0,0479	+ 0,0835	+ 0,0465	10
11	30 685	0,07316	0,07269	1,0065	+ 0,0079	—	0,0057	0,0275	0,0292	0,07215	0,9925	— 0,0033	+ 0,0179	— 0,0022	11
12	33 416	0,07967	0,08870	0,8983	— 0,0466	0,0401	0,0405	0,0520	0,0369	0,08035	0,9059	— 0,0429	— 0,0420	— 0,0399	12
13	36 137	0,08616	0,09990	0,8625	— 0,0643	0,0328	0,0267	0,0194	0,0361	0,08478	0,8486	— 0,0713	— 0,0910	— 0,0669	13
14	36 604	0,08728	0,10448	0,8353	— 0,0782	0,0317	0,0242	0,0267	0,0219	0,08518	0,8152	— 0,0887	— 0,1245	— 0,0836	14
15	35 630	0,08495	0,10199	0,8330	— 0,0794	0,0296	0,0225	0,0047	0,0193	0,08186	0,8027	— 0,0954	— 0,1384	— 0,0900	15
16	32 876	0,07839	0,09334	0,8399	— 0,0758	0,0293	0,0226	0,0350	—	0,07556	0,8095	— 0,0918	— 0,1306	— 0,0865	16
17	28 464	0,06787	0,08039	0,8442	— 0,0735	0,0345	0,0272	0,0041	0,0089	0,06720	0,8359	— 0,0778	— 0,1023	— 0,0732	17
18	23 531	0,05611	0,06539	0,8580	— 0,0665	0,0462	0,0422	0,0077	0,0063	0,05775	0,8831	— 0,0540	— 0,0572	— 0,0504	18
19	19 601	0,04674	0,05039	0,9274	— 0,0327	0,0514	—	0,0372	0,0224	0,04808	0,9541	— 0,0214	— 0,0004	— 0,0183	19
20	15 886	0,03788	0,03689	1,0267	+ 0,0114	0,0706	0,0061	0,0384	0,0379	0,03888	1,0538	+ 0,0228	+ 0,0645	+ 0,0229	20
21	12 315	0,02936	0,02572	1,1415	+ 0,0575	0,0214	0,0146	0,0301	0,0337	0,03059	1,1890	+ 0,0752	+ 0,1324	+ 0,0730	21
22	9 764	0,02328	0,01712	1,360	+ 0,1335	0,0351	0,0224	0,0537	0,0414	0,02346	1,370	+ 0,1367	+ 0,2014	+ 0,1319	22
23	7 489	0,01786	0,01090	1,638	+ 0,2144	0,0394	0,0271	0,0470	0,0500	0,01757	1,612	+ 0,2074	+ 0,2707	+ 0,1993	23
24	5 503	0,01312	0,00665	1,973	+ 0,2952	0,0390	0,0306	0,0381	0,0421	0,01286	1,934	+ 0,2865	+ 0,3357	+ 0,2752	24
25	3 807	0,00908	0,00389	2,331	+ 0,3675	0,0357	0,0323	0,0270	0,0319	0,00922	2,370	+ 0,3748	+ 0,3997	+ 0,3593	25
26	2 731	0,00651	0,00219	2,97	+ 0,473	0,0372	0,0382	0,0431	0,0346	0,00648	2,96	+ 0,471	+ 0,461	+ 0,451	26
27	1 918	0,00457	0,00119	3,85	+ 0,585	0,0380	0,0453	0,0420	0,0425	0,00447	3,76	+ 0,575	+ 0,519	+ 0,552	27
28	1 215	0,00290	0,00062	4,66	+ 0,668	0,0350	0,0486	0,0228	0,0309	0,00303	4,87	+ 0,688	+ 0,575	+ 0,660	28
29	888	0,00212	0,00031	6,74	+ 0,829	0,0386	0,0648	0,0657	0,0400	0,00202	6,43	+ 0,808	+ 0,628	+ 0,775	29
30	554	0,00132	0,00015	8,63	+ 0,936	0,0369	0,0742	0,0274	0,0422	0,00132	8,64	+ 0,937	+ 0,679	+ 0,898	30
31	359	0,00086	0,00007	11,9	+ 1,075	0,0370	0,0921	0,0383	0,0459	0,00085	11,8	+ 1,072	+ 0,728	+ 1,029	31
32	235	0,00056	0,00003	17,0	+ 1,230	0,0376	0,119	0,0423	0,0403	0,00054	16,5	+ 1,218	+ 0,774	+ 1,167	32
33	81	0,00019	0,00002	12,9	+ 1,111	0,0261	0,078	—	—	0,00035	23,3	+ 1,367	+ 0,825	+ 1,312	33
34	17	0,00004	0,00001	7	+ 0,833	—	—	—	—	0,00020	33	+ 1,519	+ 0,861	+ 1,464	34
35	3	0,00001	0,00000	2	+ 0,362	—	—	—	—	0,00013	49	+ 1,690	+ 0,905	+ 1,623	35
36	0	0,00000	0,00000	—	—	—	—	—	—	0,00007	72	+ 1,857	+ 0,942	+ 1,788	36

Table 14. Measured traffic 2. Distribution function of the inter-arrival times

Counter No.	Time t in s	Time t with $1/y$ as time unit	Number of recordings	$\varphi(t)$	e^{-yt}	$D(t)$	η_0 from $D(t)$ and (128)
Sum	0	0	79 276	1,0000	1,0000	0,0000	
7	0,182	0,021	77 491	0,9775	0,9797	— 0,0022	0,113
8	0,482	0,054	74 847	0,9441	0,9472	— 0,0030	0,060
9	0,986	0,111	70 460	0,8888	0,8949	— 0,0061	0,065
10	1,481	0,167	66 533	0,8393	0,8463	— 0,0070	0,054
11	2,452	0,276	59 302	0,7481	0,7587	— 0,0106	0,059
12	3,452	0,389	52 935	0,6677	0,6778	— 0,0101	0,048
13	4,452	0,501	47 085	0,5939	0,6056	— 0,0117	0,051
14	5,452	0,614	41 932	0,5289	0,5411	— 0,0122	0,053
15	6,452	0,727	37 304	0,4706	0,4835	— 0,0129	0,058
16	7,452	0,839	33 247	0,4194	0,4320	— 0,0126	0,060
17	8,452	0,952	29 697	0,3746	0,3860	— 0,0114	0,059
18	9,452	1,065	26 545	0,3348	0,3448	— 0,0100	0,058
19	10,452	1,177	23 760	0,2997	0,3081	— 0,0084	0,056
20	11,452	1,290	21 286	0,2685	0,2752	— 0,0067	0,054
21	12,452	1,403	19 014	0,2398	0,2460	— 0,0061	0,059
22	13,452	1,515	17 010	0,2146	0,2197	— 0,0052	0,064
23	14,452	1,628	15 204	0,1918	0,1963	— 0,0046	0,076
24	15,452	1,741	13 611	0,1717	0,1754	— 0,0037	0,094
25	16,452	1,853	12 197	0,1539	0,1567	— 0,0029	0,136
26	17,452	1,966	10 942	0,1380	0,1400	— 0,0020	0,429
27	18,452	2,078	9 824	0,1239	0,1251	— 0,0012	—
28	19,452	2,191	8 918	0,1125	0,1118	+ 0,0007	0,030
29	20,452	2,304	8 031	0,1013	0,0999	+ 0,0014	0,041
30	21,452	2,416	7 246	0,0914	0,0892	+ 0,0022	0,048
31	23,452	2,642	5 869	0,0740	0,0712	+ 0,0028	0,046
32	25,452	2,867	4 753	0,0600	0,0569	+ 0,0031	0,044
33	27,452	3,092	3 881	0,0490	0,0454	+ 0,0036	0,046
34	29,452	3,178	3 176	0,0401	0,0362	+ 0,0038	0,048
35	31,452	3,543	2 587	0,0326	0,0289	+ 0,0037	0,047
36	33,452	3,768	2 143	0,0270	0,0231	+ 0,0039	0,051
37	35,452	3,993	1 742	0,0220	0,0184	+ 0,0035	0,048
38	37,452	4,219	1 428	0,0180	0,0147	+ 0,0033	0,048
39	39,452	4,444	1 181	0,0149	0,0118	+ 0,0032	0,049
40	41,452	4,669	991	0,0125	0,0094	+ 0,0031	0,053
41	46,452	5,232	623	0,0079	0,0053	+ 0,0025	0,056
42	51,452	5,796	397	0,0050	0,0030	+ 0,0020	0,059
43	56,452	6,359	245	0,0031	0,0014	+ 0,0014	0,057
44	61,452	6,922	171	0,0022	0,0010	+ 0,0012	0,070
45	66,452	7,485	116	0,0015	0,0006	+ 0,0009	0,078
46	71,452	8,048	87	0,0011	0,0003	+ 0,0008	0,100
47	76,452	8,612	57	0,0007	0,0002	+ 0,0005	0,104
48	81,452	9,175	39	0,0005	0,0001	+ 0,0004	0,114
49	86,452	9,738	31	0.0004	0,0001	+ 0,0003	0,149
50	91,452	10,301	23	0.0003	—	+ 0,0003	0,182
51	96,452	10,865	14	0,0002	—	+ 0,0002	0,174

Table 15. Measured traffic 2. Difference quotients formed from measured values of φ (t)

Table No.	Time t in s	φ (t)	First quotient times 10^6	Second quotient times 10^7	Third quotient times 10^8	Fourth quotient times 10^9	Fifth quotient times 10^{10}
Sum	0	1,00000	— 108 535	+ 62 306	— 32 839	+ 26 394	— 22 086
10	1,481	0,83926	— 87 027	+ 44 402	— 13 170	+ 5 518	— 12 349
12	3,452	0,66773	— 69 395	+ 36 538	— 8 772	— 6 795	+ 9 373
14	5,452	0,52894	— 54 780	+ 31 275	— 14 208	+ 4 452	+ 728
16	7,452	0,41938	— 42 270	+ 22 750	— 9 756	+ 5 471	— 3 956
18	9,452	0,33484	— 33 170	+ 14 945	— 3 191	— 859	+ 1 506
20	11,452	0,26850	— 24 203	+ 11 754	— 4 394	+ 1 550	— 555
24	15,452	0,17169	— 14 800	+ 6 481	— 1 914	+ 439	— 90
28	19,452	0,11249	— 9 615	+ 4 184	— 1 211	+ 258	— 40
31	23,452	0,07403	— 6 268	+ 2 731	— 798	+ 171	— 31
33	27,452	0,04896	— 4 083	+ 1 773	— 490	+ 85	— 8
35	31,452	0,03263	— 2 665	+ 1 087	— 285	+ 58	— 10
37	35,452	0,02197	— 1 578	+ 518	— 110	+ 18	— 3
40	41,452	0.01250	— 749	+ 232	— 47	+ 6	— 1
42	51,452	0,00501	— 285	+ 90	— 22	+ 5	
44	61,452	0,00216	— 106	+ 23	— 1		
46	71,452	0,00110	— 61	+ 20			
48	81,452	0,00049	— 20				
50	91,452	0,00029					

Table 16. Measured traffic 2. Ordinary differences formed from measured smoothed φ (t)-curve

Time t with $1/y$ as unit	Smothed φ (t)	$\Delta\varphi$	$\Delta^2\varphi$	$\Delta^3\varphi$	$\Delta^4\varphi$	$\Delta^5\varphi$	$\Delta^6\varphi$	$\Delta^7\varphi$	$\Delta^8\varphi$
0	1,00000	— 0,64282	+ 0,41938	— 0,27649	+ 0,18288	— 0,11966	+ 0,07493	— 0,04055	+ 0,01018
1	0,35718	— 0,22344	+ 0,14289	— 0,09361	+ 0,06322	— 0,04473	+ 0,03438	— 0,03037	+ 0,03191
2	0,13374	— 0,08055	+ 0,04928	— 0,03039	+ 0,01849	— 0,01035	+ 0,00401	+ 0,00154	— 0,00638
3	0,05319	— 0,03127	+ 0,01889	— 0,01190	+ 0,00814	— 0,00634	+ 0,00555	— 0,00484	+ 0,00311
4	0,02192	— 0,01238	+ 0,00699	— 0,00376	+ 0,00180	— 0,00079	+ 0,00071	— 0,00173	+ 0,00416
5	0,00954	— 0,00539	+ 0,00323	— 0,00196	+ 0,00101	— 0,00008	— 0,00102	+ 0,00243	
6	0,00415	— 0,00216	+ 0,00127	— 0,00095	+ 0,00093	— 0,00110	+ 0,00141		
7	0,00199	— 0,00089	+ 0,00032	— 0,00002	— 0,00017	+ 0,00031			
8	0,00110	— 0,00057	+ 0,00031	— 0,00019	+ 0,00014				
9	0,00053	— 0,00026	+ 0,00012	— 0,00005					
10	0,00027	— 0,00014	+ 0,00007						
11	0,00013	— 0,00007							
12	0,00006								

Table 17. Measured traffic 2. Adaptation of the measured difference curve $D(t)$ to normal form 1

yt	$D(t)$	η_0	$D_{\eta_0}(t)$ for $\eta_0 = 0,0550$	$D(t) - D_{\eta_0}(t)$
0,2	— 0,0082	0,056	— 0,0080	— 0,0002
0,4	— 0,0110	0,052	— 0,0116	+ 0,0006
0,6	— 0,0125	0,055	— 0,0124	— 0,0001
0,8	— 0,0127	0,061	— 0,0115	— 0,0012
1,0	— 0,0107	0,060	— 0,0098	— 0,0009
1,2	— 0,0081	0,057	— 0,0077	— 0,0004
1,4	— 0,0061	0,060	— 0,0056	— 0,0005
1,6	— 0,0046	0,071	— 0,0036	— 0,0010
1,8	— 0,0033	0,098	— 0,0018	— 0,0015
2,0	— 0,0016	—	— 0,0002	— 0,0014
2,2	+ 0,0007	0,031	+ 0,0010	— 0,0003
2,4	+ 0,0020	0,057	+ 0,0020	—
2,6	+ 0,0026	0,052	+ 0,0027	— 0,0001
2,8	+ 0,0031	0,050	+ 0,0032	— 0,0001
3,0	+ 0,0034	0,051	+ 0,0036	— 0,0002
3,5	+ 0,0039	0,054	+ 0,0040	— 0,0001
4,0	+ 0,0036	0,053	+ 0,0037	— 0,0001
4,5	+ 0,0032	0,053	+ 0,0032	—
5,0	+ 0,0028	0,056	+ 0,0027	+ 0,0001
6,0	+ 0,0017	0,053	+ 0,0017	—
7,0	+ 0,0011	0,058	+ 0,0010	+ 0,0001
8,0	+ 0,0008	0,068	+ 0,0006	+ 0,0002
9,0	+ 0,0004	0,065	+ 0,0003	+ 0,0001
10,0	+ 0,0002	0,073	+ 0,0002	—
12,0	+ 0,0001	0,055	+ 0,0001	—

Table 18. The relative moments μ_n for all cases of measured traffic

Measured traffic No.	μ_1	μ_2	μ_3	μ_4	μ_5	μ_6
1	1,0000	1,0343	1,1049	1,2167	1,3776	1,5976
2	1,0000	1,0273	1,0824	1,1686	1,2919	1,4611
3	1,0000	1,0181	1,0521	1,1009	1,1636	1,2397
4	1,0000	1,0107	1,0326	1,0668	1,1151	1,1789
5	1,0000	1,0128	1,0366	1,0709	1,1160	1,1678
6	1,0000	0,9896	0,9687	0,9354	0,8911	0,8357
7	1,0000	0,9925	0,9771	0,9533	0,9209	0,8797
8	1,0000	0,9980	0,9923	0,9804	0,9586	0,9236
9	1,0000	1,0083	1,0214	1,0365	1,0510	1,0624
10	1,0000	1,0014	1,0025	1,0015	0,9966	0,9859
11	1,0000	1,0092	1,0252	1,0458	1,0693	1,0940
12	1,0000	1,0065	1,0162	1,0254	1,0301	1,0269
13	1,0000	1,0251	1,0744	1,1491	1,2511	1,3830
14	1,0000	1,0581	1,1715	1,3443	1,5858	1,9099

Table 19. η_0 computed from μ_n/μ_{n-1} for all cases of measured traffic except 6, 7 and 8. In the last column η_0 is computed from $P_n/P_{n,0}$ in the vicinity of $n = A$.

Measured traffic No.	$n = 2$	$n = 3$	$n = 4$	$n = 5$	$n = 6$	η_0 from $P_n : P_{n,0}$
1	0,0343	0,0341	0,0337	0,0331	0,0319	0,0305
2	0,0273	0,0268	0,0266	0,0264	0,0262	0,0270
3	0,0181	0,0167	0,0155	0,0143	0,0131	0,0180
4	0,0107	0,0108	0,0111	0,0113	0,0114	0,0061
5	0,0128	0,0118	0,0110	0,0105	0,0093	0,0154
9	0,0083	0,0065	0,0049	0,0035	0,0022	0,0119
10	0,0014	0,0006	—	—	—	0,0059
11	0,0092	0,0079	0,0067	0,0056	0,0046	0,0141
12	0,0065	0,0048	0,0030	0,0012	—	0,0138
13	0,0251	0,0241	0,0232	0,0222	0,0211	0,0236
14	0,0581	0,0536	0,0492	0,0449	0,0409	0,0761

Table 20. The relative moments μ_n for different cases of measured traffic, computed from the combinations of occupations (B) in comparison with certain moments (Z) computed from the states

Measured traffic No.	μ_3		μ_4		μ_6		μ_9	μ_{12}
	B	Z	B	Z	B	Z	B	B
1	1,102	1,105	1,216	1,217	1,795	1,598	3,133	4,88
2	1,079	1,082	1,162	1,169	1,714	1,461	2,659	1,05
3	1,211	1,052	1,267	1,101	1,732	1,240	2,457	2,00
4	1,123	1,033	1,168	1,067	1,374	1,179	1,611	1,33
5	1,226	1,037	1,252	1,071	1,752	1,168	2,564	1,90
6	1,049	0,969	1,037	0,935	0,961	0,836	0,591	0,68
7	1,062	0,977	1,066	0,953	1,101	0,880	0,887	0,63
8	1,223	0,992	1,306	0,980	1,724	0,924	1,347	—
9	1,151	1,021	1,264	1,037	1,812	1,062	1,157	1,83
10	1,103	1,003	1,179	1,002	1,580	0,986	1,484	0,85
11	1,216	1,025	1,200	1,046	1,642	1,094	2,173	0,66
12	1,225	1,016	1,441	1,025	1,683	1,027	1,730	0,58
13	1,094	1,074	1,163	1,149	1,663	1,383	2,485	2,13
14	1,299	1,172	1,449	1,344	2,150	1,910	3,366	5,38

Table 21. Comparison between the deviations for different cases of measured traffic.

Measured traffic No.	η_{0Z}	η_{0A}	$\frac{\eta_{0A}}{\eta_{0Z}}$	k_A	k_y	k_s
1	0,0343	—	—	20	34	4,2
2	0,0273	0,0550	2,01	8,5	24	4,2
3	0,0181	0,0470	2,60	12	15	4,7
4	0,0107	—	—	5,8	7,9	3,1
5	0,0128	0,0455	3,55	9,3	5,2	7,5
6	—	—	—	0,07	2,3	1,9
7	—	—	—	0,66	4,1	2,6
8	—	—	—	0,03	2,3	2,6
9	0,0083	0,0422	5,08	2,2	5,8	2,2
10	0,0014	0,0428	30	1,8	3,3	3,1
11	0,0092	0,0416	4,5	7,1	2,3	7,9
12	0,0065	0,0350	5,4	4,6	0,90	7,4
13	0,0251	—	—	6,2	9,2	1,5
14	0,0581	0,0800	1,38	31	42	5,2

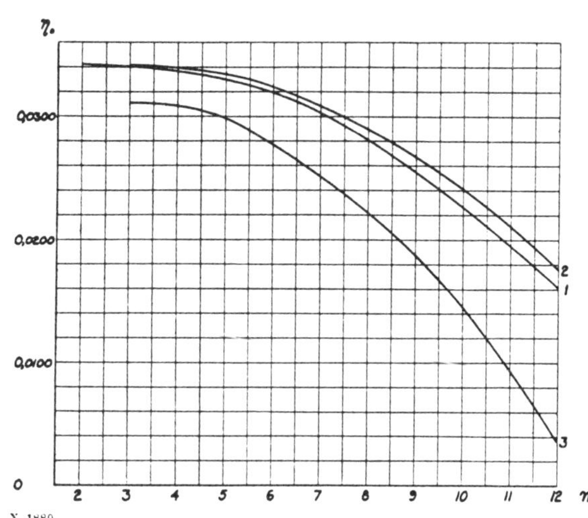

X 1880

Fig. 25. Measured traffic 1. η_0 computed from the relative moments μ_n
Curve 1: Computed from μ_n/μ_{n-1}
Curve 2: Computed from μ_n/μ_{n-2}
Curve 3: Computed from the second differences in (154 e).

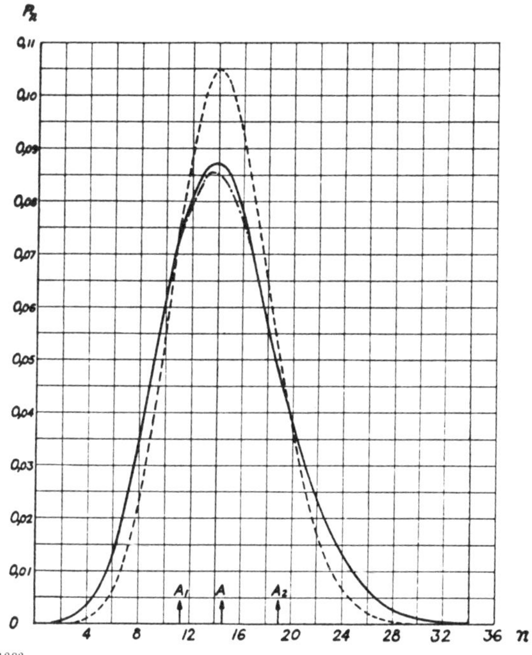

X 1883

Fig. 28. Measured traffic 1. Measured and computed states
——— measured P_n
– – – computed $P_{n,0}$ for random traffic.
– · – · computed P_n for $\eta_0 = 0.0369$, remaining parts coincide with the measured P_n

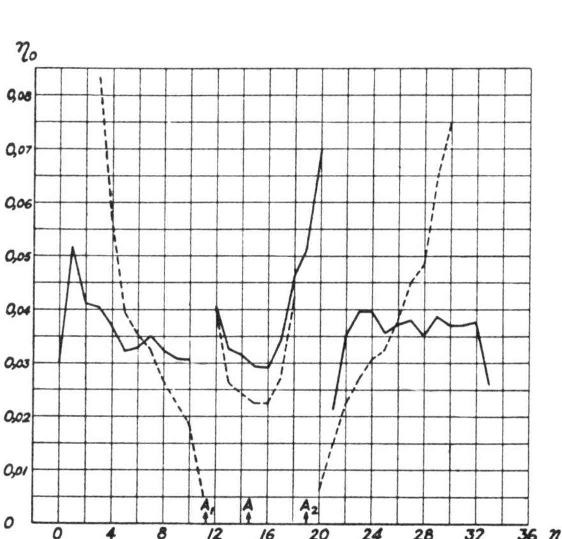

X 1881

Fig. 26. Measured traffic 1. η_0 computed from $P_n/P_{n,0}$
——— computed from (157 c)
– – – computed from (158)

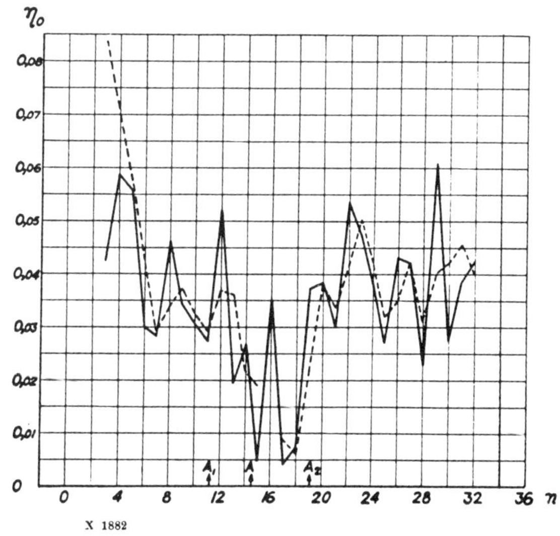

X 1882

Fig. 27. Measured traffic 1. η_0 computed from
——— P_n/P_{n-1} according to (156 a)
– – – P_n/P_{n-2} according to (156 b)

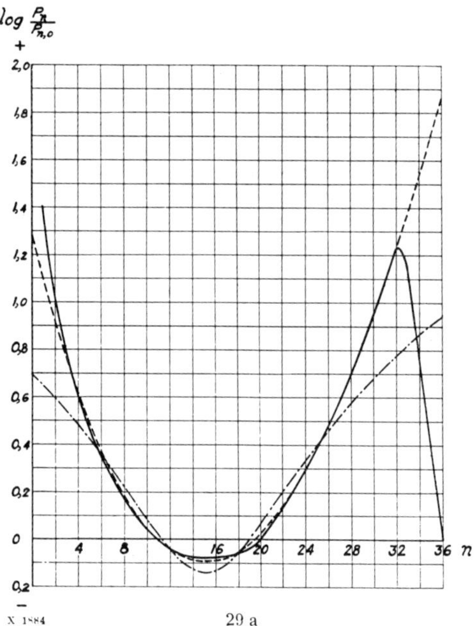

29 a

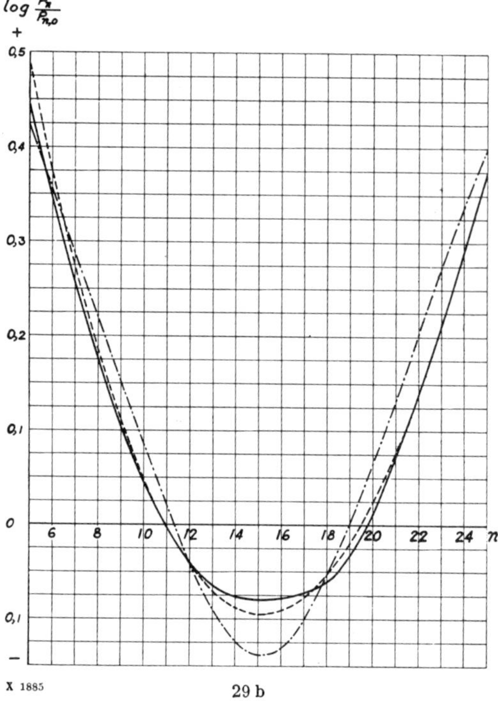

29 b

Fig. 29. Measured traffic 1. log $P_n/P_{n,0}$

Figs. 29 a and b:

——— according to the measurement results.

– – – according to normal form 1 for $\eta_0 = 0.0369$.

— · — · according to the approximate formula (158) for $\eta_0 = 0.0369$.

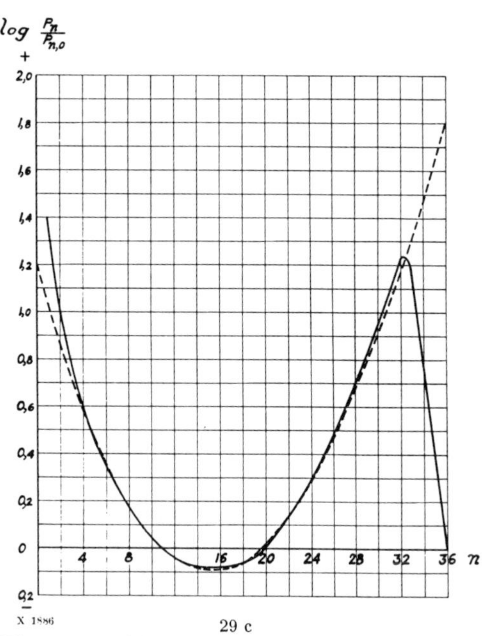

29 c

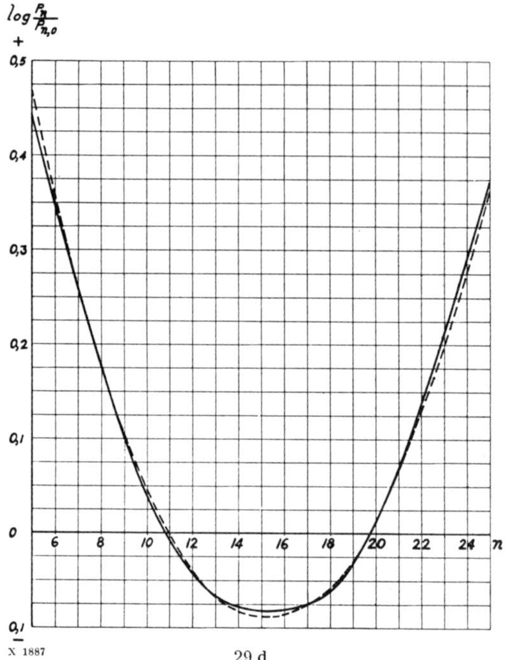

29 d

Figs. 29 c and d:

——— according to the measurement results.

– – – according to normal form 1 for $\eta_0 = 0.0343$.

Fig. 30. Measured traffic 2. η_0 computed from the measured points of the difference curve. On the abscissa the counter numbers are indicated

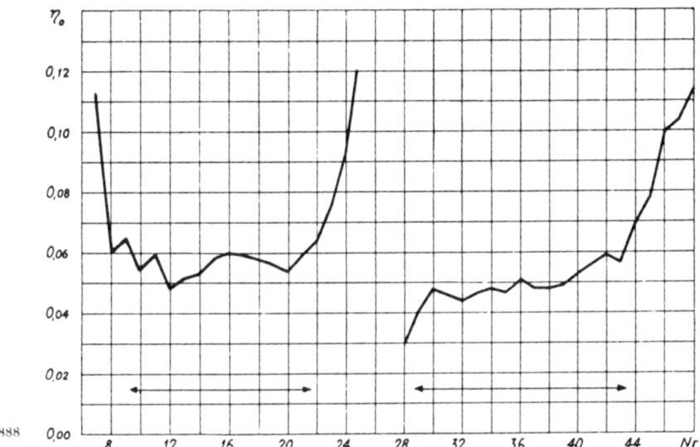

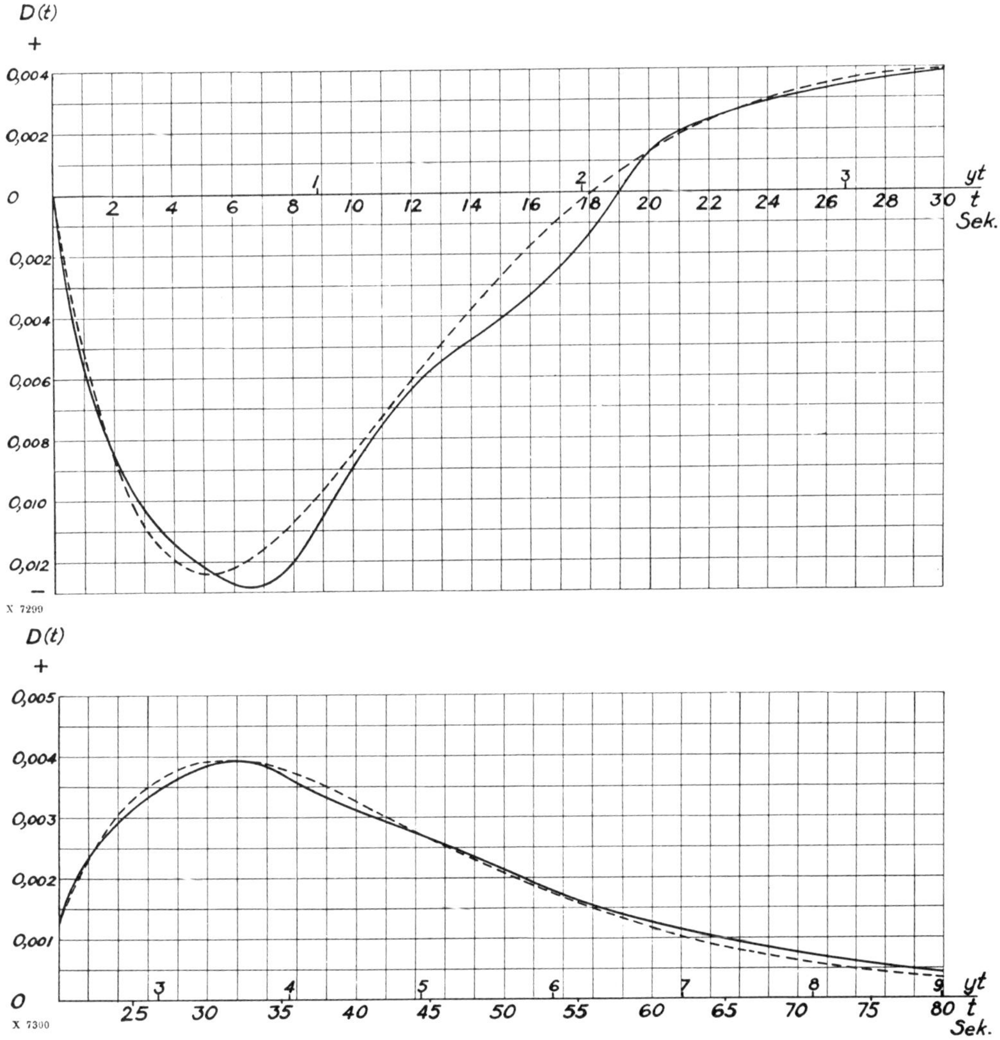

Fig. 31. Measured traffic 2. The difference curve of the distribution function of the inter-arrival times. The upper and lower figures show different domains with different scales

——— measured difference curve $D(t)$

– – – computed difference curve $D\eta_0(t)$ for normal form 1 and at $\eta_0 = 0.0550$.

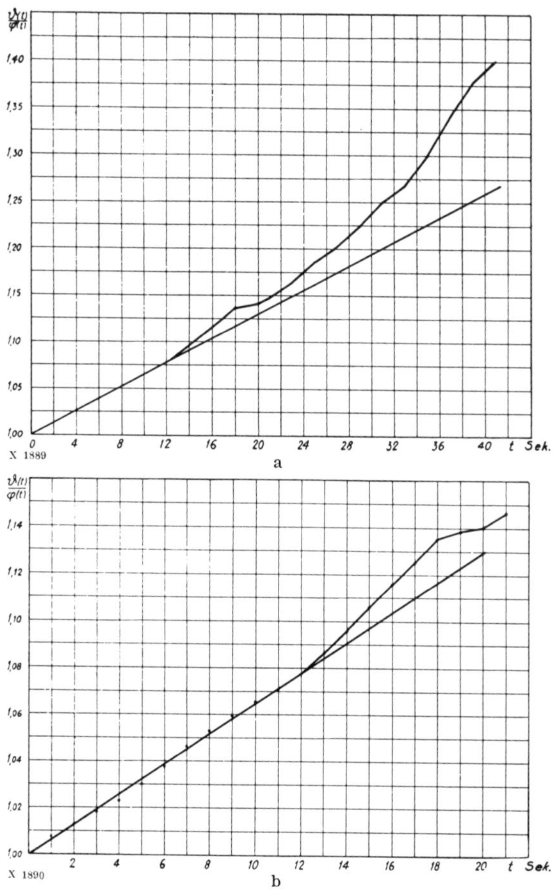

Fig. 32. Measured traffic 2. $\vartheta(t)/\varphi(t)$ as a function of t. Figs. 32 a and b show different domains with different scales. The straight line represents the asymptote from origo. The points corresponding to the measured values are partly interconnected by a broken line

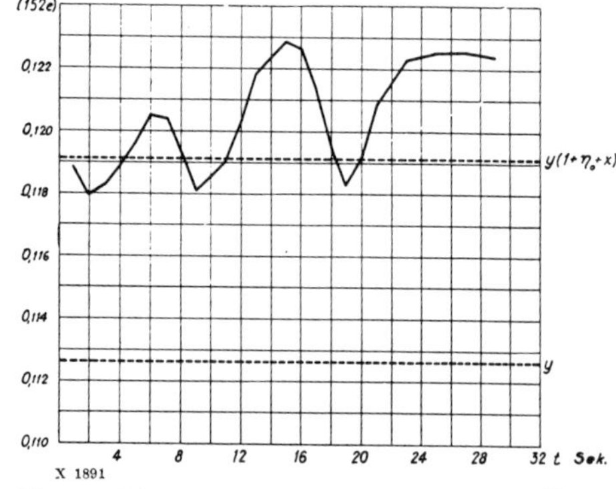

Fig. 33. Measured traffic 2. The left-hand member of the expression (152 e) as a function of t. The broken line interconnects the measured points

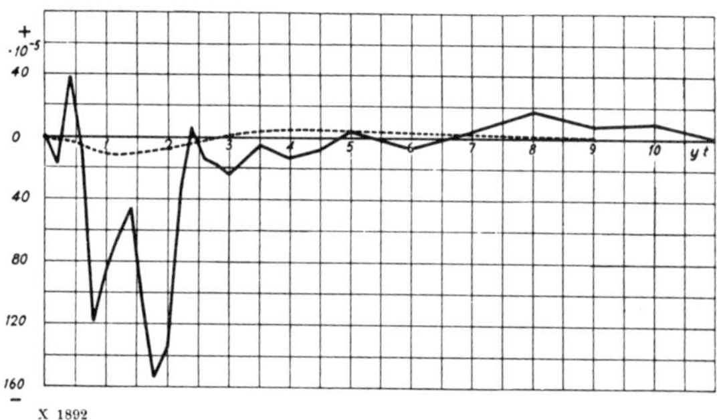

Fig. 34. Measured traffic 2. Difference between two difference curves

——— $D(t) - D\eta_0(t)$ at $\eta_0 = 0.0550$.

– – – $D_{\eta_0, \kappa}(t) - D_{\eta_0}(t)$ at $\eta_0 = \kappa = 0.0287$

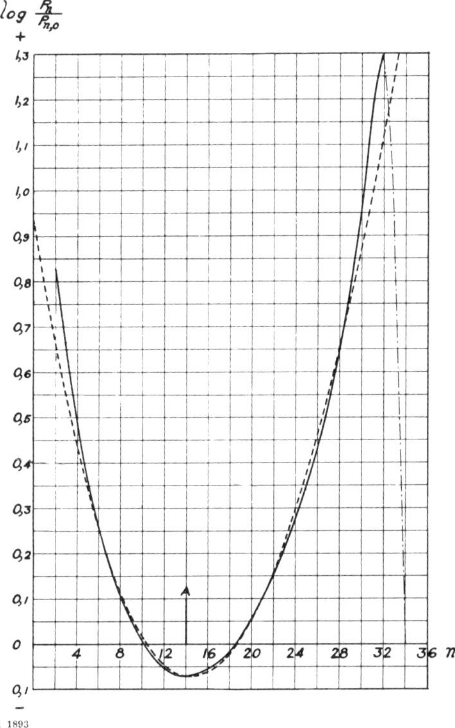

Fig. 35. Measured traffic 2. log $P_n/P_{n,0}$ as a function of n
——— measured
– – – computed according to normal form 1 for $\eta_0 = 0.0273$

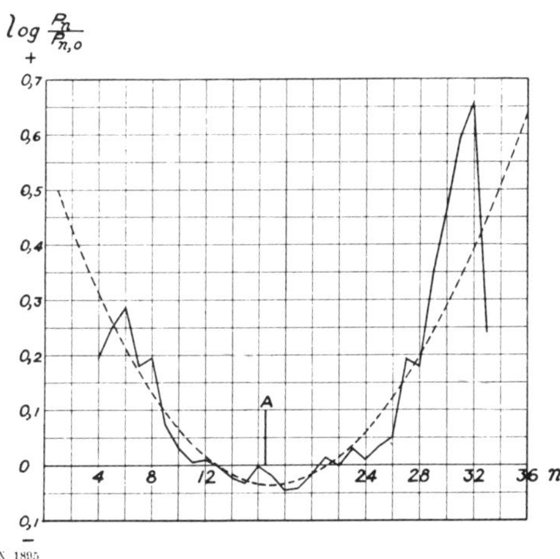

Fig. 37. Measured traffic 4. log $P_n/P_{n,0}$ as a function of n
——— measured
– – – computed according to normal form 1 for $\eta_0 = 0.0107$

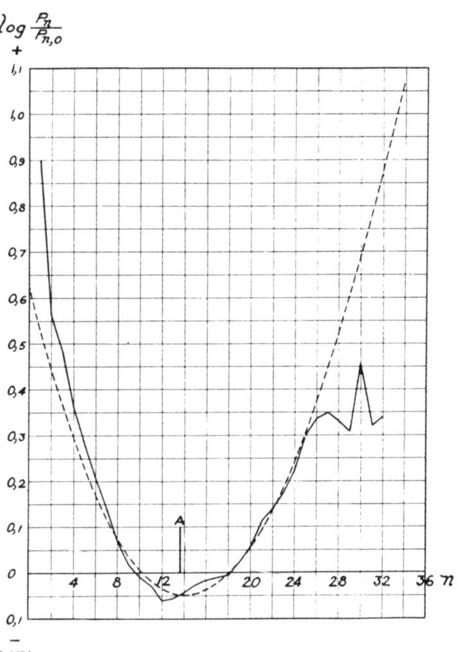

Fig. 36. Measured traffic 3. log $P_n/P_{n,0}$ as a function of n
——— measured
– – – computed according to normal form 1 for $\eta_0 = 0.0181$

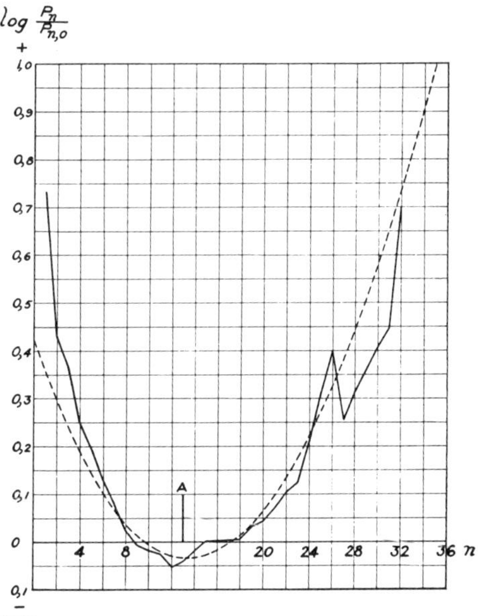

Fig. 38. Measured traffic 5. log $P_n/P_{n,0}$ as a function of n
——— measured
– – – computed according to normal form 1 for $\eta_0 = 0.0128$

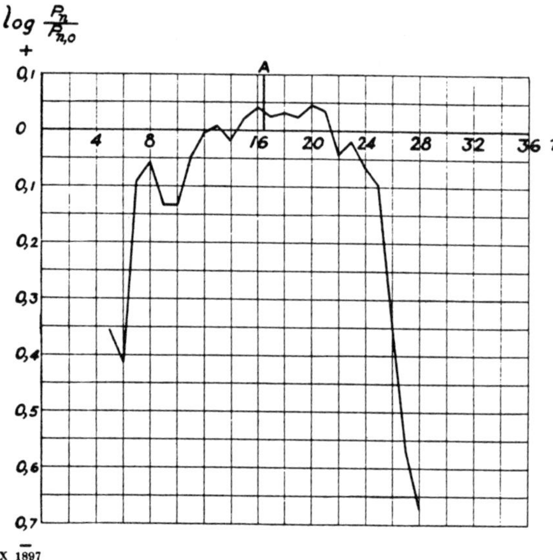

Fig. 39. Measured traffic 6. log $P_n/P_{n,\,0}$ as a function of n

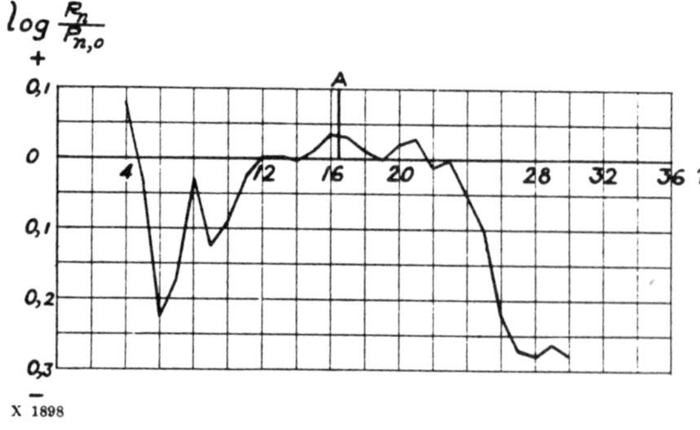

Fig. 40. Measured traffic 7. log $P_n/P_{n,\,0}$ as a function of n

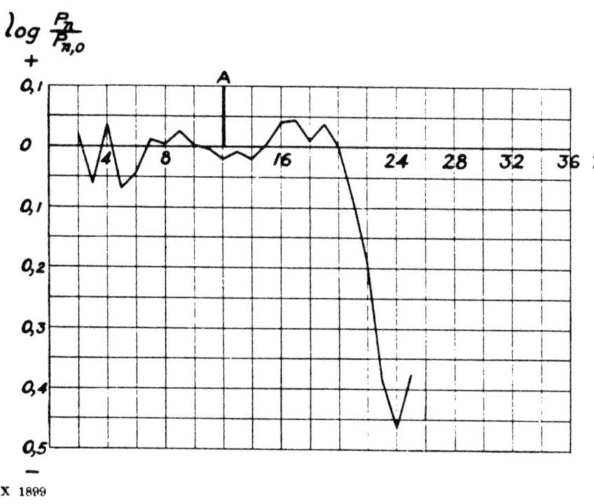

Fig. 41. Measured traffic 8. log $P_n/P_{n,\,0}$ as a function of n

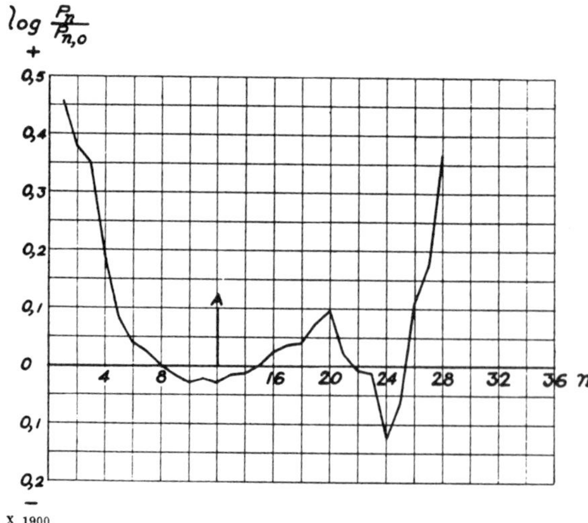

Fig. 42. Measured traffic 9. log $P_n/P_{n,\,0}$ as a function of n

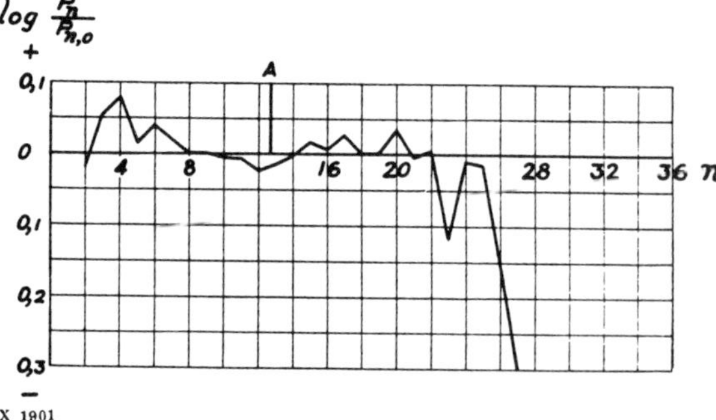

Fig. 43. Measured traffic 10. log $P_n/P_{n,\,0}$ as a function of n

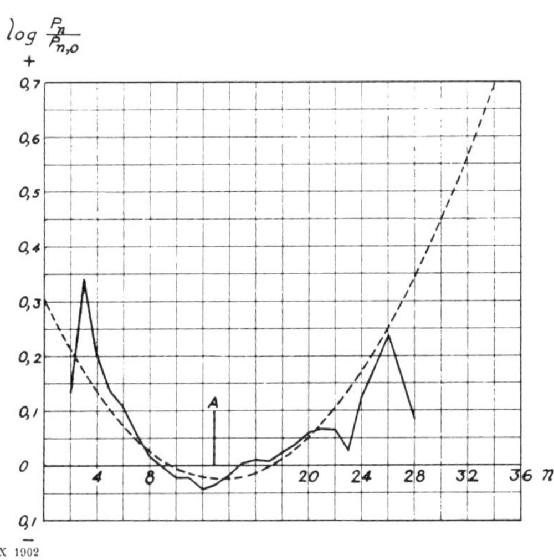

Fig. 44. Measured traffic 11. log $P_n/P_{n,\,0}$ as a function of n

——— measured

– – – computed according to normal form 1 for $\eta_0 = 0.0092$

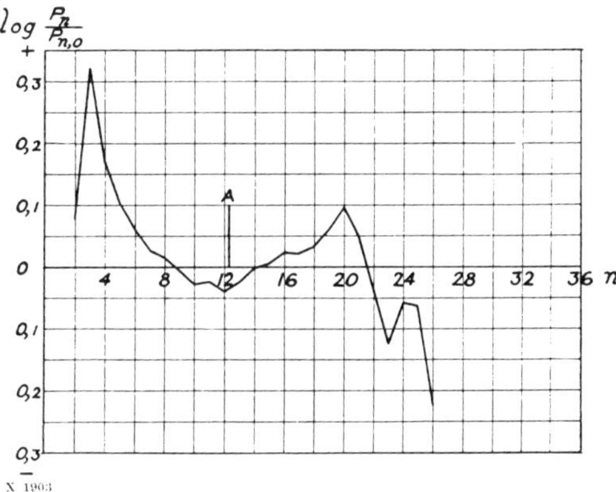

Fig. 45. Measured traffic 12. log $P_n/P_{n,\,0}$ as a function of n

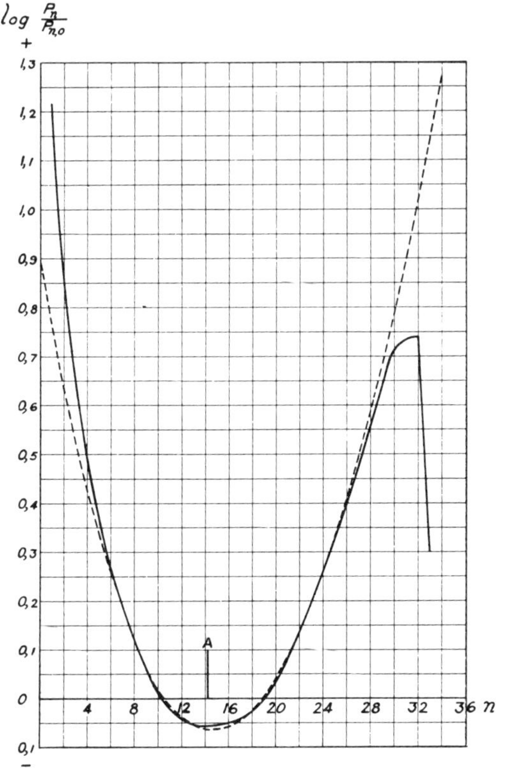

Fig. 46. Measured traffic 13. log $P_n/P_{n,\,0}$ as a function of n

——— measured

– – – computed according to normal form 1 for $\eta_0 = 0.0251$

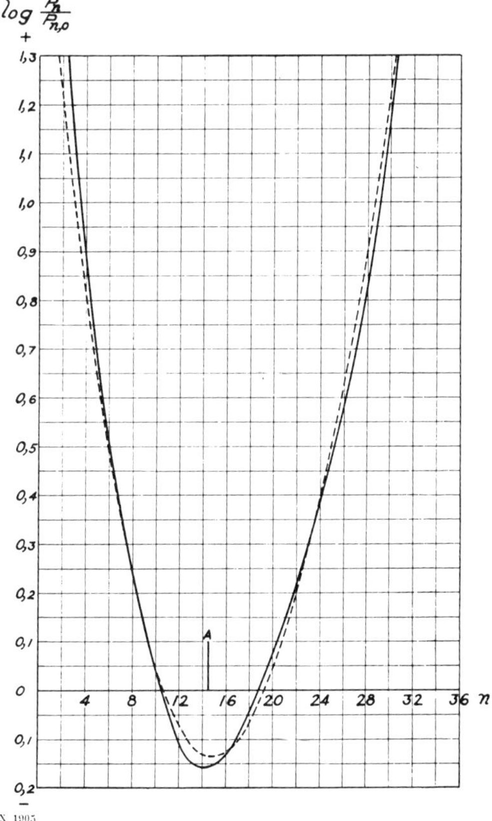

Fig. 47. Measured traffic 14. log $P_n/P_{n,\,0}$ as a function of n

——— measured

– – – computed according to normal form 1 for $\eta_0 = 0.0581$

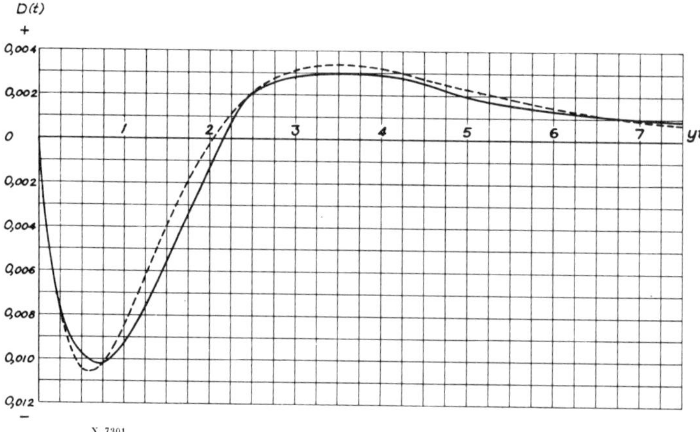

Fig. 48. Measured traffic 3. The difference curve of the inter-arrival times
——— measured
– – – computed according to normal form 1 for $\eta_0 = 0.0470$

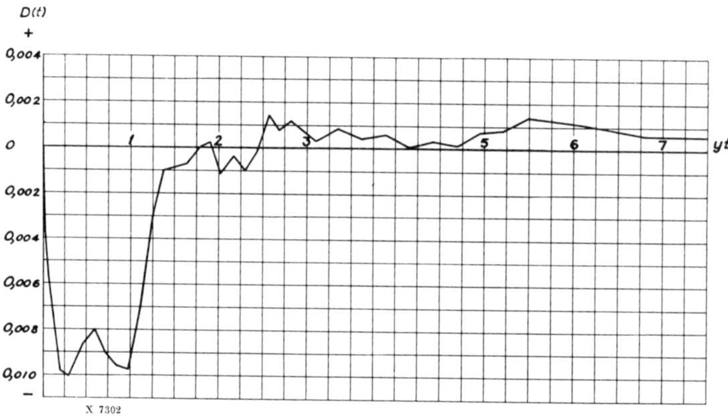

Fig. 49. Measured traffic 4. The difference curve of the inter-arrival times

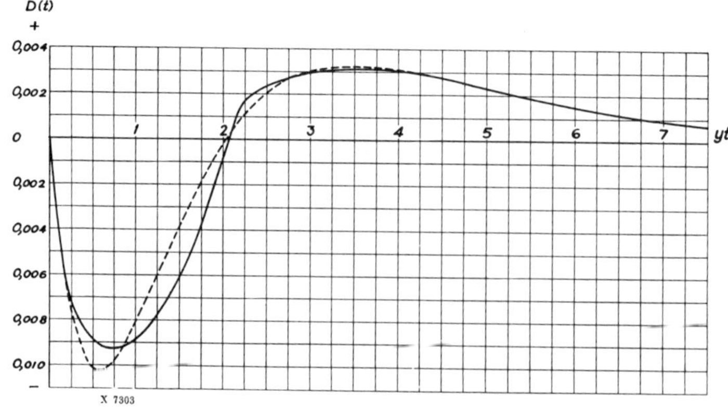

Fig. 50. Measured traffic 5. The difference curve of the inter-arrival times
——— measured
– – – computed according to normal form 1 for $\eta_0 = 0.0455$

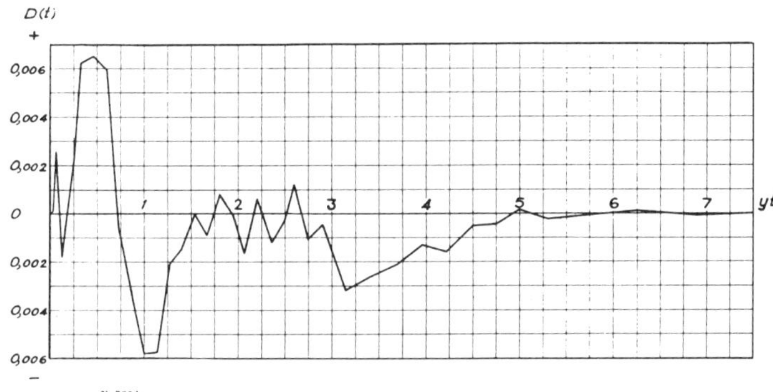

Fig. 51. Measured traffic 6. The difference curve of the inter-arrival times

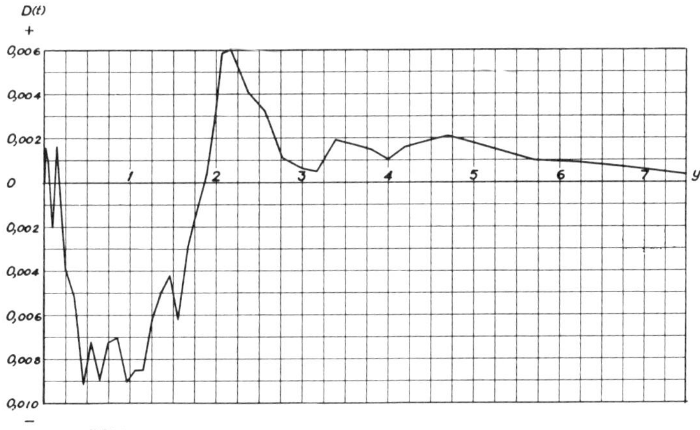

Fig. 52. Measured traffic 8. The difference curve of the inter-arrival times

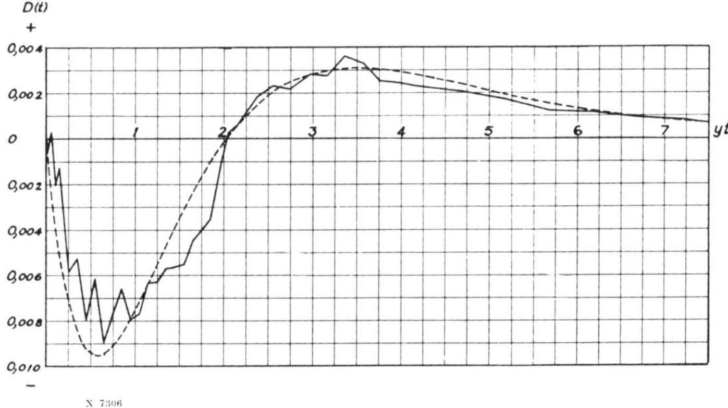

Fig. 53. Measured traffic 9. The difference curve of the inter-arrival times

——— measured

– – – computed according to normal form 1 for $\eta_0 = 0.0422$

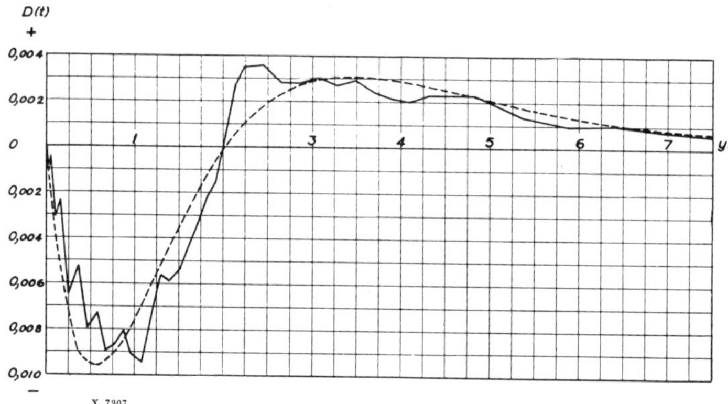

Fig. 54. Measured traffic 10. The difference curve of the inter- arrival times
——— measured
– – – computed according to normal form 1 for $\eta_0 = 0.0428$

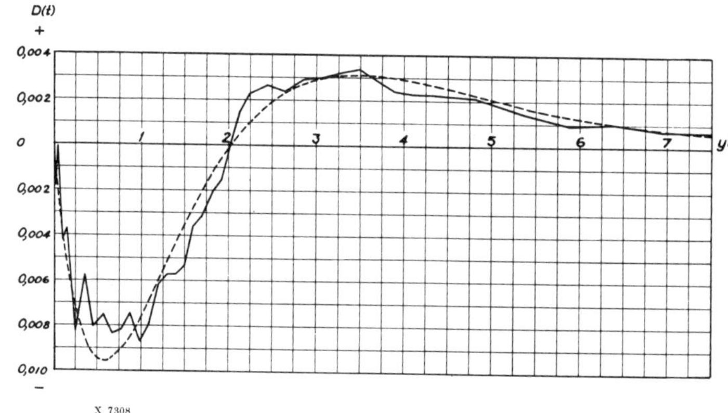

Fig. 55. Measured traffic 11. The difference curve of the inter-arrival times
——— measured
– – – computed according to normal form 1 for $\eta_0 = 0.0416$

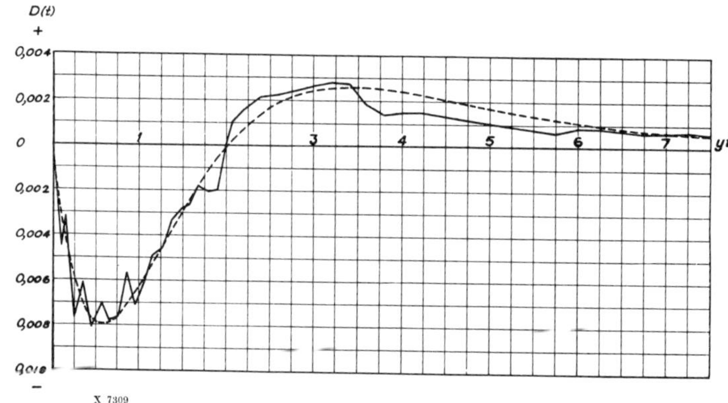

Fig. 56. Measured traffic 12. The difference curve of the inter-arrival times
——— measured
– – – computed according to normal form 1 for $\eta_0 = 0.0350$

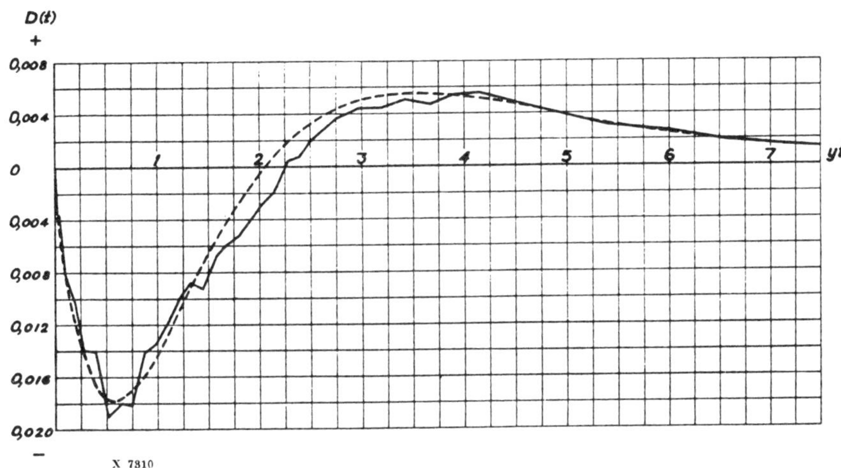

Fig. 57. Measured traffic 14. The difference curve of the inter-arrival times
——— measured
– – – computed according to normal form 1 for $\eta_0 = 0.0800$

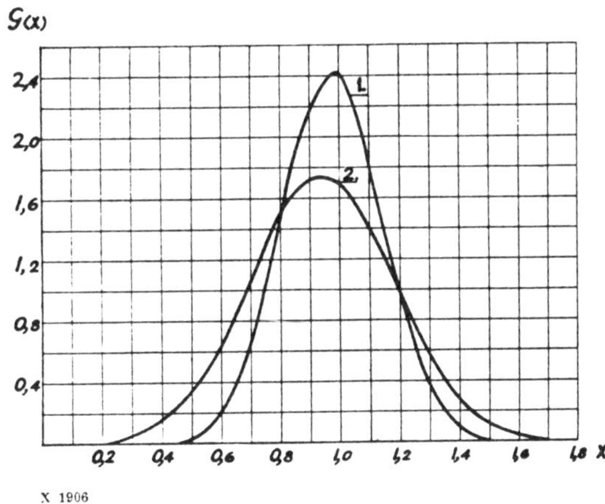

Fig. 58. Measured traffic 14. The frequency function $g(x)$ according to normal form 1

Curve 1: The intensity variation of the traffic. $\eta_0 = 0.0273$. x is measured with A used as unit

Curve 2: The intensity variation of the number of calls. $\eta_0 = 0.0550$. x is measured with y used as unit

List of symbols

Only the special notations that have been used extensively in this work are defined here. Most of the symbols used have in different connections obtained different indices and those are not presented here. Nor are ordinary mathematical symbols presented or special notations, appearing only at one or a few places.

A	Traffic intensity
b_ν	Auxiliary constant in computations according to normal form 2 as defined in (136 a)
c_ν	Auxiliary constant in computations according to normal form 2 as defined in (142 a)
$D(t)$	Difference function of the inter-arrival times
$D_\nu(t)$	Approximate value of the difference function
$D_{\eta_0}(t)$	Difference function according to normal form 1
$D_{\eta_0,\kappa}(t)$	Difference function according to normal form 2
$E_{1,n}$	Blocking according to the Erlang formula for loss systems
F, f	Functions in general, mostly frequency functions of inter-arrival times
$F_n(x_1, x_2, \ldots x_n)$	Defined in (1 b)
$f_n(T)$	Defined on p. 63
$F(T, x)$	Defined in (74)
$F_n\begin{pmatrix}x_1, x_2, \ldots x_n\\ y_1, y_2, \ldots y_n\end{pmatrix}$	Defined in (94)
$f_y(x, \xi)$	Defined on p. 81
$G(x)$	Intensity distribution function
$g(x)$	Intensity frequency function
$H(\xi)$	Defined on p. 81
K_μ	Defined on p. 34
$L_{n,\mu}$	Defined in (137 a)
M_n, m_n	Moments in most cases for intensity distributions
$m_{y,\sigma}$	Defined in (63)
P_n	State quantity for traffic with intensity variations
$P_{n,0}$	State quantity for random traffic
$P_{\mu,\nu}$	Defined on p. 33
s	Mean occupation time
T, t	Time variables
$V(t, \xi)$	Defined in (104)
$v(t)$	Defined in (112)
Y, y	Call intensities
α	Phase displacement, used only in Chapter 7
α	Ratio between the scanning interval and the mean occupation time in measurements according to the scanning method, p. 153
$\alpha_{p,n}$	Combinations of device occupations, p. 116
$\beta_{r,n}$	Characteristic constants, defined in (22)
ϵ	Form factor
η_0	Parameter of normal forms 1 and 2
$\Theta(t)$, $\vartheta(t)$	Distribution functions of the next call
κ	Parameter of normal form 2
μ_n	Relative moments
$\pi_r(t)$	Characteristic function, defined in (20)
τ_0	Parameter of normal forms 1 and 2
$\varphi(t)$	Distribution function of the inter-arrival times
$\psi_r(t)$	Characteristic function, defined in (18)
ω	Frequency in angular measure, Chapter 7.

References

The literature on telephone traffic problems is very rich. Here are only listed titles to which reference has been made in the previous text.

1. *Bernstein, S.:* "Sur les fonctions absolument monotones", Acta Mathematica, Vol. 52, 1929, pp. 1 – 66.

2. *Dubourdieu, J.:* "Les fonctions absolument monotones et la théorie mathématique de l'assurance-accidents", Comptes Rendus de l'Academie des Sciences, Paris, Vol. 206, 1938, pp. 556 – 557.

3. *Erlang, A. K.:* "Solution of some Problems in Probability Theory of Importance for Automatic Telephone Exchanges" (in Danish), Elektroteknikeren, København, 1917.

4. *Erlang, A. K.:* "Lösung einiger Probleme der Wahrscheinlichkeitsrechnung von Bedeutung für die selbsttätigen Fernsprechämter", Elektrotechn. Zeitschrift, Berlin, 1918, p. 504.

5. *Feller, W.:* "Completely Monotone Functions and Sequences", Duke Mathem. Journ., Vol. 5, No. 3, 1939.

6. *Lundberg, O.:* "On Random Processes and their Application to Sickness and Accident Statistics", Dissert. Stockholm 1940.

7. *Palm, C.:* "Calcul exact de la perte dans les groupes de circuits échelonnés", Ericsson Technics, Stockholm, Vol. 4, 1936, p. 41.

8. *Palm, C.:* "Inhomogeneous Telephone Traffic in Full-Availability Groups", Ericsson Technics, Stockholm, Vol. 5, 1937, p. 1.

9. *Palm, C.:* "Analysis of the Erlang Traffic Formulæ for Busy-Signal Arrangements", Ericsson Technics, Stockholm, Vol. 6, 1938, p. 39.

10. *Palm, C.:* "Accuracy in Measurements of Traffic by the Scanning Method" (in Swedish), Tekn. Medd. från Kungl. Telegrafstyrelsen, Stockholm, 1941, No. 7 – 9.

11. *Palm, C.:* "A Form Factor for Judging Waiting Time Distributions" (in Swedish), Tekn. Medd. från Kungl. Telegrafstyreelsen, Stockholm, 1943, No. 1 – 3.

12. *Palm, C.:* "Some Corollaries of the Erlang Formulæ" (in Swedish), Tekn. Medd. från Kungl. Telegrafstyrelsen, Stockholm, 1943, No. 1 – 3.

Mathematical handbooks

13. *Doetsch, G.:* "Theorie und Anwendung der Laplace-Transformation", Berlin, 1937.

14. *Frank, P.* and *v. Mises, R.:* "Die Differential- und Integralgleichungen der Mechanik und Physik", Vol. 1, Braunschweig, 1930.

15. *Hobson, E.:* "The Theory of Functions of a real Variable", Cambridge, 1926.

16. *Jahnke, E.* and *Emde, F.:* "Funktionentafeln mit Formeln und Kurven", Leipzig and Berlin, 1938.

17. *Kowalewski, G.:* "Determinanten-Theorie", Leipzig, 1909.

18. *Netto, E.:* "Lehrbuch der Combinatorik", Leipzig, 1901.

19. *Nörlund, N. E.:* "Vorlesungen über Differenzenrechnung", Berlin, 1924.

20. *Whittaker, E. T.* and *Watson, G.:* "A Course of modern Analysis", Cambridge, 1940.